EXPLODING WIRES

EXPLODING WIRES

VOLUME 4

Proceedings of the Fourth Conference on the Exploding Wire Phenomenon, Held at Boston, October 18-20, 1967, under the Sponsorship of the Air Force Cambridge Research Laboratories, Office of Aerospace Research, with the Cooperation of the Lowell Technological Institute Research Foundation

Edited by

WILLIAM G. CHACE

Air Force Cambridge Research Laboratories
Office of Aerospace Research
Bedford, Massachusetts

and

HOWARD K. MOORE

Lowell Technological Institute Research Foundation
Lowell, Massachusetts

SPRINGER SCIENCE+BUSINESS MEDIA, LLC 1968

ISBN 978-1-4899-7330-6 ISBN 978-1-4899-7328-3 (eBook)
DOI 10.1007/978-1-4899-7328-3

Library of Congress Catalog Card Number: 59-14822

Originally published by Plenum Press in 1968
Softcover reprint of the hardcover 1st edition 1968

Preface

This fourth volume of the Exploding Wires series, like its predecessors, contains the proceedings of a Conference on the Exploding Wire Phenomenon, the fourth, held in Boston in October of 1967. For the first time in this series, the discussion which followed the papers has been included in the proceedings. In all other respects the format remains the same.

As in previous cases, this volume contains papers on the relation of shock waves and exploding wires; or uses, particularly as exploding bridge wires; on chemical reactions induced by wire explosions; and on the general theory of the wire explosion itself. In addition, several new areas are covered. There are two papers on the resistivity–density relation in wire explosions. There are papers on the use of lasers and also on the use of x rays for the study of exploding wires. Spectroscopy as applied to wire explosions is discussed in four papers.

The Conference Committee included Dr. F.C. Bennett, Mr. M.A. Levine, Dr. C.P. Nash, and Mr. R.J. Reithel. The editors wish to thank these and all the others who gave so generously of their time and talent to make the Conference and this volume possible.

Mrs. William Watson was again the efficient conference secretary, and the art work was again done by Mr. C. V. Fish.

William G. Chace
Howard K. Moore

Contents

Vaporization Waves in Metals

F.D. Bennett and G.D. Kahl

Ballistic Research Laboratories
Aberdeen Proving Ground, Maryland

The vaporization wave hypothesis is discussed and its merits and defects are examined. The vaporizing model is visualized on thermodynamic grounds as carrying the liquid metal through a continuous succession of states either on or near the liquidus line in the two-phase region. On this line, the adiabatic sound speed for wet vapor will limit the rate of propagation of the vaporization front into the liquid. Experimental data for wire explosions of Al, Ag, Cu, Au, Pb, and Hg (frozen) are analyzed for wave speeds. While the influence of thermal expansion of the liquid can be accounted for theoretically, insufficient thermal data are available for the metals to permit correction of the wave speeds for this effect. The experimentally derived wave speeds are compared with theoretical values of the adiabatic sound speed in the wet vapor obtained from a modified van der Waals equation of state. At low velocities, the agreement is satisfactory but higher values deviate considerably from theory. Possible causes of the deviations are discussed. These include the crudity of the fluid dynamic model, neglect of thermal expansion, lack of information about the relationship between density and electrical conductivity and the approximation imposed by the van der Waals equation.

INTRODUCTION

We discuss here some recent experiments to examine further the hypothesis that vaporization waves are responsible for the anomalous resistance rise in exploding wires. While vaporization waves may be expected in all high-temperature matter on general thermodynamic grounds [1, 5], the first instance in which their existence could be recognized appears to be the exploding wire phenomenon [2]. if the vaporization of a superheated metal cylinder is limited by the speed with which the head of the vaporization wave travels from periphery to the axis, then an upper bound is set on the rate at which the conducting cross section diminishes. Resistance increases above the melting point, larger than the usual linear rise with temperature, can be related to the decreasing cross section of the conductor; thus, wave speeds can be obtained from the electrical pulse data. In our earlier paper [2], we reported wave speeds for copper wires. Here we present data for Cu, Pb, Al, Au, Ag, and Hg (frozen) wires and make comparison with predictions from theory.

From a critic's point of view, a number of limitations on the applicability of the vaporization wave hypothesis can be stated. The success of the experimental method of analysis depends on the assumption of a linear relation between resistivity and specific energy applied above the melting point up through the two-phase region. This is, at best, a fairly crude approximation but cannot be replaced until better information is available. Certain metals, such as Fe, Ni, and W, display nonlinear resistivity decreases above the melting point and, therefore, cannot be treated by the present method. In the case of tungsten, at least, we know now that these nonlinear effects are associated with conduction and arc phenomena in the peripheral vapor emitted by the wire during expansion. There is reason to hope that eventually wave phenomena may be studied in tungsten wires immersed in a dielectric liquid such as silicone vacuum pump oil. Needless to say, the effects of a dense ambient medium on the wave propagation and the electrical pulse shape are not well understood, although some preliminary unpublished experiments show that voltage breakdown and peripheral arc formation can be slightly delayed by ambient liquids of high breakdown strength.

At the relatively high densities of the exploding wire experiment, collision frequencies are high ($\sim 10^{15}$/sec), the liquid and vapor should obey continuum fluid mechanics, and local thermodynamic equilibrium may be assumed from point to point in the medium.

To discuss the fluid dynamical behavior of a superheated metal, one desires to specify a homogeneous, high-temperature medium at uniform pressure as the initial state. For the fast explosions in which we are interested, a plausible argument can be given that nearly uniform conditions obtain, at least for the early phases of the explosion. This means that, assuming that transient skin effect has damped out, kinetic pressures rapidly exceed magnetic pinch pressures and heat losses via conduction, convection, and radiation can be neglected on the time scale of the experiment. Estimates based on simplified calculations of the expected effects show that these conditions may be satisfied for events occurring within intervals of a few microseconds.

Dynamical effects themselves may be expected to lead to nonuniformities of pressure, temperature, and density throughout the medium as the flow develops. Figures 1 and 2 show that density variations occur rather early in the expansion process. What, then, is the initial state and how uniform is the expansion which follows?

X-ray and optical shadowgraphs have already shown that the later stages of the expansion proceed with large density differences between adjacent portions of the wire. Apparently, vapor and dense striations exist side by side during appreciable intervals of the expansion process. Formerly we supposed that these nonuniform densities occurred late in the expansion regime, after the vaporization wave had traversed the wire radius. Fansler and Shear [4] demonstrate by means of x-ray shadowgraphs, correlated in time with electrical pulse data, that striations appear at a much earlier

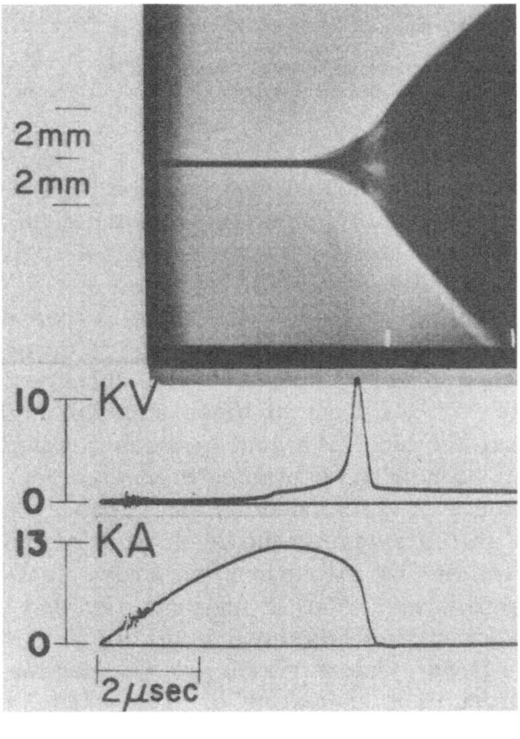

Fig. 1. Correlated electrical and streak camera data for 10-mil copper wire. (V = 3 kV, C = 32 μF.)

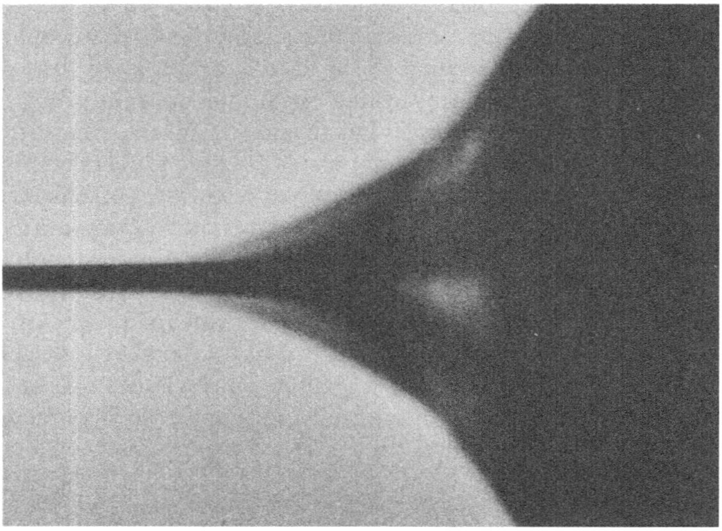

Fig. 2. Enlargement of expansion region of Fig. 1. Note vapor veil is penetrated by backlighting. Luminosity from interior is seen just after voltage peak.

stage than previously thought, and must be regarded as practically concurrent with more elementary expansion processes such as volume expansion of the liquid in the two-phase region and expansion through the head of the vaporization wave.

Light-scattering studies of the metal vapor peripheral to the wire—studies made at this laboratory and to appear soon in report form—indicate appreciable particle-size changes during the expansion process. In the light of these experimental indications, we recognize that the ideal of a homogeneous medium expanding from uniform initial conditions continues to be elusive.

The presence of visible density striations in the early expansion stages would militate against the assumption of a uniform vaporization wave diminishing the conducting cross section at a definite, though variable, speed. Rather, one may ask what conduction processes could simultaneously involve the striations and the nearby less dense regions. One may also ask whether the striations arise from nonuniformities of heating or from initial conditions in the crystalline solid wire, or whether striations form because of local condensation from a more uniform, vaporous state attained soon after melt and prior to the violent expansion.

No matter what the answers to these questions, the viability of the vaporization wave hypothesis, at least in its application to exploding wires, is clearly still open to question; nevertheless, one cannot say that it is decisively disproved by experiment.

The evidence in its favor is compelling. Our results show that experimentally derived wave speeds correlate with those calculated from a van der Waals equation of state for metals. Both the onset energy and the form of the wave speed function are correctly represented. In terms of scaling laws based on critical temperatures, the wave speed curves for several metals cluster about the single theoretical curve. These results provide an independent check of recent methods of estimating critical temperatures hitherto inaccessible to measurement. One cannot readily reject a theoretical approach with demonstrable virtues such as these. One hopes that by refinement its deficiencies can be eliminated; however, the loss of its initial fine simplicity is to be expected in the ensuing complications.

THERMODYNAMICAL MODEL

In this section, a simplified fluid model will be developed to represent the transient behavior of a material heated rapidly from the solid state up through its critical temperature. For typical metals, the heat energy for melting is small compared to that required to vaporize the system; therefore, we concentrate on the mechanism of the liquid–vapor transition. For such a two-phase condensing fluid, there are a variety of state equations. The most famous is the van der Waals equation, which typifies the essential features of a condensing fluid and which will herein be used in a slightly modified form.

Consider unit mass of material, and let P, V, and T be pressure, specific volume, and temperature, respectively. In the single phase (either liquid or vapor), the pressure is assumed given by the van der Waals function, P_w,

$$P_w(V, T) = \hat{R}T/(V - b) - a/V^2 \tag{1}$$

with inequalities $T > 0$, $V \geq b$, where a and b are substantive constants for each material and $\hat{R}$ is the gas constant. For subcritical temperatures, there is a two-phase region where both the liquid and vapor phases coexist at the same pressure and temperature; in coexistence, the pressure is not that given by Eq. (1) above; instead it is the vapor pressure function $P_A(T)$. For the van der Waals system, this latter function is not explicitly represented by an analytic function and must be found by computation. To compute $P_A(T)$ as well as the $V_3(T)$ and $V_1(T)$ loci of the saturated liquid and vapor lines, a generalization of the Maxwell criterion [5] is used, so that at fixed $T < T_c$ (subscript c always denotes critical conditions),

$$\int_{V_3(T)}^{V_1(T)} P_w dV = P_A(T)\,[V_1(T) - V_3(T)] + \phi(T) \tag{2}$$

where

$$\phi(T) = \int_T^{T_c} - [C_V(\tau)(\text{vapor}) - C_V(\tau)(\text{liq})]\, d\tau + T \int_T^{T_c} [C_V(\tau)(\text{vapor}) - C_V(\tau)(\text{liq})] \frac{d\tau}{\tau} \tag{3}$$

Here C_V is the specific heat at constant volume of the single phases. For either phase, $C_V(T)$ is a function of temperature only, and is chosen to be consistent with the van der Waals system. If $C_V(T)$ (vapor) is identical with $C_V(T)$ (liq), then $\phi(T)$ is zero and (2) reduces to the usual "equal area" rule first given by Maxwell. We note that the equal-area rule is restrictive in demanding identical specific heat functions for the liquid and vapor phases. This restriction is lifted when the generalized rule above is used, but its use requires some information about specific heats. The constants a and b for the material are conveniently replaced in terms of the critical values of the system by the well-known relations:

$$a = 3 P_c V_c^2; \qquad b = V_c/3; \qquad R\,T_c = (8/3) P_c V_c \tag{4}$$

Knowledge of the critical constants as well as of the specific heats completes the description of the equilibrium state of the system. One can now find all the pertinent thermodynamical quantities, including those of the coexistence state. These will be explicitly exhibited as needed.

We use this model to analyze the behavior of materials heated from comparatively low temperatures. In particular, we consider a system in the molten state just above the melting temperature and allow it to be heated to critical conditions. The usual adiabatic speed of small amplitude waves (sound speed) in the system is given by $c^2 = (dp/d\rho)_{ad}$ which becomes, with $\rho = 1/V$,

$$c^2 = -V^2\left(\frac{dP}{dV}\right)_{\text{adiabatic}} = V^2\left[\frac{T}{C_V}\left(\frac{\delta P}{\delta T}\right)_V^2 - \left(\frac{\delta P}{\delta V}\right)_T\right] \quad (5)$$

For the condensing fluid, we note that the pressure and specific heat functions are different for the single phase regime and for the two-phase coexistence state. In the single-phase regime, we use the van der Waals functions, whereas in coexistence we use the vapor pressure function $P_A(T)$ and the specific heat $C_{VA}(V, T)$ of coexistence. Since P_A is independent of volume, the wave speed in coexistence reduces to c_w, where

$$c_w = V(T/C_{VA})^{1/2}\left(\frac{dP_A}{dT}\right) \quad (6)$$

At the saturated liquid line, there are two very different values of wave speed possible; which one applies depends on whether the disturbing wave tends to change the system to the single-phase (all liquid) state, or to the two-phase state.

The specific heat $C_{VA}(V, T)$ for the coexistence state must also be found numerically. This is done straight forwardly by finding the internal energy E_A of coexistence, and using $C_{VA} = (dE_A/dT_V)$. The internal energy of coexistence is found by using E_A and P_A for E and P in the general thermodynamical relation

$$\left(\frac{\delta E}{\delta V}\right)_T = T\left(\frac{\delta P}{\delta T}\right)_V - P \quad (7)$$

and noting from Eq. (2) that $M(T) = T(\delta P_A/\delta T) - P_A$ is independent of V; then direct integration yields

$$E_A(V, T) = E_w(V_3, T) + M(T)[V - V_3(T)] \quad (8)$$

where E_w is the internal energy of the van der Waals function. One then finds

$$C_{VA}(V, T) = \left[\frac{dE_w(V_3, T)}{dT}\right]_V + \frac{dM(T)}{dT}[V - V_3(T)] - M(T)\frac{dV_3(T)}{dT} \quad (9)$$

noting that

$$\left[\frac{dE_w(V_3, T)}{dT}\right]_V = \frac{\delta E_w}{\delta V_3}\frac{dV_3(T)}{dT} + \frac{\delta E_w}{\delta T} \quad (10)$$

For example, at the saturated liquid line, the specific heat of coexistence simplifies to

$$C_{VA}(V_3, T) = \left[\frac{\delta E_w}{\delta V_3} - M(T)\right] \frac{dV_3(T)}{dT} + \frac{\delta E_w}{\delta T} \tag{11}$$

The last term on the right is the constant-volume specific heat of the liquid; $\delta E_w/\delta V_3 = a/V_3^2$, as may be seen by using E_w and P_w in Eq. (7). With $P_A(T)$ and $V_3(T)$ found from Eqs. (2) and (3), we may now evaluate the two possible wave speeds on $V_3(T)$. These are exhibited in nondimensional form in Fig. 2 as functions of T/T_c for the example where C_V(liq) = C_V(vap) = $(3/2)\ \hat{R}$.

The wave speed in the liquid is seen from Fig. 3 to be very much larger than the vaporizing wave speed of the two-phase system. The vaporizing wave speed is very small until temperatures exceeding one-half critical are attained.

It is characteristic of the van der Waals system (and, indeed, of most physical examples) that both the isotherms and adiabatics in the all-liquid phase are very steep compared to those of either the coexistence or all-vapor states, except very near the critical

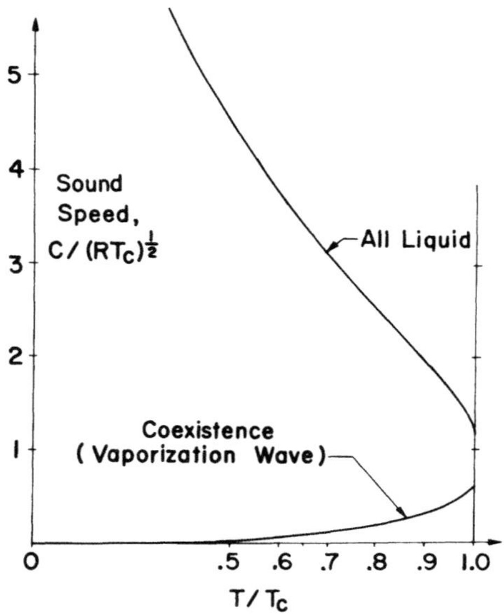

Fig. 3. Double-valued sound speeds on liquidus, van der Waals–Maxwell fluid, [C_V (liq) = C_V (vap) = (3/2) $\hat{R}$].

point. The magnitudes of these adiabatic slopes are directly proportional to the squared sound speed, so one expects the liquid sound speed to be usually much higher than the vapor or vaporizing wave speeds.

This fact suggests a simplification of our model for the system undergoing transient heating from an initially molten state at low temperatures. We assume that the heating moves the thermodynamic state along the saturated liquid line; any tendency to drive the system into the all-liquid state would be rapidly counteracted by liquid thermal expansion, which occurs at a relatively fast rate corresponding to liquid sound speed. This liquid thermal expansion lowers the pressure to the vapor pressure $P_A(T)$ where any further expansion must be accompanied by partial vaporization. This latter expansion is governed by the relatively slow vaporizing wave speed of the coexistence state. We note that due to the steepness of the all-liquid adiabatics, only a small volume increase is needed to maintain the system at the saturated liquid state.

We therefore assume the fluid state of the exploding wire in the pre-burst stage can be approximated by a molten cylinder from whose surface partial vaporization is occurring. The speed of the leading edge of this partially vaporizing wave is assumed to move with the vaporizing sound speed of the saturated liquid at temperature T, and is given by Eq. (6).

It is of interest to correlate the vaporizing wave speed with the added heat content q per unit mass rather than with temperature, because the additional heat energy can usually be found operationally by energy balance, whereas temperature is a more elusive quantity. In particular, for the modified van der Waals system employed here, one can compute the equilibrium heat content per unit mass along the saturated liquid line by using Eq. (2) and the assumed known critical constants and liquid specific heat. One integrates

$$dq = dE + PdV \tag{12}$$

along the saturated liquid line $V_3(T)$ from T_M to T, where T_M is melting temperature, using computed values of $E_w(V_3, T)$ and $P_A(T)$ from Eqs, (2) and (3). At any T between T_M and T_c, a quantity $\Delta q(T)$ is obtained, thus giving a correspondence between $\Delta q(T)$ and T. For a given material, one then assumes the heat content along the saturated liquid line to be

$$q(T) = q(T_M) + \Delta q(T) \tag{13}$$

where $q(T_M)$ is the assumed known heat content of the liquid at melt, including the latent heat of melting. It is apparent that $q(T)/\hat{R}T_c$ depends on the specific heats of the liquid and vapor phases through Eqs. (1) and (2), as well as on $q(T_M)/\hat{R}T_c$. This latter quantity is found to have nearly the constant value of 0.60 for a variety of metals, so plots of $c_w/(\hat{R}T_c)^{1/2}$ against $q/(\hat{R}T_c)$, where c_w is the vaporizing wave speed on $V_3(T)$, usually depend only on the single-phase specific heats as parameters. We exhibit such plots in Fig. 4 for several examples of specific heats.

EXPERIMENTAL

We outline here the procedure for calculating vaporization wave speeds from the electrical data. This method differs somewhat from that given previously [2], in that a means of accounting for thermal expansion of the liquid metal is incorporated in the present scheme.

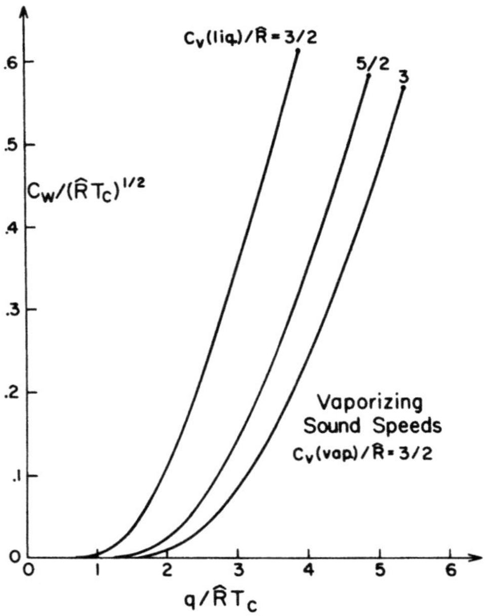

Fig. 4. Vaporizing sound speed on liquidus. Modified van der Waals equation used with liquid specific heat as parameter. Curves terminate at heat energy of critical point.

Data Reduction for Wave Speed

The voltage across the wire and the current through it are obtained from oscilloscope traces as previously described [6]. After accounting for the inductive voltage on the voltage probe, one then has the resistive voltage V_R as a function of time. Typical $V_R(t)$ and $I(t)$ measurements are seen in Fig. 1. We assume the rapid increase of voltage to its peak is due to an increasing resistance caused primarily by a reduction of the conducting cross section of the molten wire; this reduced cross section is attributed to the partial vaporization proceeding inward from the wire surface.

Pre-vaporization Heating of the Liquid Metal

As heat is added to the molten metal, a slight thermal expansion will occur. Since the length-to-radius ratio of the wire is large, expansion waves traveling at the local speed of sound will not reach the wire ends during the short time interval of the explosion. We therefore expect the thermal expansion to be effective in increasing the radial dimension only, and account for it by assuming the entire mass is contained within a cylinder of length of that of the original wire l and average radius $r_1(t)$. It is clear that $r_1(t)$ must increase slightly with time, according to the heat addition and consequent average density decrease of the liquid.

In order to describe the assumed model of the wire explosion, we wish to map certain boundary, wave path, and particle trajectories in an r, t plane. If m_s represents the mass of the solid wire and m the mass of the wire within radius r before vaporization, we can write $m = \pi r^2 l d(t)$, where $d(t)$ is the uniform density of the metal, a density which varies only with time. For m constant, it follows that $r \equiv r(m, t)$ describes the variation with time, due to thermal expansion, of the representative mass particle at the radius which encloses mass m. In this mapping, m is a parameter and t the variable of interest. Accordingly, $r_1(t) = r(m_s, t)$ gives the curve in the r, t plane of the expanding outer boundary of the liquid wire. Because of the assumption of uniform density, all other particle trajectories corresponding to $m \leq m_s$ are similar curves.

The trajectory of the vaporization wave, which we denote by $r_2(t)$, cuts across the particle trajectories in passing from larger to smaller radii. With the above definitions, we could write $r_2(t) = r[m_2(t), t]$, where $m_2(t)$ represents the variable mass ahead of the vaporization wave at time t. Here the parameter m changes from point to point of the r_2 curve; however, m_2 is not known *a priori* but must be determined from experiment.

Vaporization Model

With increased heat content of the liquid, the threshold of vaporization will be exceeded and a wave will start. We assume that the cylindrical interface of average radius $r_2(t)$ separates the intact molten core from the outer, nonconducting, partially vaporized mixture. The local velocity of this interface in a reference frame fixed with a given set of fluid particles is denoted by c_T. In a laboratory fixed reference frame, the interface velocity is $dr_2(t)/dt$. The velocity of the liquid particles at $r = r_2(t)$ (which is the same as the velocity of the reference frame attached to these particles) is $(dr/dt)_{m_2}$. Therefore,

$$c_T = \left(\frac{dr(t)}{dt}\right)_{m_2} - \frac{dr_2(t)}{dt} \tag{14}$$

If no vaporization occurred, the entire liquid mass m_s of the wire would be contained within the radius $r_1(t)$. Assuming the density of the liquid core to be constant with radius, the mass m ($< m_s$) contained within $r(t)$, [$\leq r_2(t) \leq r_1(t)$] is related to the total mass by $r(t) = (m/m_s)^{1/2}\, r_1(t)$. Differentiating this expression, and noting that both m and m_s are time-independent, we find

$$\frac{dr(t)}{dt} = (m/m_s)^{1/2}\, \frac{dr_1(t)}{dt} \tag{15}$$

The mass, $m_2(t)$, of material contained within $r_2(t)$ does change with time, since $r_2(t) = (m_2/m_s)^{1/2}\, r_1(t)$. With the foregoing relations

$$c_T = - r_1(t)\, \frac{d}{dt}\, [(m_2/m_s)^{1/2}] \tag{16}$$

The electrical resistance R_2 of the conducting cylinder of radius $r_2(t)$ is

$$R_2/R_0 = \frac{\rho(t)}{\rho_0} \left[\frac{r_0}{r_2(t)}\right]^2 \tag{17}$$

where subscript 0 denotes some suitably chosen reference conditions which identify the beginning of the wave, and ρ is the electrical resistivity. Furthermore, if $d(t)$ be the liquid density of the core, we have

$$m_2(t)/m_s = \left[\frac{r_2(t)}{r_0}\right]^2 \frac{d(t)}{d_0} \tag{18}$$

The rate at which total heat Q is added to the core is $dQ/dt = I(t)\,V_R(t) = I(t)^2 R_2(t)$, a measured quantity. Since this heat input is assumed to be deposited uniformly in the conducting cylinder by uniform current distribution, we can write $dQ/dt = m_2\,(dq/dt)$, where q is the heat content per unit mass. With the foregoing relations, we then have

$$[R_2(t)/R_0]\,\frac{dQ}{m_s} = \left\{\left[\frac{\rho(t)}{\rho_0}\right]\frac{d(t)}{d_0}\right\}dq \tag{19}$$

Using the experimental values of $V_R(t)$ and $I(t)$, we can numerically integrate the left-hand side from the reference conditions $R(Q_0)$ to the values for $Q(t)$ at any chosen time within our data record. A corresponding integration of the right-hand side, assuming we know the law relating the variation of resistivity and density with heat content, would then relate the measured heat input to the resulting heat content per unit mass of the liquid core.

Unfortunately, $\rho(q)$ has not been measured at sufficiently high temperatures for most metals. A further complication is that the model assumed thus far is an average one, in the sense that both ρ and d apply here to the assumed uniform cylinder employed to represent the actual conducting path. In order to proceed, we use an extrapolation of the values of ρd measured at the lower q values before the vaporization becomes important. For those q values between melt q_M and the onset of vaporization q(vap), we note that $m_2 = m_s$ and $r_2(t) = r_1(t)$. Using the latter relations in Eqs. (17) and (18) gives $R_2/R_0 = [\rho(t)/\rho_0][d(t)/d_0]$ for q between q_M and q(vap). Plots of R_2/R_0 versus q in this energy range indicate a linear relation between the product ρd and q. We assume that this linear relation continues to hold for $q > q$(vap); this assumption enables us to integrate the right-hand side of Eq. (19) with respect to $q(t)$. The resulting integral relation, evaluated numerically, correlates the measured input heat energy $Q(t) - Q_0$ with the heat content per unit mass $q(t)$ of the liquid core.

Using Eqs. (17) and (18), we then have

$$m_2/m_s = [\rho(t)/\rho_0][d(t)/d_0][R_0/R_2(t)] \tag{20}$$

for the range $q(t) > q$(vap). Since the right-hand side is now known as a function of time, numerical differentiation with respect to t can be done for use in Eq. (16). Moreover, if one knows how $r_1[q(t)]$ varies with $q(t)$, then c_T is obtained as a function of q. By hypothesis, $r_1(t)/r_0$ is $[d_0/d(t)]^{1/2}$, so one needs $d[q(t)]$ as a function of q. Since the density $d(q)$ is expected to decrease with q, it follows that $r_1(q)$ should increase with q. If one uses the modified van der Waals theory described previously, assumed the state of the liquid core to be that of the saturated liquid, and takes the reference conditions to correspond to those of the liquid at melt, one finds $d(q)/d_M$ decreases slowly from the value unity until q values near critical are approached.

If we define

$$c_{T0}(q) = -r_s \frac{d(m_2/m_s)^{1/2}}{dt} \tag{21}$$

where r_s is the original radius of the solid wire, we have from Eq. (16)

$$c_T(q) = F(q)\, c_{T0}(q) \tag{22}$$

where

$$F(q) = \left(\frac{r_0}{r_s}\right)\left(\frac{r_1(q)}{r_0}\right) \tag{23}$$

The quantity $c_{T0}(q)$ is obtained directly from the electrical data using the model just described. On the other hand, $F(q)$ depends on the thermal expansion of the liquid and cannot be obtained from our present data. Estimates of $F(q)$ indicate it may be replaced by a constant a little larger than unity for temperatures less than 0.9 critical. $F(q)$ will be described more completely below. For the present, we show experimental values of c_{T0} vs. q.

Experimental Conditions

Using the above procedure, we have obtained experimental values of c_{T0} vs. q from tests on copper, lead, aluminum, gold, silver, and mercury wires. The copper data have been reported previously [2], although not in the scaled form given here. We first mention briefly the experimental conditions and give tables of pertinent quantities.

A capacitor of 31.5 μF was used for all tests except those on copper, and the circuit ringing frequency was 43 kc/sec. Voltage and current traces were photographed on a type 555, double-beam Tektronix oscilloscope. From measurements of these traces we obtain $V_R(t)$ and $I(t)$; numerical integration gives $\Delta Q(t) = \int_0^t V_R I d\tau$, and we have $R_2(t) = V_R/I$. We list in Table I the conditions for all wires except copper; copper test conditions are given in [2]. The explo-

Table I. Test Conditions

Metal	Initial wire diameter		Length, cm	Initial capacitor voltage, kV	No. of tests
	mils	cm			
Lead	5.1	.0130	1.1	1.2	1
Lead	10	.0254	1.0	1.0	1
Lead	10	.0254	1.0	2.0	1
Lead	10	.0254	1.0	3.0	1
Aluminum	5.6	.0142	1.1	2.0	1
Aluminum	10	.0254	1.1	3.0	2
Gold	10	.0254	2.0	3.0	2
Silver	10	.0254	2.0	3.0	1
Mercury	20	.0508	2.0	2.5	1
Mercury	20	.0508	2.0	3.0	1

sions occurred in air at atmospheric pressure, and the wires were maintained at room temperature before current switch-on, except for the frozen mercury wires; they were kept at dry ice–acetone temperatures.

From the measurements, the scaled functions $S_2(t) = R_2 \pi r_s^2/l$ and $U(t) = \Delta Q(t)/m_s$ were plotted as ordinate and abscissa, respectively, for each metal; here r_s and m_s are the initial radius and mass of the wires. For a given metal we find that the S_2 vs. U plots for wires of different radii and initial capacitor voltage coincide with each other until a fixed value of U is exceeded. For $U > U_V$, the plots diverge from each other, according to initial wire diameter and capacitor voltage [2].*

For U between a typical low value U_M and the higher value U_V, the experimental functions S_2 vs. U can be approximated by a straight-line segment for each metal. $S_2 = S_0[1 + \beta(U - U_0)]$; on this segment then the ratio $S_2/S_0 = R_2/R_0$ as defined by Eq. (17), and we take our reference conditions (S_0, V_0) at the largest value of $U = U_V$ where the experimental curves depart significantly from linearity. Table II gives the experimental quantities typifying these straight-line segments for the different metals. The last column, Δq_1, is the computed heat content per unit mass of metal prior to electrical heat addition.

*The notation of [2] is slightly different from that used here. A factor π was inadvertently omitted from the ordinate of Fig. 4 reproduced there.

Table II. Resistive and Heat Input Measurements

Metal	ρ_M, μΩ-cm	S_M, μΩ-cm	U_M, kJ/gm	$(dS/dU)_{\text{linear}}$, μΩ-cm/(kJ/gm)	S_0, μΩ-cm	U_0, kJ/gm	Δq_1, kJ/gm
Copper	21.1	29	1.10	8.40	40.5	2.4	.083
Lead	95	102	0.085	144	146	0.39	.035
Aluminum	24.2	26	1.4	5.1	48	5.7	.20
Gold	31.2	35	0.35	66	60	0.72	.031
Silver	17.2	19	0.75	22.6	28	1.14	.063
Mercury	91	108	0.045	700	125	0.07	.030

*First column gives tabulated values of liquid resistivity at melt [7]. Last column gives heat content prior to test, estimated from tabulated specific heats.

The value $S_M = S_2(U_M)$ should be comparable with the resistivity of the liquid metal at melt, ρ_M. Previously tabulated values of ρ_M are given in the first column [7]. Assuming the wire length constant, one expects $\rho_M = S_M(d_s/d_M)$ where d_s/d_M is the ratio of solid density to the liquid density at melt, a factor about 7 to 12 per cent greater than unity for these metals. Thus, one expects ρ_M to be several per cent larger than our measured S_M, whereas we find the converse in these experiments.

Extra care was taken with the frozen mercury wires,* both to prevent premature melting and to contain the metal vapor. A special electrode cell was constructed and cooled to the dry ice–acetone temperature. The wire was attached to the electrodes and the electrodes were enclosed by a 2-in.-diameter lucite cylinder, sealed with O rings. This cell was placed in the circuit and energized within seconds after removal from the cold box. Even with these precautions, erratic electrical behavior was noted on a number of trials with the frozen mercury wires. In at least one case, the wire melted before the current pulse was applied. In another, examination of unused portions of the wire disclosed remnants of the glass capillary not completely dissolved away. Only two of the mercury tests were judged worthy of data reduction.

Experimental Wave Speeds

We show in Fig. 5 the experimental results in deducing c_{T0} and q from the data. The q values contain the computed heat content

*We are particularly indebted to A. V. Grosse, J. Cahill, and Mrs. L. Streng of the Research Institute of Temple University for supplying the frozen mercury wires. The wires tested thus far were made by freezing mercury in capillary tubes and dissolving the capillary in acid.

of the solid metal just before the test, as given by Δq_1 in Table II. Data points connected by a given line were obtained from a single explosion.

We attempt to correlate the data from the different metals by dividing c_{T0} by $(\hat{R}T_c)^{1/2}$ and q by $\hat{R}T_c$ to give $\bar{c}_{T0}$ and $\bar{q}$; the resulting plot is shown in Fig. 6. Only the bounds of the data are indicated for copper, lead, and aluminum. The critical temperatures T_c used here are those estimated by A.V. Grosse and his associates [8, 9, 10]. These and other constants are given in Table III.

The experimental curves $\bar{c}_{T0}$ for the different metals (excepting mercury) are seen to coincide only at the lower values of $\bar{q}$ where the experimental wave speeds begin to be detectable. This scaled heat content of wave onset is about 2.1 ± 0.2. Taking this fact as an experimental test of the critical temperatures used for the scaling, then the values used for T_c appear to be satisfactory within the ± 10% data scatter.

At heat contents higher than that of wave onset, the scaled curves for the different metals depart from each other. In particular, those for the monetary metals, copper and silver, bend to

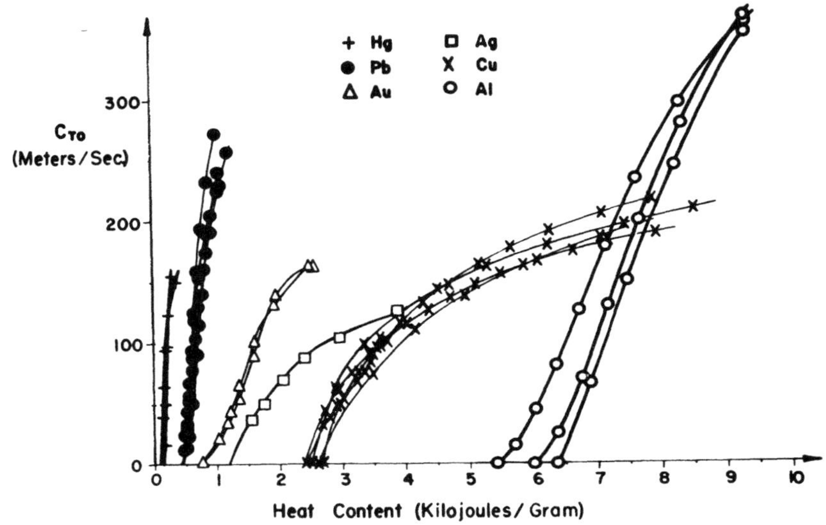

Fig. 5. Experimental wave speeds. Thermal expansion of core has been neglected.

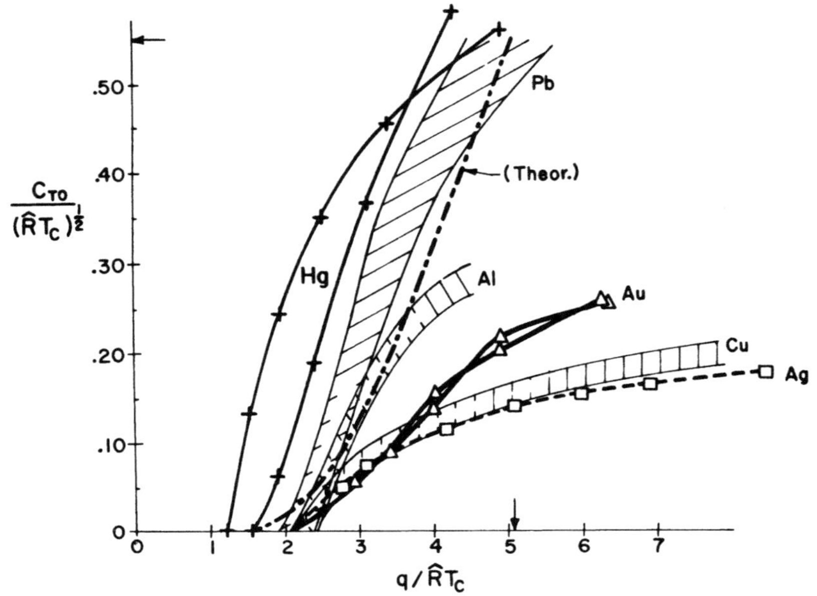

Fig. 6. Scaled wave speeds. Theoretical curve is vaporizing sound speed [Fig. 3 center curve adjusted 0.1 units right for $q(T_M)$]. Arrows show critical point limits for theoretical curve.

the right and go to scaled heat contents considerably higher than the other metals. We note c_{T0} should be multiplied by the factor $F(q/\hat{R}T_c)$ defined in Eq. (23) to obtain the scaled local velocity of the deconducting wave front, $\bar{c}_T$. If this factor were the same for the different metals, its use would not improve the correlation of $\bar{c}_T$ over that of the $\bar{c}_{T0}$ quantities. It is possible to compute a theoretical $F(q/\hat{R}T_c)$ curve using the modified van der Waals model described above. Aside from insignificant differences caused by assuming different values of specific heat $C_V/\hat{R}$ of the liquid this theoretical curve is identical for all the metals. Moreover, the theoretical curve is not defined for $\bar{q}$ values larger than that corresponding to the critical point. We show this curve in Fig. 7 for the example where $C_V(\text{liq}) = (5/2)\hat{R}$, and $C_V(\text{vap}) = (3/2)\hat{R}$. For this same example, we also show on Fig. 6 the scaled theoretical vaporizing sound speed, $\bar{c}_w = c_w/(RT_c)^{1/2}$. This curve terminates at $\bar{q} \simeq 5$. We expect the speed of the wave front, c_T, to be closely related to the vaporizing sound speed c_w along the saturated liquid line of the metal. Note that the experimental data for copper and silver greatly exceed the limiting heat content of the theoretical curve. The simplifying as-

Table III. Scaling Constants

Metal	T_c, °K	T(melt), °K	M, gm/mol	$\hat{R}T_c$, kJ/gm	$(\hat{R}T_c)^{1/2}$, m/sec
Copper	8500	1356	63.5	1.088	1043
Lead	5400	601	207.2	0.217	465
Aluminum	8650	933	27.0	2.67	1633
Gold	9500	1336	197.2	0.401	630
Silver	7460	1234	107.9	0.575	758
Mercury *	1733	234	200.6	0.0718	268

* T_c is estimated critical temperature [8–10].

sumption of a linear extrapolation of ρd to the higher heat energies may be questionable for these metals.

None of the wave speed data shown here were obtained from electrical data at times larger than that corresponding to a 10% descent of the voltage past its peak value. Two reasons for the choice of this limiting time are: a) in those tests where initial capacitor voltages are "well matched" [6] to the wire, the very rapid decrease of both voltage and current with time after the voltage peak makes the resistance measurements inaccurate; and b) in those instances where the capacitor voltage is not "well matched," the measured resistance reaches a maximum and then decreases, suggesting the occurrence of arc breakdown. In a given test, when the curvature of the R_2 vs. U curve changes from concave upward, no further data from that test are used.

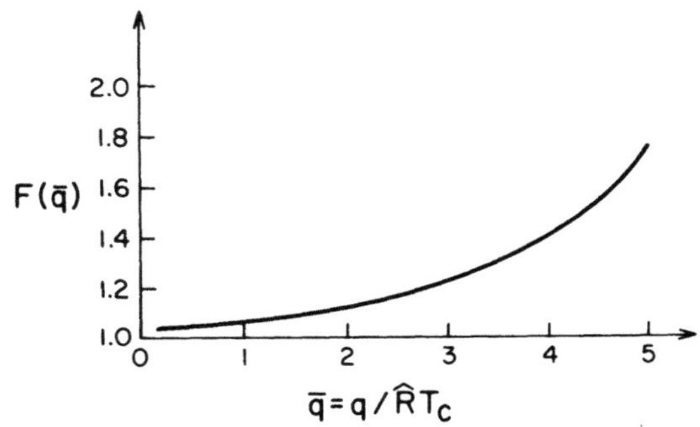

Fig. 7. Factor accounting for thermal expansion of liquid. Modified van der Waals model [C_V(liq) = (5/2) $\hat{R}$. C_V (vap) = (3/2) $\hat{R}$] is used.

DISCUSSION

With respect to the question of the vaporization wave hypothesis, the present experiments are somewhat ambiguous. If there were no agreement between the experimental findings and the theoretical implications of the wave speed and thermodynamic models, one could reject the hypothesis as incorrect. On the other hand, the agreement between experimental and theoretical wave speeds is sufficiently good so that the wave speed model continues to survive as a partial explanation of the expansion process.

The assumption of an incoming, cylindrical wave front separating the conducting core from the nonconducting wet vapor enables one to deduce front velocities which are comparable with the vaporizing sound speed of the saturated liquid. No other signal speed, either suggested or measured from thermodynamic data, is of the same order of magnitude as the vaporizing sound speed.

When different metals are considered, the use of a similarity law suggested by the theory permits correlation of the lower temperature wave speeds within a reasonably narrow band. This convenient scaling, which harmonizes the data for the various metals, permits detailed comparisons which reveal significant differences between them. For example, the effects of adjusting critical temperatures to improve the correlation can easily be seen.

The method of data analysis identifies the radius–time locus of that front which by hypothesis abruptly separates the finite conductivity of the intact liquid core from zero conductivity vapor. Such a locus probably does not exist physically; yet it is in a sense analogous to the boundary layer displacement thickness of a viscous fluid flow and plays a similarly useful role in treating a complicated problem. From physical grounds, one can maintain that the electrical conductivity, on the average, varies as a continuous function of radius, decreasing from that of the liquid metal core at the radius of the leading edge of the vaporizing expansion wave to zero at some larger radius. If this argument is admitted, then the average radius of the front which abruptly separates finite from zero conductivity must be larger than that of the leading edge of the vaporizing expansion wave. The reasoning proceeds as follows: The transition region of conductivity fans out with time, as all finite expansion waves do, and the abrupt change in conductivity must fall relatively further behind the head of the expansion wave; hence, the speed of this assumed front would be lower than that of the vaporizing sound speed in the molten metal. The experimental values for the noble metals appear to lie below the theoretical curve even after estimated corrections for dilatation; i.e., multiplication by $F(\bar{q})$, are applied.

The good agreement of wave onset for the different metals at the lower specific energies probably reflects both the fact that the assumed linear behavior of ρd with q cannot be greatly in error here and also the likelihood that the early trajectories of both the average conductivity wave front and the vaporization wave front are closely coincidental. The divergence of the curves for the different metals at higher specific energies and the fact that finite wave-front velocities for copper and silver are obtained at energies greatly in excess of those expected to drive the metal past critical temperature cannot be explained merely by revising the linear extrapolation of ρd.

The abundant data for copper [2]* exhibit large and increasing resistances at specific energies sufficiently high to heat the original mass of copper well beyond critical temperatures. This fact is independent of any assumption about the behavior of the product ρd for the saturated liquid. One is forced to conclude that the supercritical copper must have a finite resistivity both in the core and in the denser parts of the expanding flow. Thus, it follows that the electrical resistivity of the metal vapor is an important parameter in fixing the position of the wave front which shuts off conduction. With this added parameter, whose effects cannot yet be accounted for, one should not expect perfect correlation of the conduction-front wave speeds for the different metals when using a scaling law based on mechanical and thermodynamical arguments alone. The present scaling law should apply to the electrical wave fronts only when the vaporization and electrical conduction waves are nearly coincident, as they are expected to be at wave onset.

The physical properties of supercritical materials at high densities are not well known, but there are some theoretical reasons [3, 11] for thinking that electrical conductivity may persist or even be enhanced in materials at temperatures above critical and at normal liquid densities. According to Rouse's calculations made with a modified Saha equation, copper at normal densities and temperatures near 1 eV will have up to 15% concentration of the first ion present on account of pressure ionization. Presumably the free electrons will participate in conduction and ohmic heating processes as usual. Thus, arguments based on the presumed zero

*See Fig. 4 of this reference where resistance data for several wires are plotted vs. specific energy obtained by dividing the total deposited energy at a given time by the initial wire mass.

conductivity of condensing materials above the critical temperature can be seriously in error. If this is the case, the indicated high temperatures may be real and the temperature-multiplying capabilities of the exploding wire experiment are realized in practice. Since the argument of an earlier paragraph shows that the most conservative method of calculating specific energies for copper leads to the conclusion that supercritical temperatures ($T_c \sim 8500°K$ for Cu) are indeed attained, the conclusion seems inescapable that some process capable of increasing the number of current carriers; e.g., pressure ionization of the neutral atoms is an important factor at the higher temperatures and densities encountered in the wire cores.

A further anomaly in the thermal behavior of the metals studied should be noted. Measurement of input heat energy of the wires always makes it appear larger than expected. For example, the experimental heat content of the liquid at melt is always larger than that computed by using tabulated specific heats and heats of fusion. For the metals listed in Table II, these excesses are 55, 20, 30, 80, 100, and 90 per cent, respectively. The point corresponding to the liquid at melt is taken where the S_2 vs. U curves first show an abrupt decrease of slope to the linear portion; this point also corresponds to the first abrupt slope decrease easily visible on the voltage–time curve. The slope of the linear portion of the S_2, U curve can be related to tabulated values of the temperature coefficient of liquid resistivity. Even after accounting for the expected volume change and using the best estimated values for liquid specific heat, the experimental slope is smaller than expected. Conversely, for a given liquid resistivity, the corresponding heat content from the present data is much higher than that obtained using handbook values for the temperature coefficient of liquid resistivity.

The presence of the striations at an early stage, as discovered by Fansler and Shear [4], at first appears to be a strong argument against the existence of a vaporization wave phenomenon. The striations do, in fact, rule out the possibility that the vaporization wave proceeds uniformly into the material on a front with cylindrical symmetry about the wire axis. Rather, one must argue that as vaporization begins at temperatures far below critical and with specific energies too small to produce complete vaporization, the striations may be the physical evidence of the random statistical way in which vaporization takes place at favored surface sites during passage of the wave. The shape and number of striations may be indications of the manner in which small regions are cooled by local thermal transport processes to support the vaporization of adjacent regions. If so, this new hypothesis would force a reconsideration of our primitive ideas of the symmetry of the process of vaporization, but not necessarily a revision of its fundamental basis. Clearly, such a view of the striation problem provides an alternative, or a supplement, to theories based on assumption of the prior existence of elastic or plastic waves, buckling phenomena, and the like.

SUMMARY

The energy-dependent expansion of a cylinder of superheated liquid is offered as an idealized physical model of flow phenomena occurring in exploding wire events. Magnified streak shadowgraphs of copper wires show complex radial motions accompanying two major stages; viz., 1) a linear boundary expansion, during which a vapor veil proceeds ahead of a denser expanding core and, 2) a parabolic boundary expansion characterized by violent acceleration of the inner core as a result of sudden energy addition. Correlated electrical data show that the main deposit of electrical energy occurs near the end of the linear expansion, during which the vaporization wave is presumed to initiate the expansion.

Vaporization wave speeds have been obtained from experiments on wires of Al, Pb, Ag, Cu, and Au and from initial preliminary experiments on solid Hg wires. The wave-speed curves for these metals show similar behavior, although the initial specific energies differ by a factor up to 10^2. A thermodynamic model of the expansion is obtained by assuming a modified van der Waals equation of state together with a state path lying on the liquidus line in the $P-V$ plane. Values of the adiabatic sound speed calculated from the thermodynamic model and representing expansions from the liquidus line into the two-phase region agree well with the initial portions of the experimental wave speed curves. When speed is scaled by $(RT_c)^{1/2}$ and specific energy by RT_c, a plot of corresponding states is obtained wherein for all the metals the experimental data initially overlap the theoretical curve but deviate in different ways at larger values of the scaled variables.

Discussion of these deviations shows that to some extent they can be attributed to the approximations inherent in the fluid dynamic and thermodynamic models and to assumptions made necessary by the lack of electrical and thermal data for the elements.

ACKNOWLEDGMENTS

The authors are grateful to D.D. Shear and T. Long for performing the experiments; to H.S. Burden, who assisted with the data measurement and reduction; and to D.C. Mylin, who made many of the numerical computations.

REFERENCES

1. F.D. Bennett, Phys. Fluids 8:1425 (1965).
2. F.D. Bennett, G.D. Kahl, and E.H. Wedemeyer, in Exploding Wires, Vol. 3, W.G. Chace and H.K. Moore, eds., Plenum Press, New York (1964), p. 65.
3. F.D. Bennett, Phys. Fluids 8:1106 (1965).
4. K.S. Fansler and D.D. Shear, this volume, p. 185.
5. G.D. Kahl, Phys. Rev. 155:78 (1967).
6. F.D. Bennett, H.S. Burden, and D.D. Shear, Phys. Fluids 5:102 (1962).
7. N.E. Cusack, Rep. Prog. in Physics, XXVI, The Institute of Physics and The Physical Society, Stonebridge Press, Bristol, England (1963), p. 361.
8. A.V. Grosse, J. Inorg. Nucl. Chem. 22:23 (1961).
9. J.A. Cahill and A.D. Kirshenbaum, J. Phys. Chem. 66:1050 (1962).
10. A.V. Grosse, Rev. Hautes Temper. et Refract. 3(2):115-146 (1966).
11. C.A. Rouse, Astrophys. J. 139:339 (1964).

DISCUSSION

Comment by O.H. Zinke, University of Arkansas: We have been calculating vaporization wave velocities from resistance using your energy-corrected technique and with Au, Cu, and Al we obtain velocities similar to yours. The vaporization wave seems to travel for 0.50 (±0.25) microseconds before breakdown across the wire. Breakdown occurs in our case in the vapor from the wire because we explode wires in vacuum.

Question by M. Levine, Air Force Cambridge Research Laboratory, Bedford, Mass.: Do you believe the fact that the sound velocity in the liquid is large compared with the vaporization wave will allow the wire to expand uniformly rather than expand by reason of a vaporization wave?

Answer by Author: No. The large sound speed in the liquid phase ensures that compressions of the liquid will be quickly relieved so that the thermodynamic path stays on or near the liquidus line. The liquid will indeed expand thermally to a slight extent, but the vaporization process will be dominated by the much smaller sound speed characteristic of the coexistence region.

Question by B. Ya'akobi, Princeton Plasma Physics Laboratory, Princeton, N.J.: If the deviations from the theoretical curve in the scaled vaporization speed are indeed caused by electrical conductivity, then one would expect a higher deviation for lower ionization potential. Scaling the speed further by $\exp(-Ei/kT_c)$ should then bring the graphs for the different metals closer.

Answer by Author: The suggested scaling would appear to work the other way. (Ionization potentials were furnished orally by another person.) At present, the elements with higher ionization potentials appear to agree better with the thermodynamic sound speed.

Nevertheless, the effect of ionization potentials is reflected in the modified Saha equation calculations of Rouse, Astrophys. J. 139:339 (1964), where possible conditions for pressure ionization are developed.

Question by W.G. Chace, Air Force Cambridge Research Laboratory, Bedford, Mass.: Where did you get the van der Waals a and b of the metals?

Answer by Author: The van der Waals equation is $P = \hat{R}T/(v - b) + a/v^2$. The a, b, and $\hat{R}$ parameters of van der Waals'equation can be expressed in terms of the critical values of pressure P, volume V, and temperature T (i.e., P_c, V_c, and T_c by $b = V_c/3$, $a = 3P_cV_c^2$, and $\hat{R}T_c/P_cV_c = 8/3$. Using these relations, we find that the quantity we need for comparing vaporizing sound speed as a function of the dimensionless variables T/T_c or $q/\hat{R}T_c$ (q is the heat content per unit mass) is just T_c. For the metals we examine, estimates of T_c have been made by A. V. Grosse and his colleagues at the Research Institute of Temple University.

Initial Behavior of an Exploding Wire

T. S. Chang and M. L. Chang

North Carolina State University
Raleigh, North Carolina

The nonlinear circuit equations describing the initial phase of an exploding wire are considered. It is shown that the solution for any arbitrary variation of the wire resistance with the internal energy satisfying certain smoothness requirements can be represented by a single integral curve in the resistance–charge plane. Maclaurin series expansions of the current function and the logarithm of the resistance function for the case of linear variation of the wire resistance with the internal energy are presented in powers of the elapsed time after switch-on. The coefficients which appear in these expansions are obtained exactly and recurrently, using the basic circuit equations and the initial conditions during switch-on. The present results are compared with the approximate expressions described by F.D. Bennett.

INTRODUCTION

The initial phase of the exploding wire phenomenon has been considered analytically by several investigators. Nash and Olsen [1] reported an analysis in which the resistance of the wire was assumed to vary linearly with time. Hobson and Manka [2] presented a perturbation solution, using the classical RLC solution as the zeroth-order solution.

Actually, the circuit equations for an exploding wire are intrinsically nonlinear due to the variation of the resistance with the internal energy of the wire. Noting this nonlinearity, Bennett [3, 4] obtained some approximate solutions to the basic set of circuit equations, treating the wire resistance as a linear function of the internal energy or the temperature of the wire. These equations are also amenable to numerical integrations [5, 6].

In the present analysis, it will be demonstrated that the solution to the nonlinear circuit equations can be represented by an integral curve in the resistance–charge plane, provided that the wire resistance is a smoooth function of the internal energy of the wire. In addition, Maclaurin series expansions in powers of the elapsed time after switch-on for the current function and the logarithm of the resistance function are deduced. By truncating these series, rational approximations of the early time behavior of the exploding wire phenomenon can be obtained.

THE NONLINEAR CIRCUIT EQUATIONS

It will be assumed in the present analysis that during the initial phase of the exploding wire phenomenon, the wire resistance R_w and the specific internal energy of the wire U can be related by a piecewise continuous function $f(r)$ such that

$$u = \int_1^r d\xi/f(\xi) \tag{1}$$

is defined, where $u = mUC/Q_o^2$ is the dimensionless internal energy, $r = R_w/R_{wo}$ is the dimensionless wire resistance, m is the mass of the wire, C is the capacitance of the condenser, Q_o is the initial charge on the condenser, and R_{wo} is the initial wire resistance.

For an RLC circuit, it is known that

$$di/dt = -A(r_c + r)i - Bq \tag{2}$$

$$dq/dt = i \tag{3}$$

where $q = Q/Q_o$, $i = IT_o/Q_o$, $t = T/T_o$, $r_c = R_c/R_{wo}$, $A = R_{wo}T_o/L$, $B = T_o^2/CL$, Q is the charge on the condenser, I is the current in the circuit, T is the time after switch-on, T_o is a characteristic time, R_c is the circuit resistance, and L is the circuit inductance. In addition, the rate of the increase of the internal energy can be calculated by the following expression:

$$du/dt = Di^2r$$

where

$$D = CR_{wo}/T_o \tag{4}$$

Combining Eqs. (1) and (4), it is found that

$$d\ln r/dt = Di^2f(r) \tag{5}$$

Equations (2), (3), and (5) form the basic set of nonlinear circuit equations governing the initial behavior of an exploding wire. The initial conditions are: $i = 0$, $q = -1$, and $\ln r = 0$, at $t = 0$.

INTEGRAL CURVE IN THE LN $r - q$ PLANE

Multiplying Eq. (2) by i, combining with Eqs. (3) and (5), and integrating, it is found that

$$i^2 + B(q^2 - 1) + (2A/D)\,v = 0 \tag{6}$$

where

$$v(r) = u(r) + r_c \int_1^r d\xi/[\xi f(\xi)] \tag{7}$$

For linear variation of the wire resistance with the internal energy, f is a constant, and Eq. (7) reduces to the expression given by Bennett [4].

Combining Eqs. (3), (5), and (6), it will not be difficult to demonstrate that

$$d \ln r/dq = g(\ln r, q) \tag{8}$$

where

$$g(x, y) = Df(e^x)[B(1 - y^2) - (2A/D)\,v(e^x)]^{1/2} \tag{9}$$

The integral curve in the $\ln r - q$ plane,

$$q = G(\ln r) \tag{10}$$

of Eq. (8), passing through the initial point (0, -1), represents the nonlinear solution of the basic exploding wire circuit equations, Eqs. (2), (3), (5). Since $g(\ln r, q) = D'i$ is zero at ($\ln r = 0$, $q = -1$) this integral curve will be generated more conveniently with an asymptotic expression for $d \ln r/dq$ near the point (0, −1).

As an example, consider the case for the linear variation of r with u; i.e., for $f = f_o$ = constant. Equation (9) becomes

$$g(x, y) = D'[B(1 - y^2) - A'(r_c \ln x + e^x - 1)]^{1/2} \tag{11}$$

where $A' = 2A/D'$ and $D' = Df_o$. The integral curve for Eq. (11), passing through the point (0, -1), can be calculated by starting the integration with the asymptotic expressions developed in the next section. Using this integral curve, the circuit functions can be obtained successively, by first integrating Eq. (5) and then combining with Eqs. (6) and (10). To start the integration for $\ln r$, it will be again more convenient to use the asymptotic expression near $t = 0$ as developed in the next section.

MACLAURIN SERIES REPRESENTATION OF THE CIRCUIT FUNCTIONS FOR $f = f_o$

For linear variation of the wire resistance with the internal energy, Eq. (5) may be written as

$$dr/dt = D'i^2r \tag{12}$$

By differentiating Eq. (13) successively with respect to t, it will not be difficult to demonstrate that for $m \geq 2$,

$$\frac{d^m r}{dt^m} = 2D' \sum_{n=0}^{m-1} a_{m,n} \frac{d^n r}{dt^n} \sum_{k=0}^{K-1} (a_{m-n,k} + \bar{a}_{m-n,k}) \frac{d^k i}{dt^k} \cdot \frac{d^{m-n-k-1} i}{dt^{m-n-k-1}} \tag{13}$$

where

$$K = \begin{cases} (m-n+1)/2, \text{ if } (m-n) \text{ is odd} \\ (m-n)/2, \text{ if } (m-n) \text{ is even} \end{cases} \tag{14}$$

The coefficients $a_{m,n}$ and $\bar{a}_{j,k}$ satisfy the following conditions and recurrence relation:

$$a_{m,o} = a_{2,1} = a_{m,K} = 0$$

$$a_{m+1,n} = a_{m,n-1} + a_{m,n}$$

$$\bar{a}_{j,k} = \begin{cases} -a_{1,0}/2, \text{ if } j = 1 \text{ and } k = 0 \\ a_{j-1,k-1}, \text{ if } k = K-1 \text{ and } j \text{ is even} \\ 0, \text{ otherwise} \end{cases} \tag{15}$$

Evaluating the expression, Eq. (13), at $t = 0$ and writing out the coefficients $a_{m,n}$ and $\bar{a}_{j,k}$ explicitly, it is found that, for $m \geq 2$,

$$R_m = 2D' \sum_{n=0}^{m-1} \binom{m-1}{n} R_n \sum_{k=1}^{K-1} A_{m-n,k} \binom{m-n-1}{k} I_k I_{m-n-k-1} \tag{16}$$

where

$$R_m = (d^m r/dt^m)_{t=0} \quad \text{and} \quad I_m = (d^m i/dt^m)_{t=0} \tag{17}$$

$$A_{j,k} = \begin{cases} 1/2, \text{ if } k = K \text{ and } j \text{ is even} \\ 1, \text{ otherwise} \end{cases} \tag{18}$$

and $\binom{a}{b}$ is the binomial coefficient.

Differentiating Eq. (3) successively and using the same procedure as above, it can be shown that, for $m > 2$,

$$I_m = -Ar_c I_{m-1} - BI_{m-2} - A \sum_{n=1}^{m-1} \binom{m-1}{n} I_n R_{m-n-1} \tag{19}$$

Starting from $I_o = 0, R_o = 1, R_1 = 0, I_1 = B$, Eqs. (16) and (19) can be used to solve for I_m and R_m recurrently. The Maclaurin series expansion for i about $t = 0$ is then

$$i = \sum_{m=0}^{\infty} I_m t^m / m! \tag{20}$$

Since the wire resistance r varies exponentially with t^3 for small t, it will be more convenient to represent ln r (instead of r) in a Maclaurin series expansion. This can be accomplished by differentiating Eq. (5) successively for $f = f_o$. The result for $m \geq 2$ is

$$L_m = \left(\frac{d^m \ln r}{dt^m}\right)_{t=0} = 2D' \sum_{k=1}^{K-1} A_{m,k} \binom{m-1}{k} I_k I_{m-k-1} \tag{21}$$

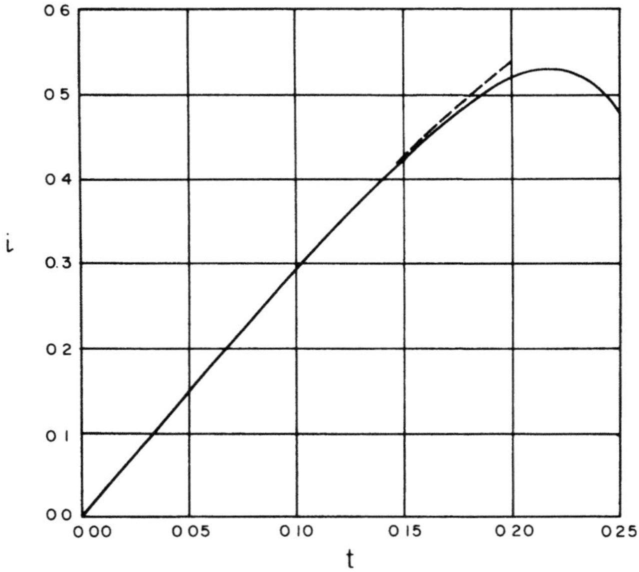

Fig. 1. i vs. t for $A = 0.4$, $B = 3.0$, $D' = 100$, $r_c = 0.1$, $f = f_o$. Asymptotic expansion ($M = 9$), – – –; numerical result, ———.

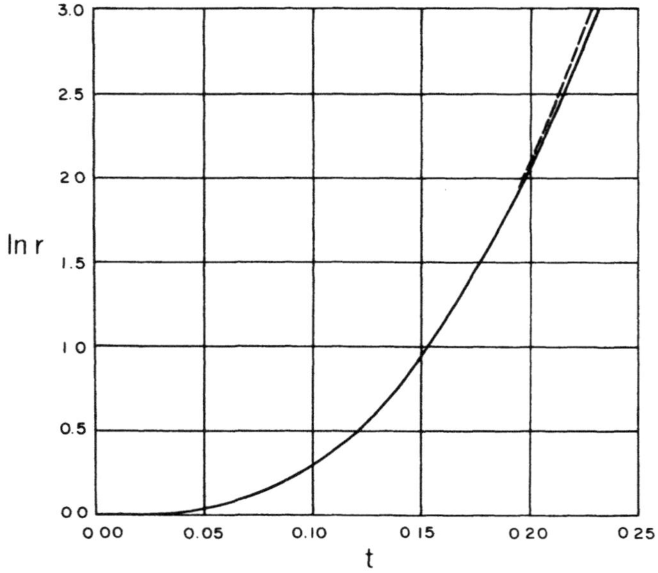

Fig. 2. ln r vs. t for $A = 0.4$, $B = 3.0$, $D' = 100$, $r_c = 0.1$, $f = f_o$. Asymptotic expansion ($M = 9$), – – –; numerical result, ———.

Since the l_m's are known, the values of L_m can be obtained recurrently from Eq. (21) with $L_o = L_1 = 0$. The Maclaurin series expansion for ln r is then

$$\ln r = \sum_{m=o}^{\infty} L_m t^m/m! \tag{22}$$

Summarizing, Eqs. (17), (19), and (21) represent the recurrence formulae from which the coefficients of the Maclaurin series expansions in Eqs. (20) and (21) for the circuit functions, i and ln r, can be generated exactly. Truncating these series at M (= a positive integer) terms, rational approximations of i and ln r for small time t are obtained. As examples, the truncated expressions for i and ln r, for $M = 6$, are given below:

$$i = B\{t - A(r_c + 1)t^2/2 + [A^2(r_c + 1)^2 - B]\, t^3/6 - A(r_c + 1)[A^2(r_c + 1)^2 - 2B]\, t^4/24 + [A^4(r_c + 1)^4 - 3A^2B(r_c + 1)^2 + B^2(1 - 8D')]\, t^5/120 - A(r_c + 1)[A^4(r_c + 1)^4 - 4A^2B(r_c + 1)^2 + 3B^2 - 58B^2D']\, t^6/720\} \tag{23}$$

$$\ln r = 2B^2D't^3\{1/6 - A(r_c + 1)\,t/8 + [7A^2(r_c + 1)^2 - 4B]\,t^2/120 - A(r_c + 1)[3A^2(r_c + 1)^2 - 4B]t^3/144\} \qquad (24)$$

Comparing these expressions, Eqs. (23) and (24), with those given by Bennett [4], it is noted that the differences set in "at and after" the "t^4" and "t^5" terms, for i and $\ln r$, respectively. As examples, the truncated expressions for i and $\ln r$ for the case of $A = 0.4$, $B = 3.0$, $D' = 100$, $r_c = 0.1$, and $M = 9$, are calculated and given in Figs. 1 and 2.

Using the truncated Maclaurin series as valid asymptotic expansions near $t = 0$, the integral curve $q = G(\ln r)$ defined by Eqs. (8), (10), and (11) can be calculated. In terms of $q = G(\ln r)$, the expression $\ln r$ as a function of t can be obtained by integrating Eq. (5) for $f = f_o$ in the form of a quadrature.

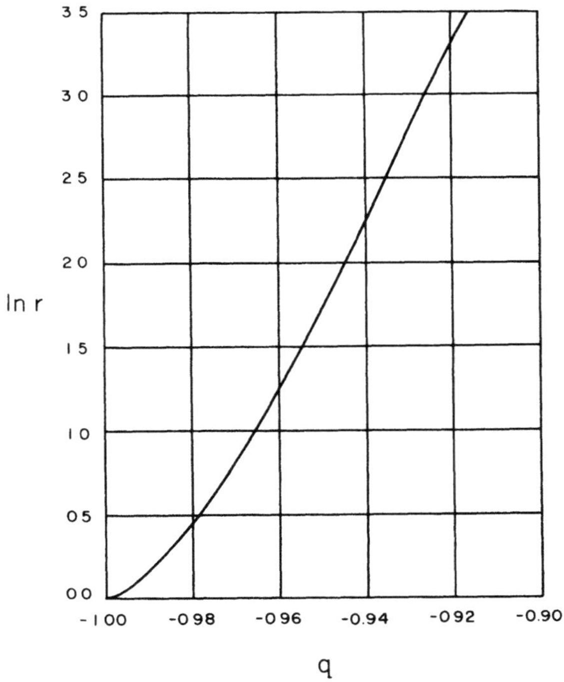

Fig. 3. $\ln r$ vs. q for $A = 0.4$, $B = 3.0$, $D' = 100$, $r_c = 0.1$, $f = f_o$.

$$t - t_1 = \int_{\ln r_1}^{\ln r} dx/\{BD'[1 - G^2(x)] - 2A(r_c \ln x + e^x - 1)\} \tag{25}$$

where (t_1, ln r_1) represents a point near $t = 0$ given by the asymptotic expansion for ln r. Using Eqs. (6), (25), and $q = G$ (ln r), the expressions $i(t)$ and $q(t)$ can be obtained algebraically. These circuit functions can, of course, be integrated numerically using the basic nonlinear differential equations without using the integral curve in the ln $r - q$ plane. To demonstrate the validity of the asymptotic expansions based on the truncated Maclaurin series, the numerically integrated results for i and ln r for $A = 0.4$, $B = 3.0$, $D' = 100$, $r_c = 0.01$, and $f = f_o$, are compared with the asymptotic results in Figs. 1 and 2. The corresponding curve relating ln r and q is given in Fig. 3.

REFERENCES

1. C.P. Nash and C.W. Olsen, Phys. Fluids 7:209 (1964).
2. A. Hobson and C.K. Manka, J. Appl. Phys. 37:1897 (1966).
3. F.D. Bennett, Phys. Fluids 7:147 (1964).
4. F.D. Bennett, Phys. Fluids 9:471 (1966).
5. G.W. Anderson and F.W. Nielsen, in Exploding Wires, Vol. 1, W.G. Chace and H.K. Moore, eds., Plenum Press, New York (1959), p. 97.
6. E. David, in Exploding Wires, Vol. 1, W.G. Chace and H.K. Moore, eds., Plenum Press, New York (1959), p. 271.

DISCUSSION

Question by F.D. Bennett, Ballistics Research Laboratory, Aberdeen Proving Ground, Md.: Is it fair to say that these approximate solutions will not be useful much beyond the linear portion of the current curve? In other words, can we go beyond i_{max} with the approximate solution?

Answer by Author: I feel that the Maclaurin series representations of the circuit functions are valid from $t = 0$ to the time at which the first maximum of the current occurs. The quadrature solution, however, is not limited by i_{max} .

Similarities in Spark and Exploding Wire Discharges

Ihor M. Vitkovitsky

Naval Research Laboratory
Washington, D. C.

Study of underwater sparks, the aim of which was better understanding of the breakdown phenomena in water and the development of parallel discharges, has underscored the similarity with exploding wire discharges. Energy conversion, initial dynamics, and pause phenomena in spark discharges are some of the many aspects that are common to both types of discharges. The electrical parameters that govern the behavior of spark discharges such as in the experiments with parallel discharges are relevant to studies of the exploding wires. Some of the applications of parallel discharges which extend the utility of the exploding wires (e.g., generation of strong simultaneous pressure pulses) will be indicated.

Study of underwater spark discharges, the aim of which is better understanding of the breakdown phenomena in liquids, has underscored the similarity with the exploding wire discharge phenomena. Such work is useful in the understanding of exploding wire physics as well as in the design of exploding wire experiments.

The most evident similarities are the effects of such discharges; i.e., dynamics of pressure pulse and generation of light. But more important are, for instance, the pause phenomena which occur in both types of discharges, although they are of a somewhat different type. These occur in exploding wires as the result of the change of the conductor into a dielectric (i.e., wires stop conducting [1]). In spark discharges these occur when the dielectric changes to a conductor [2].

Traditionally, the two fields—spark and exploding wire discharges—have remained separate. An attempt at crossbreeding of concepts from both fields is made here by applying some of the physical results of the high-voltage dielectric breakdown. These are also helpful in the design of the exploding wire experiments.

The "pause" in spark discharges is defined as the time between the application of breakdown voltage across the dielectric and the development of full discharge (i.e., discharge channel resistance becomes much smaller than the driving impedance of the voltage source). The pause consists of an inductive component τ_L, given by the ratio of the channel inductance and circuit impedance, and the resistive component τ_R, defined [2] as

$$\tau_R = \frac{5}{Z_o^{1/3} F^{4/3}} \text{ (nsec)} \tag{1}$$

Equation (1) is defined in convenient engineering units: Z_o, the impedance of the generator, is in ohms; F, the electric field at the surface, is in MV/cm. It holds for liquids and solids and has been extended to gases by use of the root of the density ratio as a multiplying factor. This equation can be derived assuming that the power input into the discharge channel determines the expansion rate and the channel temperature through the ideal gas law and Spitzer resistivity.

Since similar gross considerations govern the explosions of conducting wires, it is instructive to apply Eq. (1) to the estimate of the length of the time needed for the full development of the discharge.

F. W. Webb et al. [3] have shown in their work that the discharge of sufficient energy into the wire initially draws full current from the storage system. As energy is deposited into the wire, the current is limited as if the wire were changing from conductor to a nonconductor. At this time, the state of the exploding wire conduction is analogous to the limited current flow in the dielectric breakdown, before the full current flow develops. This period is given by Eq. (1). The field strength F and the impedance of the generator in the experiments described in [3] are approximately 0.024 MV/cm and 15 ohms, respectively. Using these values in Eq. (1) to predict the duration of the limited current flow period, a 3×10^{-7} sec period is obtained. This compares with an observed period of about 6×10^{-7} sec measured in Al and Pt wire explosions. The discrepancy is due to the high density of the exploded wires relative to the density of solid dielectrics for which Eq. (1) has been developed.

Similarly good predictions are made on the duration of the limited current flow in the experiments where much higher voltage and stored energy were used, as in the NRL exploding wire study

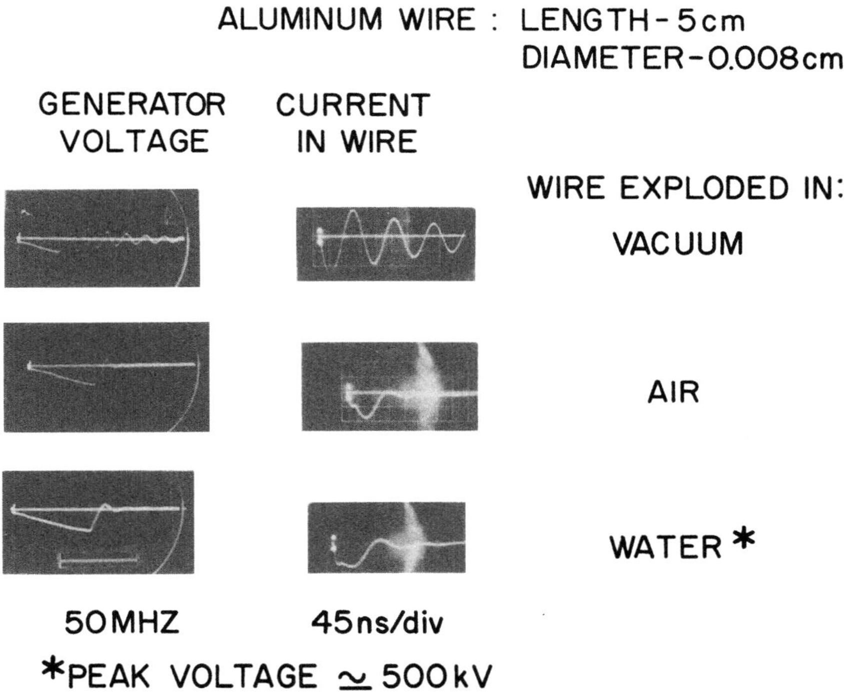

Fig. 1. Experimental current and voltage waveforms characteristic of wire explosions utilizing very high voltage energy storage system. Time scale of 50 MHz shown in bottom frame.

[4]. Figure 1 shows voltage and current waveforms of wire explosions occurring in vacuum, air, and water. The undamped waveform is a result of discharges around the wire when explosion chamber wall flashover occurs. Explosions in air show a damped discharge, indicating that most of the current flows through the freely expanding wire. The last case is that of the underwater explosion. Here, water restrains the free expansion of the discharge and the exploding wire serves as the discharge channel. This is very similar to channels in water breakdown, whose behavior is described by Eq. (1). The electric field across the wire is 0.1 MV/cm and the source impedance is 1.3 ohms. This indicates that τ_L is again negligible, while τ_R is $\sim$100 nsec. The observed discharge decays in less than 100 nsec with current flow of about one-third of the short-circuit current; i.e., this is a case where, due to energy storage limitation, full breakdown of the channel never occurs.

These examples show that a scaling law developed in the electric breakdown studies is also applicable to the exploding wire discharges. Other such laws can be transformed into useful criteria in the exploding wire field. In the study of liquid breakdown it has been established [2] that breakdown strength has a time-dependent form

$$F_{\max} t^{1/n} = \text{constant} \tag{2}$$

It determines the maximum voltage that can be applied across a liquid for t seconds. The value of the exponent depends on the liquid and, in the case of water, $n = 3$. The constant has a weak dependence on the electrode area. The constant is 0.3 for water if F is in MV/cm and the stressed area is small (10^{-2}cm^2). The scaling law given by Eq. (2) can be applied to exploding wires when they undergo the nonconducting phase ("pause phenomena"), as in studies of C. P. Nash and C. W. Olsen [1]; i.e., where "pausing" wire can be thought of as a dielectric with the voltage applied across it. In Fig. 3 of their work, the effect of an electric field on time-to-burst is given. The mean change in time is, fortuitously, between 1.2 and 1.3, which closely corresponds to that predicted for water by Eq. (2) for the change of factor of two in an applied field. The value of the constant is considerably less than that for water; i.e., the exploding wire during pause, perhaps not surprisingly, is not a good "dielectric" (since any decrease in local current density leads to cooling which reverts sections of the wire back to the conducting state).

The question arises whether other aspects of "pausing" wire breakdown are similar to that in dielectric breakdowns. Figure 2 shows several stages of water breakdowns. It has been shown by many authors, specifically in [2], that streamers originating at either electrode surface are the initiators of the breakdown. Some that have been arrested by nearby discharge are seen on the rounded electrode in Fig. 2 (frame 3). That, at least qualitatively, similar behavior exists in exploding wires even when they have not undergone a drastic resistance change can be seen in the study of Webb et al. (Fig. 19 of [5]). Other similarities between the phenomena are the dynamics of the discharge channels, as again shown in Fig. 2.

It has been shown here that some scaling laws of the dielectric breakdown apply to certain aspects of the exploding wire phenomena; e.g., it is possible to predict the history of the currents in the wire.

These same laws can be used to optimize the design of the exploding wire experiments. In attempts to transfer large electrical power into the wire, storage system switching, which also is governed by scaling laws like that of Eq. (1), must be compatible with the desired circuit performance. A hypothetical 10 kV storage

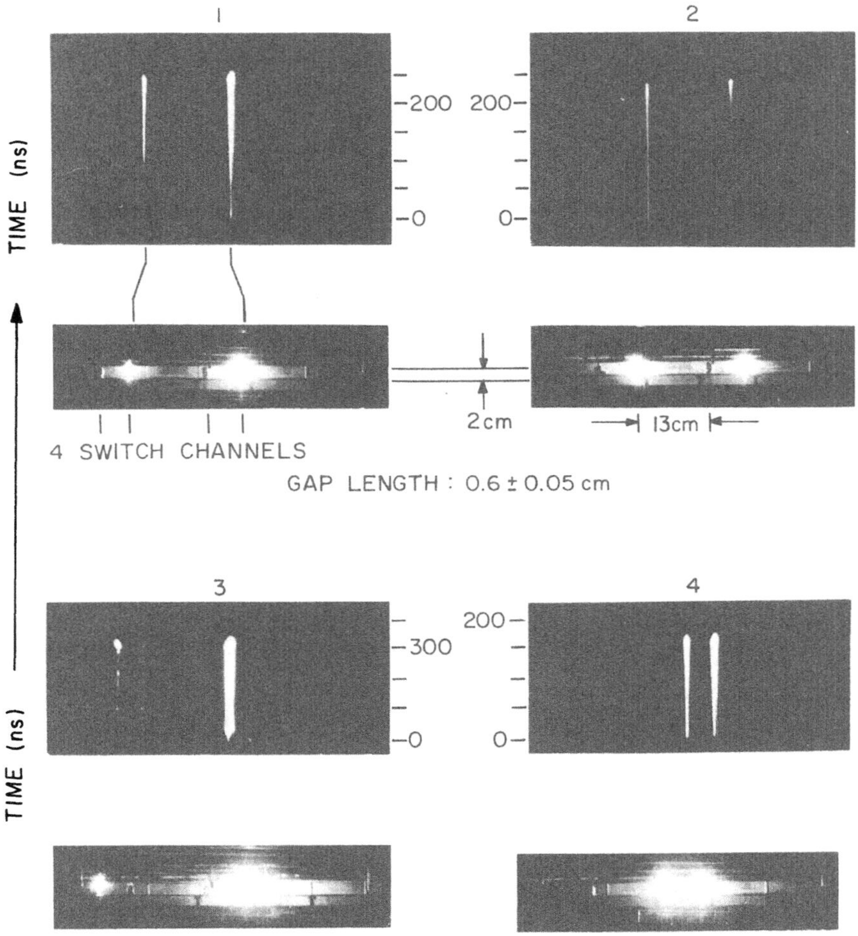

Fig. 2. Time-integrated and streak photographs of dielectric breakdown channels. Non-simultaneous (frames 1 and 2) and simultaneous (frame 3) parallel breakdowns are evident. Partial closing (streamers on the knob electrode) and evident in frame 3.

system consisting of a 1-ohm generator according to Eq. (1) would have its current rise limited to a minimum of 25, 12, or 5×10^{-8} sec when the switching uses air (1 atm), oil, or polyethelene, respectively.

However, broader study of exploding wires and their distinct characteristics is critical in determining the extent of the application of the above scaling law examples and formulation of new relations. Such work would lead to more systematic reporting of experiments for which no guidelines seem to exist at present.

REFERENCES

1. C.P. Nash and C.W. Olsen, "Factors Affecting the Time to Burst in Exploding Wires," Exploding Wires, Vol. 2, W.G. Chace and H.K. Moore, eds., Plenum Press, New York, 1962.
2. J.C. Martin, Cornell University Seminar on High Voltage Technology, Nov. 1966.
3. F.H. Webb, N. Chase, M. Ernstene, and A.V. Tollestrup, "Submicrosecond Wire Explosion Studies at Electro-Optical Systems, Inc.," Exploding Wires, Vol. 1, W.G. Chace and H.K. Moore, eds., Plenum Press, New York, 1959.
4. I.M. Vitkovitsky, P.P. Bey, W.R. Faust, R. Fulper, Jr., G.E. Leavitt, and J.D. Shipman, Jr., "Exploding Wires as Source of X-rays," Exploding Wires, Vol. 2, W.G. Chace and H.K. Moore, eds., Plenum Press, New York, 1962.
5. F.H. Webb, Jr., H.H. Hilton, P.H. Levine, and A.V. Tollestrup, "The Electrical and Optical Properties of Electrically Exploded Wires," Exploding Wires, Vol. 2, W.G. Chace and H.K. Moore, eds., Plenum Press, New York, 1962.

Exploding Wires as a Light Source for Quantitative Spectroscopy

H. Jäger and W. Lochte-Holtgreven

Kiel University
West Germany

Time-resolved investigations of the light emitted and absorbed during the explosion of wires reveal in most cases a plasma cloud which is optically thick and which has no radial symmetry. It is shown that two special types of wire explosion can be found in which the cloud is optically thin and of radial symmetry: an axial discharge within a widely expanded metal vapor cloud and a peripheral discharge evaporating only little of the wire's surface. Only in these cases can the methods of quantitative spectroscopy be applied and a determination of the radial distribution of the plasma parameters be made using Abel's integral equation.

INTRODUCTION

The spectroscopic evaluation of atomic constants such as transition probabilities or absorption cross sections requires a light source in L.T.E. with a known symmetrical temperature and density distribution which emits radiation out of an optically thin layer. One may use electric arcs for this purpose, which answer the above requirements fairly well. However, the temperatures are restricted to 10,000 to 20,000°K. Besides, an arc plasma of definite composition can generally be obtained only by feeding gaseous substances into the arc. A considerable number of elements, e.g., the alkalies, the alkaline earths, and many of the other metals have no suitable simple volatile or gaseous compounds. For these, an investigation using the arc technique is almost impossible. However, some of these latter substances can easily be drawn into thin wires. Thus, studies were made about the optical properties of exploding wire plasmas with respect to the above requirements.

From the different types of wire explosions only those which produce a plasma column symmetrical with respect to the axis of the wire allow application of the methods of quantitative spectroscopical analysis. But even in these cases the radial symmetry is observed only in the early stages of the explosion, when the optical thickness of the plasma's column is higher than 5. Thus, measurements of line intensities and line shapes are considerably handicapped. In the later stages, when the plasma cloud has expanded sufficiently to become transparent, the symmetry is not well reproducible [1]. The lack of symmetry results from the oscillating discharge, and it may be stated that, using an oscillating circuit, symmetry of the plasma cloud can be expected at best until current reverse. Thus it is necessary to evaporate the wire's material sufficiently before current reversal so that the plasma obtained is optically thin.

The different types of wire explosion were studied carefully. For our investigations we used circuits of rather slow ringing frequency (T = 100 to 200 μsec, C = 21 μF, V_o = 20 kV) because they offer the possibility of photographing different stages of the explosion under nearly constant conditions. The exposure time (5 to 15 μsec) was determined by the necessity of comparing the obtained spectra with the standard carbon arc. With low ringing frequency, on the other hand, it is difficult to obtain radial symmetry even in the first stages of the explosion.

PROPERTIES OF WIRE EXPLOSIONS WITH LOW RINGING FREQUENCY

Two special types of wire explosion seemed to respond to our requirements: the explosion with reignition along the axis and the explosion starting with symmetrical breakdown along the wire's surface leading to a peripheral discharge. Before going into detail, some remarks should be made about these two types of discharges.

Let us assume a thin wire of good electrical conductivity and an electric energy which is a multiple of the energy required for evaporation. With these conditions, the first stage of the expanding plasma cloud usually is symmetrical when the relation between the applied voltage and the length of the wire is sufficiently high. We assume further a surrounding gas of relatively high breakdown voltage, e.g., air at normal conditions. In case the wire is long enough so that no breakdown can occur, the evaporated material of the wire expands to a volume depending mainly on the pressure of the surrounding gas. The expanded cloud lasts for a time of more than 100 μsec at constant diameter. Short-time photography of this cloud using back light shows strong absorption but a zone of lower pressure along the axis.

If the length of the wire is shortened, with conditions otherwise constant, breakdown along the axis is obtained, if the conductivity of the wire is sufficiently high. With pressure of the surrounding gas about 1 atm, the length of the wire needed to obtain internal breakdown is very critical indeed (Δl = a few tenths of a millimeter, or even less). For good reproducibility, wires with very clean surfaces and pure dry gases have to be used. In any case, the plasma column is optically thick and subject to rapid alterations of the symmetry.

A further decrease in the length of the wire renders an absolutely different type of discharge, starting along the surface of the

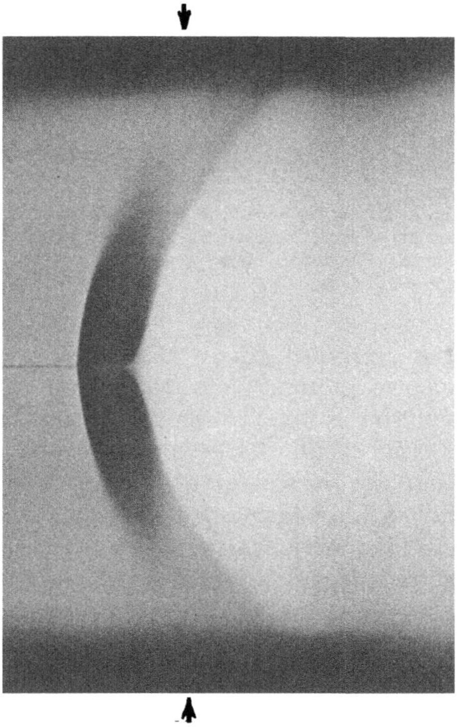

Fig. 1. Streak photograph of wire explosion showing axial breakdown in a widely expanded vapor cloud. Back light: xenon flash lamp. Cu wire: diameter 50 μ, length 10 cm. Circuit: C = 21 μF, U_o = 20 kV, T = 120 μsec. Surrounding gas: air, 266 torr.

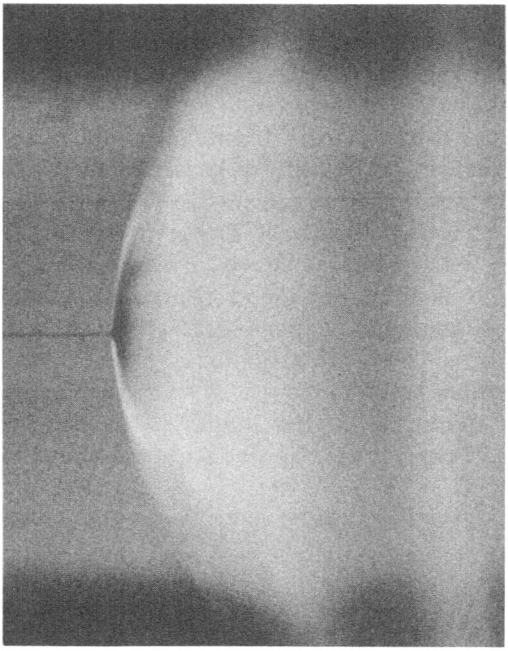

Fig. 2. Streak photograph of wire explosion in transition stage from internal discharge to external discharge. Conditions same as in Fig. 1, except pressure = 190 torr.

not yet noticeably expanded vapor cloud. A streak exposure reveals the well-known picture from explosions of short wires when small-time resolution is used. Shortly after the beginning of the explosion, the plasma column is no longer stable.

It appears that the pressure of the surrounding gas is too high to allow symmetrical discharges for an extended time. This holds in case of axial breakdown just as in the case of surface discharges.

Similar results can be obtained with a well-defined constant length of wire, the electrical conditions chosen as before, but with systematic variation of the pressure of the surrounding gas. Using, for example, a copper wire 10 cm in length and 50 μ in diameter and electrical conditions as indicated above, we obtain at 1-atm pressure of the surrounding gas an absorbing vapor cloud showing no breakdown.* With decreasing external pressure, the diameter

*Only a weak glow discharge may be observed in the axial channel.

of the vapor cloud increases rapidly. The axial zone of diminished pressure is likewise broadened. At a very definite pressure of the surrounding air (266 torr), internal breakdown is observed within the zone of diminished pressure (Fig. 1). With a further decrease of pressure, the breakdown starts at an earlier time.

However, with decreasing external pressure, the breakdown potential for surface discharges decreases likewise. This discharge may be caused partially by lower pressure at the surface of the expanding cloud and partially by properties of the wire's surface with respect to the emission of charge carriers. Under the conditions chosen, the internal discharge changes into the external discharge when the pressure is further decreased. Figure 2 shows the transition stage.

Lowering the pressure further, an ever-increasing part of the discharge current flows through the peripheral zone and the depth

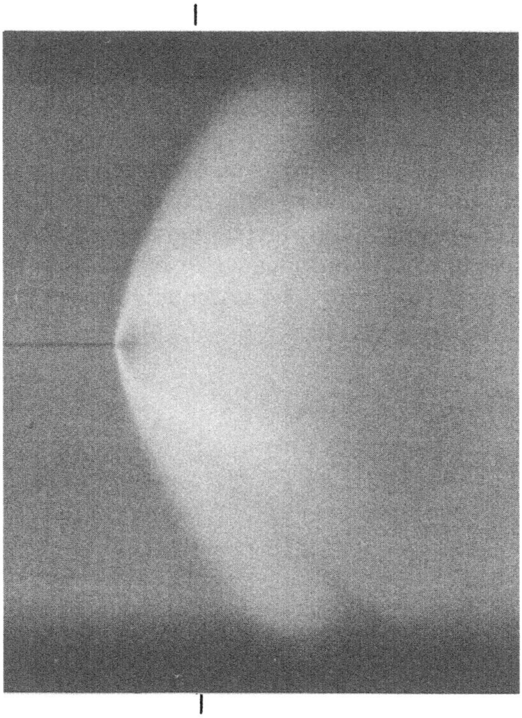

Fig. 3. Streak photograph of wire explosion with predominant external discharge. Conditions same as in Fig. 1 and Fig. 2, except surrounding gas is argon at 60 torr. In air or N_2 as surrounding gas, this type of discharge appears at a pressure of about 30 torr.

of penetration is diminished. The material of the wire itself is heated less and less (Fig. 3). Now evaporation of the wire is effected from the surface toward the axis. At pressures of about 10 torr the wire is practically unaffected, at least during the first half period, and the entire electrical energy flows within the surrounding gas.

From this, the types of explosion applicable for quantitative spectroscopy become apparent: on the one hand, the axial breakdown in a widely expanded cloud; and, on the other hand, a peripheral discharge evaporating only little of the wire's substance. In both cases we may expect radiation from an optically thin layer.

We shall now discuss the two typical cases separately.

AXIAL DISCHARGE

As was shown above, the length of the wire, the surrounding gas and its pressure, and the electrical parameters of the circuit can be chosen such that the wire is strongly evaporated while the zone of diminished pressure along the axis is still preserved. The axial discharge starts after the vapor cloud is widely expanded to about 18 mm in diameter. From thorough investigations it has been found that, in this case, the radiation is emitted from an optically thin layer. However, quantitative measurements cannot be made until the evaporated nonluminous vapor cloud surrounding the plasma column is sufficiently thin. The appropriate time can be determined

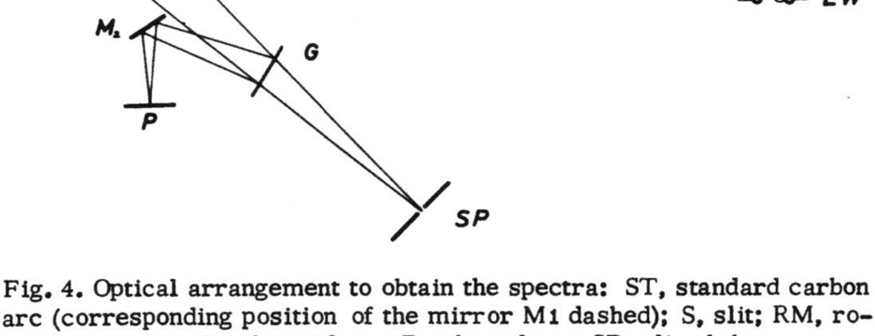

Fig. 4. Optical arrangement to obtain the spectra: ST, standard carbon arc (corresponding position of the mirror M1 dashed); S, slit; RM, rotating mirror; G, glass plate; P, photoplate; SP, slit of the spectrograph.

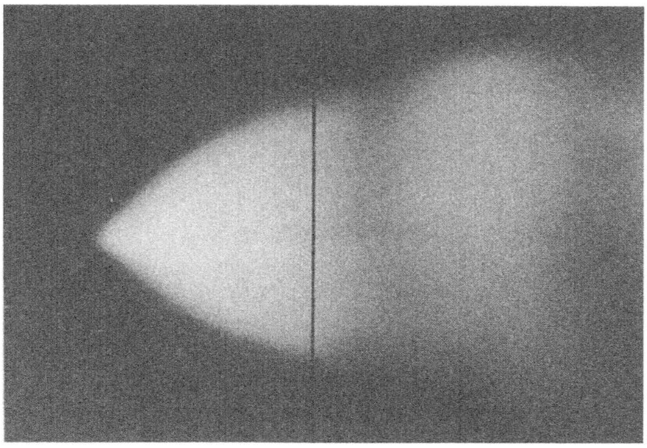

Fig. 5. Streak photograph obtained with arrangement of Fig. 4 indicating moment when exposure of a spectrum was taken. Dark line is shadow of a wire set directly before photoplate indicating position of slit of spectrograph.

from streak exposures using back light of different wavelengths controlling the absorption conditions of the vapor cloud. Later on absorption is noticeable only in the resonance lines.

In Fig. 4 we see the arrangement for obtaining spectra. Part of the light is reflected in front of the slit of the spectrograph by a glass plate. This light is used to obtain a streak photograph. A fixed mark serves to indicate the position of the slit on the streak exposure, indicating the time of the exposure on the streak photograph (Fig. 5). The explosion is triggered from a photocell using an auxiliary light beam via a second mirror attached on the revolving mirror (not shown in Fig. 5). Exposures corresponding to the conditions of Fig. 1 taken at different stages during the first light pulse rendered spectra of copper I and copper II lines. From these spectra it became evident that the light source had radial symmetry. Except for the Cu I λ 5153, 5218, and 4651 lines, all lines had top intensities of less than $^1/_{10}$ of the Planck curve corresponding to the measured radiation temperature.

Using Abel's integral equation, the temperature variation radial to the plasma column was obtained from relative intensity measurements of different pairs of Cu I lines. In addition, from the width of the Cu I λ 4540 line the radial distribution of the electron density was obtained (Fig. 6). In previous experiments [1] using exploding Cu/Mg wires, the width of this line dependent upon the electron density was measured against magnesium line widths calculated by Griem [2]. The example given in Fig. 6 shows

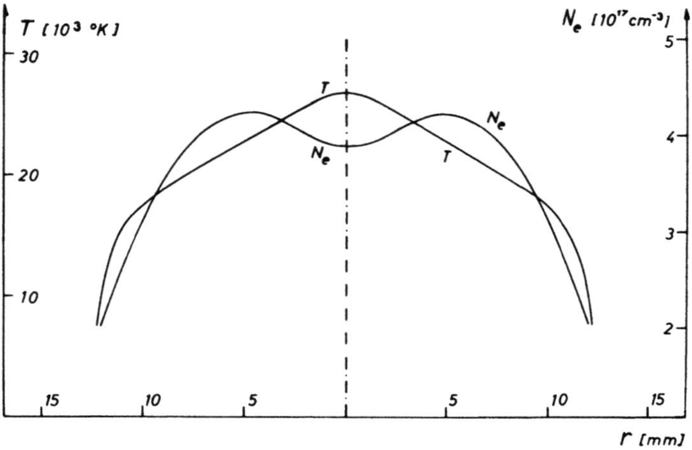

Fig. 6. Radial distribution of temperature and electron density in the plasma column of an axial discharge in a widely expanded metal vapor cylinder. The time of exposure is marked on the streak photograph (Fig. 1).

that the electron density decreases noticeably toward the center, indicating a lower pressure at the axis even at a late time.

It should be borne in mind that only the axial discharge is suited for quantitative investigation under the conditions indicated above.

Under different conditions another type of internal discharge has been observed, starting along cylindrical casings of lower pressure within the vapor cloud. In our streak photographs the zones did show up as areas of less absorption before current breakdown. The internal discharges within these cylinder casings sometimes have helical structure and are not suited for quantitative spectroscopy because of lack of symmetry.

EXTERNAL DISCHARGE

As mentioned before, the condition of the surface is important for the time of surface breakdown. Some oxide layer or a surface layer of an easily evaporated substance facilitates the appearance of charge carriers. Likewise, a small heat of evaporation of the wire's material facilitates formation of a surface discharge which is predominant if the electrical resistance of the wire is relatively high.

This is demonstrated impressively by comparing the explosion of a copper wire with an explosion of a mercury jet of exactly the same geometrical dimensions; e.g., diameter 0.3 mm and length 4 cm in a circuit of $T = 22$ μsec, $U_0 = 24$ kV, $C = 21$ μF ($E \approx 6$ kJ). The energy necessary for complete evaporation of the copper wire is 152 J, while for the mercury jet only 13 J are needed. In spite of this, the copper wire is entirely evaporated; the mercury jet not entirely. A solution of lithium in liquid ammonia has about the same resistivity as mercury, but the vapor pressure is much higher; at room temperature it is about 100 torr. Accordingly, an exploded jet from this substance leads to an even more pronounced surface discharge. It is possible to increase the resistance and the vapor pressure of this jet by lowering the concentration of lithium in the liquid. Accordingly, less and less energy is consumed by the jet and more energy is passed along the surface without ever affecting an appreciable amount of the liquid [3].

The surface discharges are generally of good radial symmetry up to the time of current reversal. This holds for substances of low vapor pressure and small pressure of the surrounding gas but just as well for substances with high vapor pressure and correspondingly high breakdown potential of the surrounding gas.

The spectrum of an explosion with predominant external discharge shows—distinctly separated—an outer plasma cylinder emitting lines of the surrounding gas and an inner cylinder emitting lines of the wire's material. (See the two different plasma cones in Fig. 3.) Temperature and electron density are below the values measured in the case of an axial discharge.

The radial distribution of temperature and electron density shows, without doubt, that the heating mechanism works from outside. The axial zone is distinctly colder (Fig. 7). The electron density and, thus, the pressure increase from without toward the surface of the unevaporated material. Optical investigation inside a certain radius is practically impossible because of absorption.

Similar distributions of T and n_e are found with poor electrical conductors of high vapor pressure. Here the cold zone is almost of the dimension of the original jet. This can be proved easily with jets of electrolytes. In this case, the distribution of n_e and T at later stages is almost rectangular, which seems important from the spectroscopic point of view since the measured line intensities and line shapes can be evaluated without the troublesome procedure of the Abel inversion.

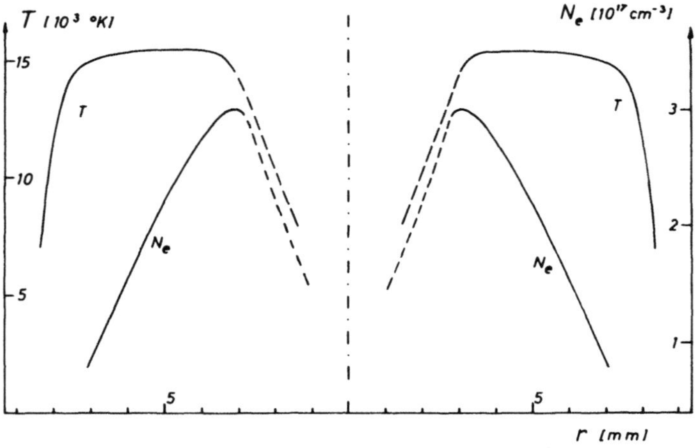

Fig. 7. Radial distribution of temperature and electron density in case of explosion with predominant external discharge. Conditions same as in Fig. 3. The time of exposure is marked there.

At an earlier time, the temperature increases toward the axis while the electron density distribution is still unchanged. This offers the possibility of investigating the plasma at different temperatures but at the same value of the electron density.

In conclusion, it seems possible to use exploding wire technique successfully in quantitative plasma analysis. Therewith a large number of elements is available for investigation, and it even appears possible that the range of temperatures known from arc research can be extended considerably.

REFERENCES

1. R. Krüger, Ph.D. Thesis, Kiel University, 1966.
2. H. Griem, Plasma Spectroscopy, McGraw-Hill Book Co., New York, 1964.
3. H. Jäger and W. Lochte-Holtgreven, Z. Physik 198:351 (1967).

An "Electrical Equation of State" of Metals Determined by an Exploding Wire Technique

W. G. Chace,* M. A. Levine, and C. V. Fish

Air Force Cambridge Research Laboratories
Bedford, Massachusetts

Using pulsed x rays and sheath suppressing methods, quantitative measurements have been made of density, conductivity, energy content, and metal vapor velocity in wire explosions. From these data an "electrical equation of state" can be calculated. This relation between physical density and electrical parameters such as conductivity can be extended through the critical region. The methods used and the calculations involved are described and typical results are presented for copper and tin.

INTRODUCTION

Several authors have established theoretical relations between resistivity and state variables for high-temperature metallic conductors. These predictions extend into the region of temperature and pressure to be found in exploding wires [1–6]. When quantitative measurements testing these theories were carried out on wire explosions, however, the results were not in good agreement with the theory. Examination of these experimental studies shows that in each case one or more possible problems existed, either in the conditions under which the wires were exploded, or in the way in which the state variables were measured. In the present work, a new set of measurements of resistivity versus density are presented which have been taken under improved conditions. Thus, in the experiments reported here, care was taken to explode the wires rapidly so that instabilities [7] were not a problem, and sheath [8, 9] formation was repressed by encapsulation [10]. To improve the accuracy of measurement of the state variables, x rays were used for all determinations of density data.

*Present address: Chace Associates, Punta Gorda, Florida.

These techniques seem to have been successful, for the results show both self-consistency and agreement with previous static results and with theory.

EXPERIMENTAL METHOD

The experimental method fell into three categories: a) x-ray methods, b) methods to eliminate the effect of the sheath, and c) measurements.

X-ray Methods

The equipment delivered a 30-nanosecond pulse from a 1-mm source with accelerating voltages variable between 75 and 105 kV. The x-ray intensity within this range of voltages was sufficient to produce good exposure of high-speed screen-type x-ray film. The cassette placed the film 5.5 cm from the wire. The x-ray source was 36 cm from the wire (Fig. 1). Although wire explosions are highly reproducible, it was felt that the very small time intervals required to follow the wire during the burst could not be recorded successfully on successive shots. For this reason, two x-ray tubes were used, placed at approximately a 45° angle. This gave two shadowgraphs on a single film (Fig. 2). The x-ray equipment allowed these two shots of the same event to be spaced at any time interval from several microseconds down to about 50 nanoseconds and the exact time of each x-ray shot to be recorded on the oscilloscope traces.

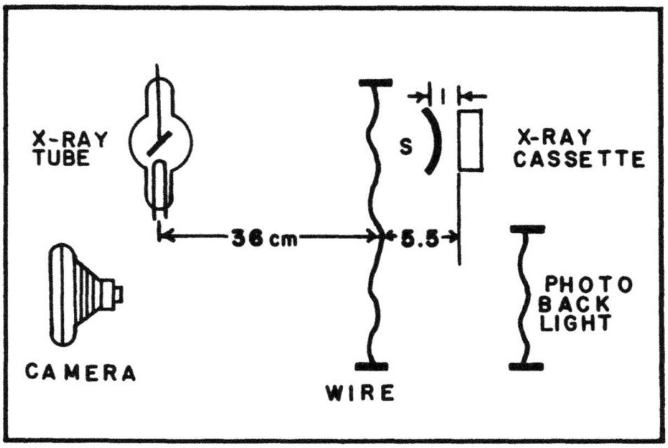

Fig. 1. Schematic equipment arrangement for x-ray study. "S" is blast shield of polyethylene to protect film from shock wave.

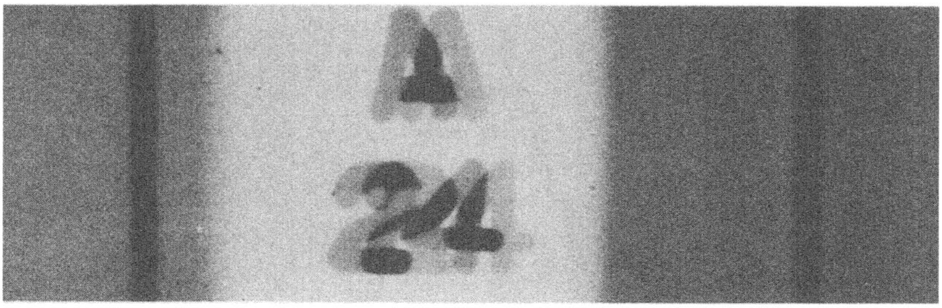

Fig. 2. Two-gun x ray. Right-hand shot taken 0.1 μsec later than left-hand.

The shadow was scanned with a recording microdensitometer and an arbitrary figure l, proportional to diameter, arrived at as shown in Fig. 3.

Encapsulation

Having established by comparison of x-ray studies and high-speed photographs that a luminous sheath existed around an exploding wire [10], it became necessary to examine the effect of such a sheath in exploding wire study.

Since the sheath becomes completely opaque to any back-lighting source known to the authors, and since the rate of expansion of the sheath differs from that of the wire material, it is necessary to make all determinations of wire shape and size by x rays.

Further, it seemed most probable that such a luminous discharge would be conducting. To eliminate this possibility, some means must be found either to eliminate the sheath or to measure its conductivity. A means of accomplishing the first of these alternatives was found in the encapsulation of the wire in a plastic of high dielectric strength (Fig. 4). Details of the method have been described in previous publications [10, 11].

Measurements

The primary electrical measurements were di/dt and initial voltage on the capacitor bank. The former was measured by a pick-up coil method as previously described [12], and the latter with a digital voltmeter connected to the capacitors through a 10:1 resistor voltage divider. This voltmeter is equipped with a locking device so that the voltage recorded at the instant of firing is locked in and may be read with an accuracy of better than ½%. Current was determined from the di/dt by an RC integrator. All other electrical quantities were calculated from these measurements and

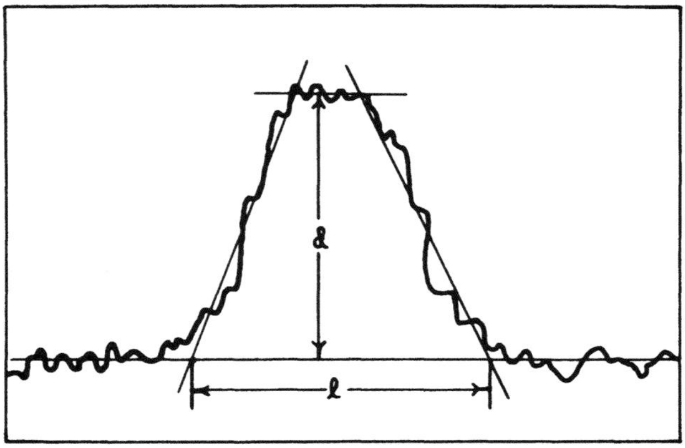

Fig. 3. Method of determining "l" dimension from microdensitometer tracing.

the constants of the discharge system: capacitance = 87.5 μF; inductance, not including the sample, = 9.6×10^{-8} H; resistance, not including the wire, 3.3×10^{-4} ohms; ringing frequency = 54 kHz.

Direct measurement of the voltage across an exploding wire is extremely difficult to accomplish and equally difficult to defend. Its calculation to any degree of accuracy is, however, a simple matter with a computer.

The equivalent circuit of Fig. 5 was used to set up a Kirchhoff's law voltage equation

$$E_w = E_c - I R_{eq} - L\,(dI/dt) - I(dL/dt) \tag{1}$$

E_w is the required voltage across the wire; E_c, the voltage across the capacitor, is

$$E_c = E_o - \left(\frac{1}{C}\right)\int I dt \tag{2}$$

E_o is the measured value of the original capacitor voltage, C the capacity of the bank and I the current. L, the total inductance is

$$L = L_{eq} + L_w \tag{3}$$

where L_{eq} is inductance of the circuit and L_w is the calculated self-inductance of the wire. From this the voltage at any time may be calculated.

Measurements made on the wire were (1) original length, (2) original diameter, (3) "l" dimension, original, and on each x ray. From these and the handbook value of the density, all wire characteristics were calculated.

CALCULATIONS

The calculations are based on the following assumptions: a) the length of the wire does not change during the explosion; b) the diameter of the wire is uniform along the axis during the time of interest, i.e., until after the restrike; c) the density remains constant along the radius during the time of interest; and d) skin effect is negligible.

Photographs of all but the most slowly exploded wires indicates that (a) is a reasonable assumption.

A large number of x rays of fast [7] exploding wire have shown only a uniform cylinder, indicating that (b) is a reasonable assumption in our experiment. Likewise, all our x-ray readings indicate that, to a first approximation, (c) is true.

Since the wires are 0.05 cm in diameter or less, a 90% penetration occurs in less than 1 μsec in copper as calculated by the formula of Langworthy et al. [13].

Resistance of the wire was calculated from Ohm's law. The current was read at the time each x ray was taken. Voltage was

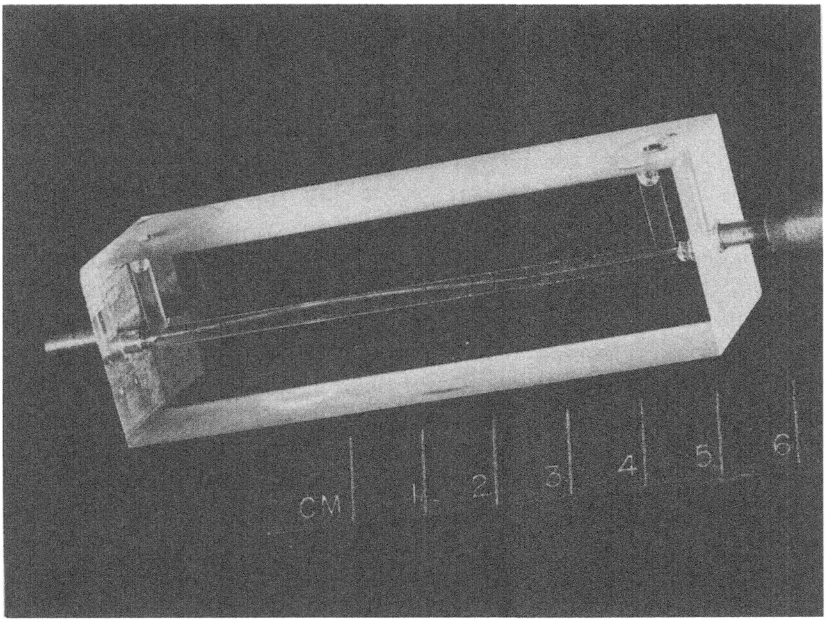

Fig. 4. Wire encapsulated in plastic. The block is Plexiglas; epoxy resin fills hole around wire.

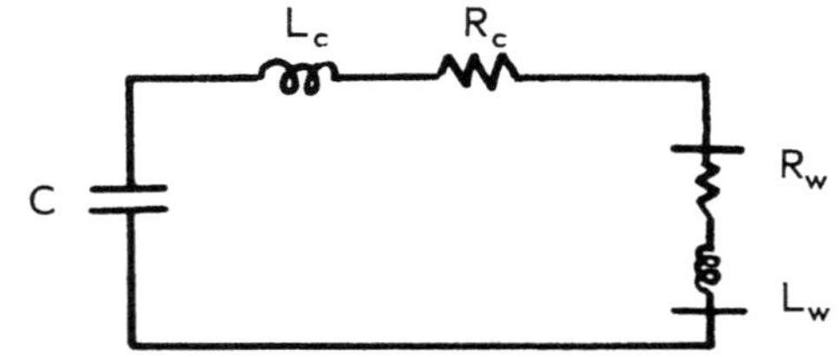

Fig. 5. Equivalent circuit of wire explosion. L_c is the capacitor bank, R is the total inductance of discharge circuit except wire, R_c is the total resistance of discharge circuit except wire, R_w is the resistance of wire sample being exploded, and L_w is the inductance of wire sample.

determined as indicated above, and from these were calculated the resistance.

Density was determined by determining change in volume from change in "l" dimension. Using the handbook value of density as ρ_o, $\rho(t)$ was calculated by simple ratio of volumes at t_o and t.

RESULTS

The results of the application of the method described above to several situations are shown in the graphs below. Figure 6, which is for a #22 copper wire 7.5 cm long and exploded at 12 kV, shows the density change of the wire material with time. Here there are three, possibly four, regions, identifiable as solid (I), liquid (II), gas (III), and dwell (IV). The very slight break between liquid and gas leads one to question whether the material is not actually near or at the "critical point" in its phase diagram.

Figure 7 is a resistivity curve for the same series of copper wire explosions and again shows the four regions. Comparison with handbook figures shows excellent agreement between region I and data for solid copper. Figures for the liquid are available for only a very small range of densities; however, for the limited range given in Landolt–Börnstein Tabellen [14] the agreement in value and especially in slope is excellent.

As pointed out in [4], it is in Region III that the results most sharply disagree with theory. In this region, the absolute values of resistivity approach those of a Lorentz gas only in the low-density region. The higher-density material with resistivity in the 10^{-4} to

10^{-3} ohm-cm is orders of magnitude removed from the value calculated from Spitzer's formula [19].

The triangle is the handbook value of resistivity at density of the liquid at the melting point, and the square is the point at which Herzfeld's molecular refractivity relation [6] would indicate a change from metallic to nonmetallic properties for copper.

Further evidence that the curve truly represents the resistivity–density relation is given by a comparison between the results of this experiment in the liquid region and a static experiment reported by Kikoin et al. [9] for mercury (Fig. 8).

Figures 9 and 10 are results of a series of explosions of encapsulated tin (0.102 cm in diameter and 7.5 cm long) also exploded at 12 kV. Here there are only two obvious regions in the resistivity

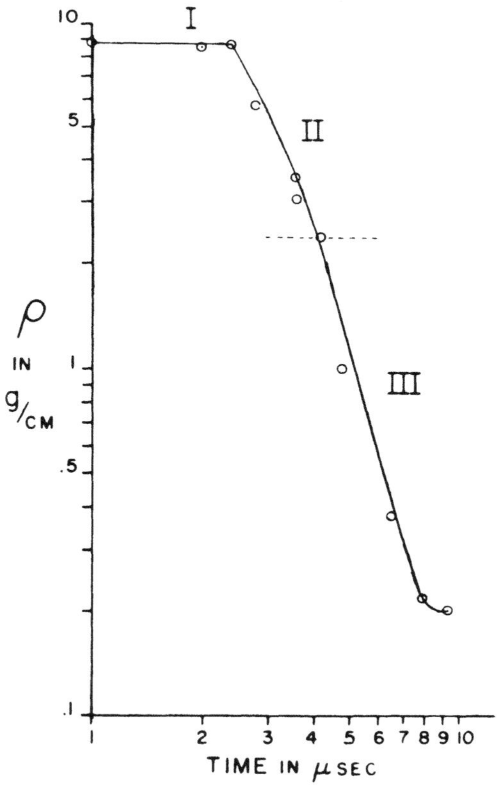

Fig. 6. Density–time relation of 0.0644-cm copper wire 7.5 cm long exploded with 12 kV. (I) Solid region, (II) "liquid" region, (III) "gas" region.

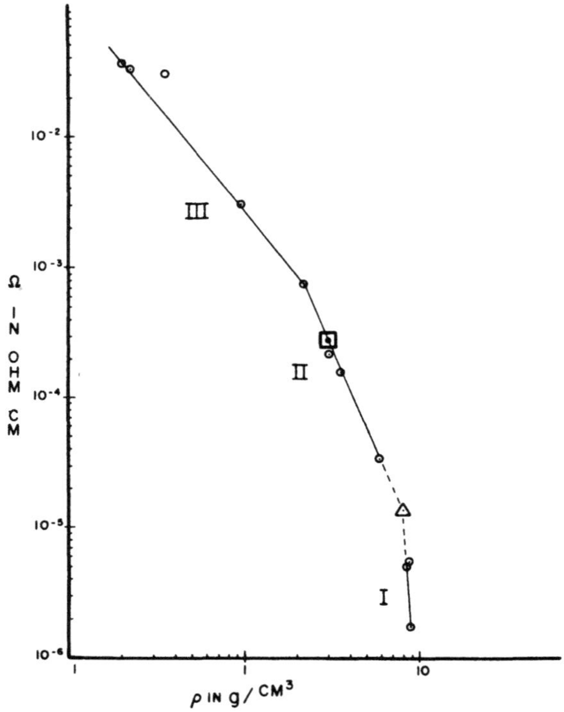

Fig. 7. Resistivity–density relation of 0.0644 cm. Copper wire 7.5 cm long exploded with 12 kV. (I) solid region, (II) "liquid" region, (III) "gas" region. △ Handbook data at m.p. ◻ Herzfeld point (see text).

plot, although the ρ vs. t plot shows breaks, which fall logically in with the current behavior. For the first 2.2 μsec, the material appears to be a solid and, indeed, a very slight break occurs in the di/dt curve at this time. This break in the di/dt curve had previously been designated as the point of melting. Another change in slope occurs at 3.5 μsec and again it fits with the current traces as a logical place for vaporization to occur. The last break in the curve, at 4.8 μsec, corresponds with the onset of dwell. Here, of course, the rate of energy delivery is small, so it is logical that expansion rate should decrease.

CONCLUSIONS

Several conclusions may be drawn from this series of experiments.

Exploding wire technique gives resistance–density relations for all metals, including the most refractory. It is now possible to compare these measurements to those taken previously under static conditions. It is noted that the curves for copper match the curves for mercury and cesium taken under static conditions.

It is noted that the density resistance curves for tin and copper are very different. No explanation for this difference is proposed.

Further, the experiment indicates that although the expansion rate of the wire material varies during the explosion, there is no time when the resistance goes through any discontinuity corresponding to an "electrical phase change," i.e., a sudden change in resistance with density. It had been previously postulated that the dwell occurred when the resistivity suddenly changed from "metallic" to gaseous conduction at, for instance, the boiling point [15]. This experiment does not give any evidence of such an occurrence. Moreover, if we follow David's estimate [16] and Nash's fifth power

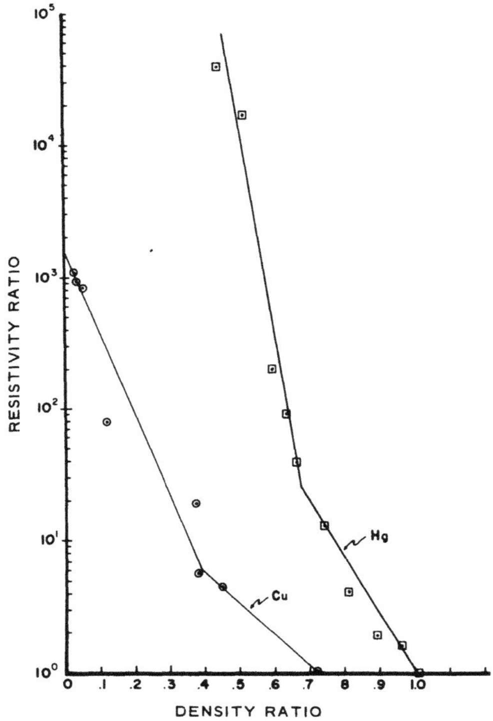

Fig. 8. Relation of resistivity vs. density for copper exploded wire and for mercury in static test.

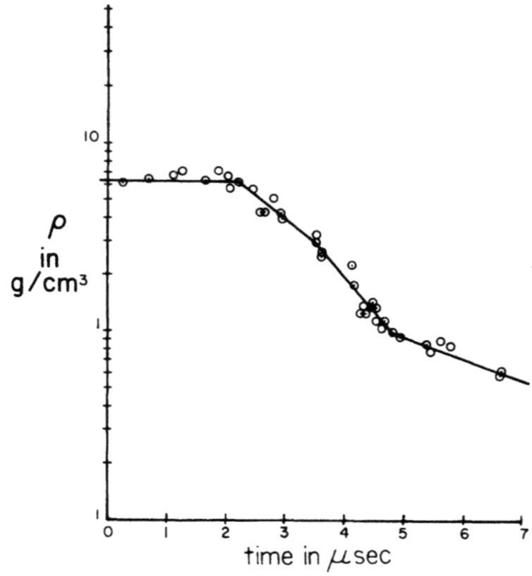

Fig. 9. Resistivity–time relation of 0.1 cm tin wire 7.5 cm long exploded with 12 kV.

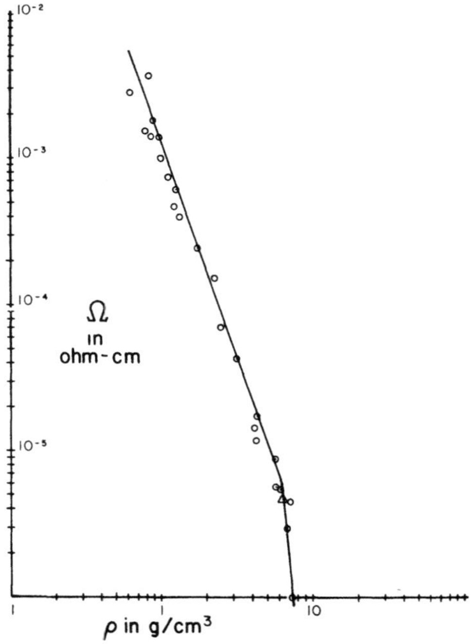

Fig. 10. Resistivity–density relation of 0.1 cm tin wire 7.5 cm long exploded with 12 kV. △ Handbook data at m.p.

law [1] it is not necessary that some sudden change from "metallic" to gaseous condition occur in order to account for the dwell. Such hypotheses as the dwell occurring at the boiling point or as a result of "breaking" of a superheated condition seem to be ruled out by these results.

A quantitative relation between the results and existing theory has not been found. This is in part due to the lack of temperature measurements. Should methods of determining temperature be found, several checks between theory and experiment would become possible.

REFERENCES

1. C.P. Nash and W.G. McMillan, Phys. Fluids 4:911 (1961).
2. S.V. Lebedev and S.E. Karkin, Zh. Eksper. i Teor. Fiz. 26:629 (1954).
3. R.J. Reithel and J.H. Blackburn, in Exploding Wires, Vol. 2, W.G. Chace and H.K. Moore, eds., Plenum Press, New York (1962).
4. F.H. Webb, H.H. Hilton, P.H. Levine, and A.V. Tollestrup, in Exploding Wires, Vol. 2, W.G. Chace and H.K. Moore, eds., Plenum Press, New York (1962), p. 37.
5. J.K. Krumhansl, Physics of Solids at High Pressures, C.T. Tomizuka and R.M. Emrick, eds., Academic Press, New York (1965).
6. K.F. Herzfeld, Phys. Rev. 29:701 (1927).
7. W.G. Chace and M.A. Levine, J. Appl. Phys. 31:1298 (1960).
8. T. Korneff, W.G. Chace, and H.D. Edelson, Proc. Intern. Conf. Ionization Phenom. in Gases, Vol. 2, Paris (1963), p. 627.
9. I.K. Kikoin, A.P. Senchenkov, E.V. Gel'man, M.M. Karsunskii, and S.P. Naurzakov, Zh. Eksper. i Teor. Fiz. 49:124 (1965). Trans. in Soviet Physics, JETP 22:89 (1966).
10. W.G. Chace, M.A. Levine, and C.V. Fish, Proc. Seventh Intern. Conf. Ionization Phenom. in Gases, Vol. 1, Belgrade (1965), p. 792.
11. W.G. Chace, Prz. Elektroteck 43:134 (1967).
12. W.G. Chace and M.A. Levine, Rep. on 21st Ann. Conf. on Physical Electronics, MIT (1961), p. 298.
13. J.B. Langworthy et al., NRL Report 5489 (DASA-1174), U.S. Naval Res. Lab. (1961), p. 74.
14. Landolt–Börnstein, Physikalisch-Chemische Tabellen, Springer Verlag, Berlin (1923).
15. B. Eiselt, Z. Physik 132:54 (1952).
16. E. David, Z. Physik 150:162 (1958).

DISCUSSION

Question by B. Ya'akobi, Princeton Plasma Physics Laboratory: Would not a study of the dependence of resistivity on temperature be more significant? If one performs the experiment with different energies, then the wire undergoes the same density changes but different temperatures and the truly significant parameter could be determined.

Answer by Author: This is certainly true and the experiment has been planned to do just that, but has not yet been carried out.

Question by E. Oktay, The University of Michigan: 1) Do you observe differences in the voltage and current traces as a result of encapsulating the wire in the Plexiglas? 2) Does the resistance calculated from these traces differ for wires exploded in air and in the Plexiglas?

Answer by Author: 1) Yes, the trailing edge of the first current pulse is less steep; in other words, the speed of burst is slowed by the heavy encapsulating material. 2) This experiment has been tried and is now at the stage of data reduction. We hope to publish the results soon.

Question by B. Markow, Institute for Exploratory Research, USAECOM, Ft. Monmouth, N.J.: What x-ray equipment was used?

Answer by Author: Field emission x ray with smallest spot size.

Question by F.D. Bennett, Ballistic Research Laboratories, Aberdeen Proving Ground, Md.: 1) How valid is the assumption of uniform density? 2) Do you see any striations in your wires?

Answer by Author: 1) Along the axis we see no variation at all. This means that the density varies less than 20%, perhaps less than 10%, which is the accuracy of our x-ray determinations. Across the radius we are less confident since we are hampered by the restrictions mentioned in the paper; i.e., our x rays are neither from a point nor parallel. From an empirical test using plastic and aluminum (ratio of about 10:1) we know we are much better than that. 2) We have been unable to observe any striations in our wires in our fast explosions (i.e., first pulse $<3.5\ \mu$sec). However, in slow explosions (first pulse $>10\ \mu$sec) they are very evident and look like those shown in your paper.

Effect of Applied Magnetic Field on the Exploding Wire Phenomenon—II

A. Sakurai

U. S. Army Engineer Waterways Experiment Station
Vicksburg, Mississippi

and

T. Takao and T. Taira

Tokyo Electrical Engineering College
Kanda, Tokyo, Japan

INTRODUCTION

This paper is a continuation of the previous work [1] in which emphasis was placed on the particular role of the effect of an applied axial magnetic field on the exploding wire phenomenon and effects on the resulting shock waves were studied theoretically as well as experimentally.

The purpose of this work is to investigate the direct interaction effect between the magnetic field and the exploding wire itself.

As a preliminary approach to this, systematic tests were conducted to compare the discharging currents as well as the voltages between results obtained with and without a magnetic field. First, the magnitude of the magnetic field necessary to cause any noticeable effects in the current and voltage was examined. Secondly, the effects of the magnetic field on the appearance of the individual wave forms of current and voltage was investigated.

EXPERIMENTAL

A block diagram of the experimental system is shown in Fig. 1 and the general view of the setup is pictured in Fig. 2.

Copper (Cu), tungsten (W), molybdenum (Mo), and phosphor-bronze (P.B.) wires were used in the tests. The diameters d and lengths L of the wires used are given in the following table:

	d, in mm		
Cu	0.1	0.25	0.3
W	0.05	0.1	0.15
Mo	0.1	0.15	
P. B.	0.1	0.2	

$L = 20, 28, 51\,\text{mm}$

First, a magnetic field was not applied and the tests were conducted for each diameter and length of wires. The voltage (V) on the condensor was systematically changed for each size wire to

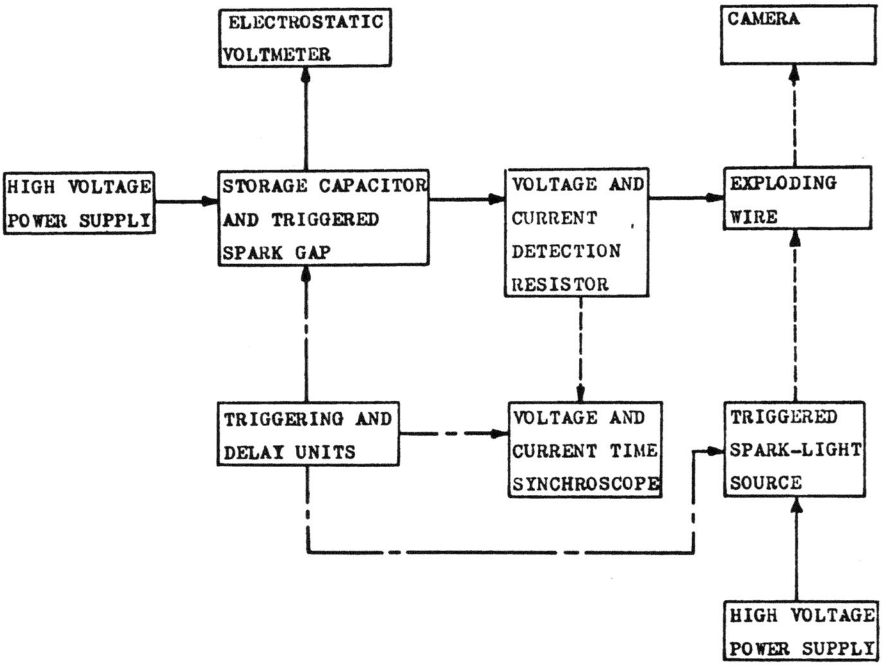

Fig. 1. Block diagram of apparatus.

Fig. 2. General view of the experimental setup.

produce the various typical explosions as designated in cases A, B, C,. . . by Chace [2]. Each of the various combinations of wire diameters and lengths for various materials was tested in this manner. The voltage (V) ranged from about 0.5 to 4 kV and the capacitance was 8 μF.

Secondly, an axial magnetic field which ranged from 3 to 8 kG was applied, and the tests were repeated for each of the individual cases.

The discharging current and voltage were recorded and photographs of the shock waves were taken.

RESULTS

Typical examples of the test results are shown in Figs. 3-9. Figure 3 shows the photograph of the cylindrical shock wave. The rest of the figures exhibit a comparison of the various cases with and without the applied magnetic field.

For these particular tests, the effect of the magnetic field becomes noticeable for B values beyond about 4 kG. In general, the effect is prominent in the "pause" period but its characteristics are dependent on the nature of the wire material. For W and Mo, as

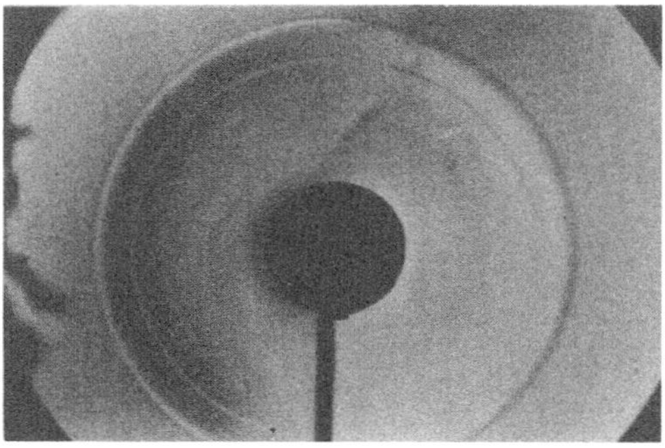

Fig. 3. The photograph of the shock wave.
Cu, $d = 0.1$ mm, $L = 20$ mm, $B = 0$.

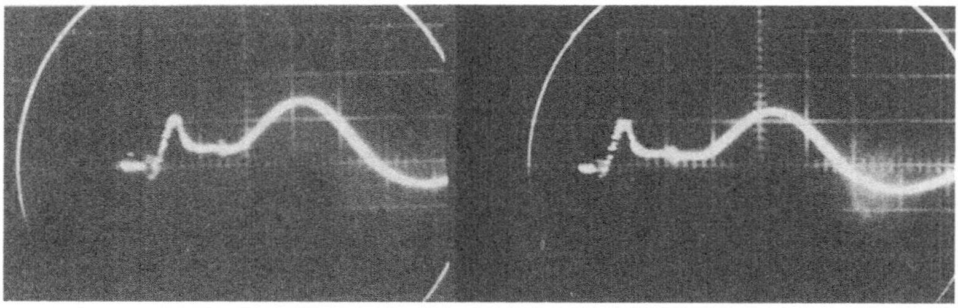

(a) (b)

Fig. 4. Current (1 kA/div) vs. time (5 μ sec/div) for W; $d = 0.1$ mm, $L = 28$ mm, $V = 1.7$ kV. (a) $B = 4.75$ kG, (b) $B = 0$.

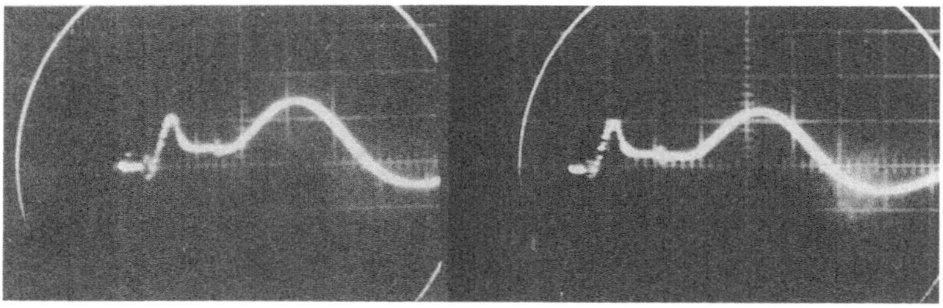

(a) (b)

Fig. 5. Current (1 kA/div) vs. time (5 μ sec/div) for Mo; $d = 0.1$ mm, $L = 28$ mm, $V = 1.8$ kV. (a) $B = 4.75$ kG, (b) $B = 0$.

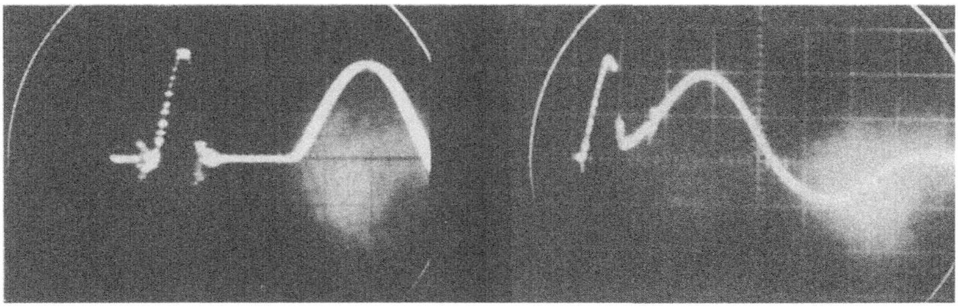

(a)

Fig. 6. Current (1 kA/div) vs. time (5 μsec/div) for Cu; d= 0.1 mm, L= 28 mm, V= 2.5 kV. (a) B= 4.75 kG, (b) B= 0.

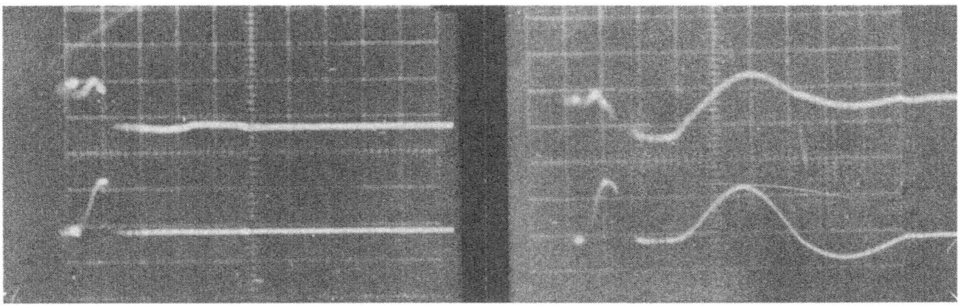

(a) (b)

Fig. 7. Current (1 kA/div) (lower trace) and voltage (0.8 kV/div) (upper trace) vs. time (5 μsec/div) for P.B. (phosphorbronze); d= 0.1 mm, L= 20 mm, V= 1.7 kV. (a) B= 7 kG, (b) B = 0.

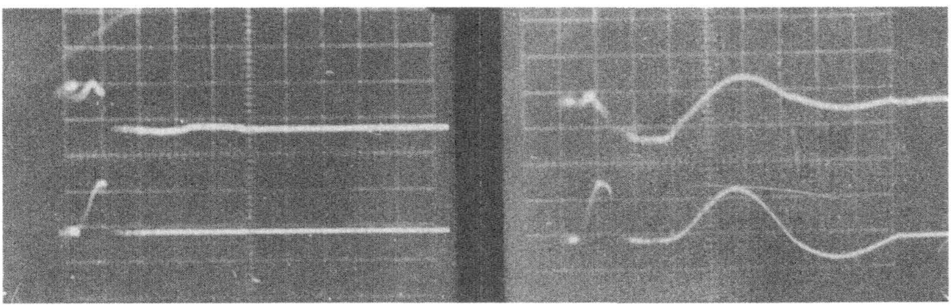

(a) (b)

Fig. 8. Current (0.2 kA/div) (lower trace) and voltage (0.4 kV/div) (upper trace) vs. time (10 μsec/div) for W; d = 0.05 mm, d = 20 mm, V = 0.8 kV. (a) B= 7 kG, (b) B= 0.

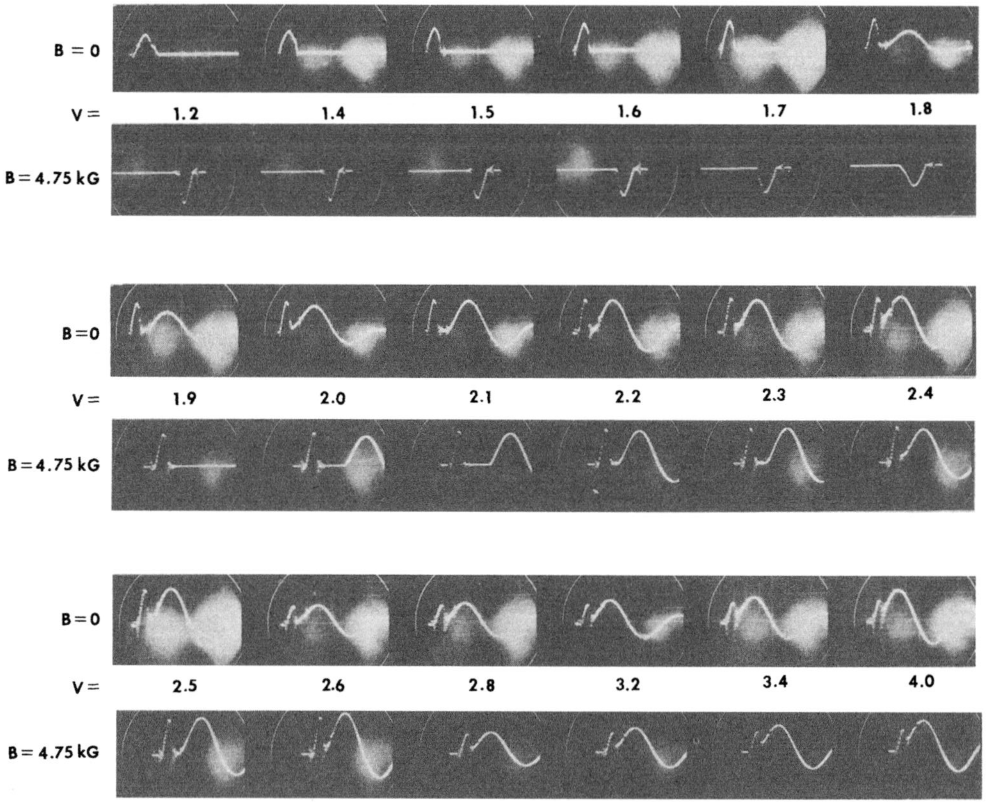

Fig. 9. A set of current (1 kA/div) vs. time (5 μsec/div) records for Cu; $d = 0.1$ mm, $L = 28$ mm as voltage (V) varies from 1.2 to 4.0 kV with and without magnetic field of $B = 4.75$ kG.

seen in Figs. 4, 5, and 8, the pause period becomes shorter and the voltage for the reignition becomes lower when the magnetic field is applied. The test results from using copper show tendencies which are just the reverse, as can be seen in Figs. 6 and 9.

Interpretation of these results cannot be made easily because they are directly related to the understanding of the real nature of the pause phenomenon, for which there are diverse views as summarized by Chace [3]. Nevertheless, it will be possible to utilize these results for a better understanding of the mechanism of the "pause" effect.

This may be done because the different test cases show the variation of the "pause" as caused by the magnetic field and, therefore, the various available theories can be checked to see if they explain the changes which were observed.

REFERENCES

1. A. Sakurai and T. Takao, in Exploding Wires, Vol. 3, W.G. Chace and H.K. Moore, eds., Plenum Press, New York (1964), p. 247.
2. W.G. Chace, in Exploding Wires, Vol. 1, W.G. Chace and H.K. Moore, eds., Plenum Press, New York (1959), p. 7.
3. W.G. Chace, in Exploding Wires, Vol. 3, W.G. Chace and H.K. Moore, eds., Plenum Press, New York (1964), p. 1.

New Evidence for Standing Waves in Electrically Exploded Iron Wires

Moody L. Coffman

Department of Physics
Oklahoma City University
Oklahoma City, Oklahoma

Data from the measurement of over 6000 interstriation distances are presented in graphical form. Data were taken from deposits of electrically exploded iron wires on glass plates, clearly revealing the striations measured by a measuring microscope. The data are fit to a frequency distribution function by least squares. Three sets of data constitute knowledge of wires exploded by the sudden discharge of a 238 μF capacitor bank at each of three voltages: 1200, 1600, and 2000 V. Evident in each of the three frequency graphs are ministriations, which are regularly spaced peaks in the frequency graph. These are submitted as further evidence of standing waves in electrically exploded iron wires. It is postulated that the standing waves responsible for the appearance of ministriations are de Broglie waves, which represent the motion of electrons from the negative terminal and holes from the positive terminal of the wire. While these waves have the properties to assure standing waves on their superposition, the currents they represent are additive to give alternate maximum and minimum values to the resulting current along the wire. This postulate allows the calculation of the electrical conductivity of iron, which is in reasonable agreement with handbook values. Finally, a brief review of properties of a metal derivable from data on striations is given.

INTRODUCTION

Striations perpendicular to an electrically exploding wire are the final state representation of the physical transformations which occur in such a small time interval that detailed analysis is impractical. Therefore, striations with their spacing relative to one another should be closely studied to determine the detailed nature of the final state of the wire and they should be a part of the explanation of any successful theory of electrically exploding wires. They have been pictured from a wire source by Arnold and Conn [1], who have analyzed the interstriation distances from explosions of copper, silver, and aluminum wires. The distribution of interstriation distances they obtained for copper wires served as a test for a microscopic model of an electrically exploding wire developed by Coffman [2].

Striations are pictured as they arise from electrically exploded foils by Woffinden [3]; and they are pictured by x-ray photography as they result from exploding wires in quartz sand by Nasilowski [4]. Nasilowski shows pictures of unduloids, which appear under certain experimental conditions at low electric field intensities, and discusses their spacing for silver and copper wires also. Although these data are of eventual interest in determining whether they are consistent with the model introduced in 1963 [2], there are possible failures in that the postulates of the theory are not purposely matched by the experimental conditions. Therefore, data on striations presented here have been taken together with some care that the postulates of the theory are not violated in the experimental design.

Happily, this new data, taken on iron wires, have revealed a new set of striations which are called ministriations because they appear with smaller separation distances than those previously studied. Furthermore, they appear to evidence the presence of de Broglie waves representing the electron motions in the wire; and they make available several new applications of exploding wire studies through the relationships developed between the interstriation distances and the electrical properties of the wire.

EXPERIMENTAL ARRANGEMENT

The wires are exploded by discharging a 238-μF capacitance bank through them. The capacitance bank may be charged to a maximum voltage of 6000 V. The switch consists of a relay-actuated parallel–series arrangement of four mercury thyratron tubes. The energy used in the explosion is so small that the capacitance bank remains practically fully charged after the explosion. The ringing frequency of the circuit was measured to be 6.4 kc/sec. The striations are captured by placing with discretion a glass microscope slide above the wire. The distance of the glass slide from the wire must be small enough so that the molten drops of metal will adhere to the glass and far enough away from the wire so that the glass will not melt appreciably. Striations are evident without the aid of magnification, but interstriation distances must be measured with care and tedium by use of a measuring microscope. An inherent problem is the definition of a striation. The use made here is simple. If there appeared to be any droplets aligned perpendicular to the orientation of the wire, a striation was present and counted. Final definition of a striation was reliably made by an experienced operator of the measuring microscope. Even so, the postulate that some striations may be predicted by the theory and not detected by this technique remains appropriate and has been retained in the theoretical explanation of the distribution of interstriation distances.

The apparatus has been used in the explosion of wires of aluminum, copper, titanium, and fuse alloys as well as the iron wires from which this data have been drawn. A Tektronix 545A oscilloscope has been used to make the usual analysis of voltage and current as a function of time in a preliminary study of the circuit characteristics. Of course, the circuit parameters do not enter directly into the theory of the origin of striations.

EXPERIMENTAL DATA

The iron wires from which the interstriation data are obtained were all $2.3\,\mu$ in diameter, and they had an electrode-to-electrode length of 2.1 cm. The data are presented in three distribution curves. Superposed on each graph is the curve which has been derived from the theoretical distribution function predicted from the microscopic model the data were to test.

Figures 1, 2, and 3 represent the distribution function, $N(x)$, of interstriation distances x taken with the initial voltage of the capacitance bank at 1200, 1600, and 2000 V, respectively. The theoretical distribution function to which the points on the left slope of each graph were fitted by least-square calculations was taken to be of the form $N = 3x^2/d_m^3$, where d_m is the interstriation distance which appears most often in the distribution and $N = 0$ for $x > d_m$, in accordance with the theory. It can be seen that the data do indeed fit the form of the distribution function on the left side. The postulate is that those interstriation distances larger than d_m are in effect the result of two or more successive interstriation distances with the intermediate striations not visible for one reason or another. When this

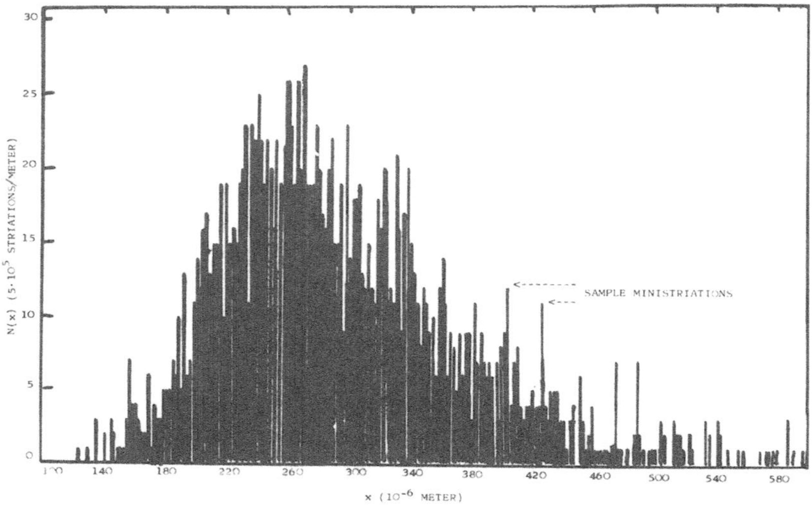

Fig. 1. Frequency distribution of interstriation distances for 2.1 cm Fe wires of 0.009-in. diameter exploded by 238-μF capacitor bank at 1200 V.

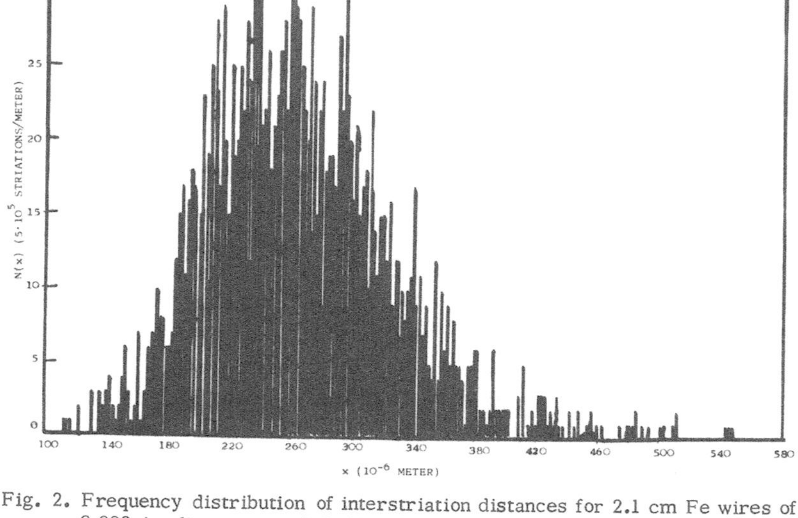

Fig. 2. Frequency distribution of interstriation distances for 2.1 cm Fe wires of 0.009-in. diameter exploded by 238-μF capacitor bank at 1600 V.

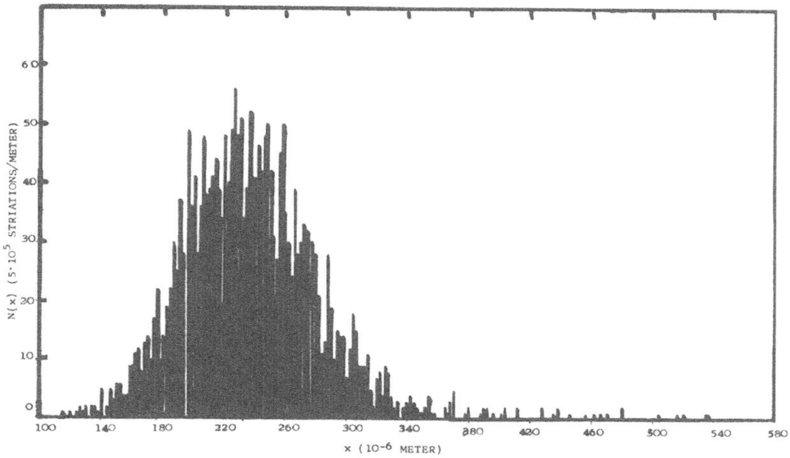

Fig. 3. Frequency distribution of interstriation distances for 2.1 cm Fe wires of 0.009-in. diameter exploded by 238-μF capacitor bank at 2000 V.

postulate is considered, the magnitude of $N(x)$ is multiplied by the ordinate appearing in the graph to yield an integral under the curve representing $N(x)$ equivalent to the total number of striations. The left side of the distribution has been interpreted to include all interstriation distances less than or equal to 260 mm.

LEAST-SQUARE FIT TO INTERSTRIATION DISTANCES

The form of the theoretical distribution function is $N(x) = Cx^2$ in the range $(0, d_m)$ and has the value zero outside this range. If the integral of $N(x)$ over the range is normalized to unity, the value of C is $3/d_m^3$. To fit the data of N striations, the value of C is N times as large.

Consider the striation distances, x_i, which have a frequency of N_i, to be fit to the theoretical distribution of the best choice of the value of C as given by the theory of least squares. Then the function, $\sum_i (N_i - Cx_i^2)^2$, is to be made a minimum value by proper choice of C. Therefore,

$$\sum_i x_i^2(N_i - Cx_i^2) = 0 \tag{1}$$

Table I. Most Probable Interstriation Distances

Figure number	Charging voltage V, V	Most probable interstriation distances $D \cdot 10^4$, m	Applied electric field intensity $E \cdot 10^{-5}$, V/m
1	1200	2.21	0.57
2	1600	2.24	0.76
3	2000	2.26	0.95

The solution of Eq. (1) for C yields

$$C = \frac{\sum_i N_i x_i^2}{\sum_i x_i^4} \tag{2}$$

Finally, the distribution normalized to the total number of striations considered in the distribution is given by Eq. (2) when it has been multiplied by the total number of striations. Therefore, d_m is determined from the relation

$$d_m^3 = \frac{3\left(\sum_i N_i\right)\left(\sum_i x_i^4\right)}{\sum_i N_i x_i^2} \tag{3}$$

The values of d_m obtained by this procedure are tabulated with the corresponding voltage across the capacitance bank in Table I, which contains the value of the electric field intensity applied to the wire also. The electric field intensity, $\vec{E}$, has a magnitude equal to the voltage divided by the length of the wire. It will be seen in the discussion of the data that the thermionic work function of the metal being exploded is related to the values of $\vec{E}$ and d_m [2].

LEAST-SQUARE FIT TO THE DISTANCE BETWEEN MINISTRIATIONS

The striking thing about the data presented in Figs. 1, 2, and 3 is the regularity with which a value of $N(x)$ peaks above its neighboring values of $N(x)$. For purposes of tabulation, a peak of this type is counted if the value of $N(x)$ at x exceeds the value of $N(x)$ at $x \pm 2 \cdot 10^{-6}$ by two. Strangely, these places in the data which represent an additional effect are most clearly seen to the right side of the peak. Some examples are marked on Fig. 1. These peaks are assumed to represent places in the wire which were encouraged to striate to a greater extent than other points in the wire. A reason for this will be given in the discussion of the data. These additional peaks are called ministriations because of their small amplitude and separation distance.

The fitting of the data for the ministriation separation distance is essentially the problem of fitting the data obtained in the measurement of the charge on an oil drop by the method of Millikan. There are places where gaps exist, but it is assumed that the apparent regularity present in the parts of the data where the ministriations are clearly in evidence carries on through the entire data range. Therefore, a frequency, or small integer, is assigned to each of the gaps which exceed the apparent mean separation distance of the ministriations. The resulting data consist of the observed ministriation separation distance y_i and the assigned frequency n_i, for each interval from the first peak which exceeds its neighbors by $2 \cdot 10^{-6}$ m to the last one.

The data is then fit by least-squares theory to obtain the best average value $\bar{x}$. Actually $\bar{x}$ is treated as a parameter in the sum $\sum_i (y_i - n_i\bar{x})^2$, which is minimized by the choice of $\bar{x}$. Thus,

$$\bar{x} = \frac{\sum_i n_i y_i}{\sum_i n_i^2} \tag{4}$$

The resulting values of $\bar{x}$ are given with their discussion in which they are related to de Broglie waves in the exploding wire.

INTERPRETATION OF DATA

From the interstriation data, the thermionic work function of iron may be determined. It is assumed that the striations occur at places in the wire where two standing waves have their antinodes superposed to provide high temperatures relative to neighboring parts of the wire [2]. These standing waves are associated with the imposed electric field with its reflection from the opposite terminal and the natural vibrations of the nuclei due to the shift in their equilibrium positions after the electric field is applied. The superposition of these effects has led to the predicted distribution of the interstriation distances expressed by the equation $N(x) = Cx^2$. Under the assumptions which led to this distribution, it can be seen

Table II. Calculated Electronic Work Functions

Charging voltage V, V	Electronic work function Φ, eV
1200	6.3
1600	8.5
2000	10.7

that the free acceleration distance of the electrons from a node to an antinode is just one-half the distance d_m, which represents the value of x at which the maximum of the interstriation distances occur. Thus, $s = d_m/2$; as d_m represents the most probable distance between standing wave antinodes. The product of this distance and the force on an electron, a conduction electron, is the energy acquired by the electron. The applied force is eE, where e represents the charge on the electron and $\bar{E}$ represents the applied electric field intensity.

The concept is that the electrons emitted by thermionic emission initiate a striation, and that an electron cannot achieve the status of thermionic emission until it has acquired enough energy to overcome the work function of the material. Therefore,

$$e\Phi = eEs = eEd_m/2 \tag{5}$$

where Φ represents the work function of the material. The values of the work function for iron obtained from exploding wire data in Table I are listed in Table II with the voltage applied in each case. It is seen that the data are higher than the values 4.04-4.76 eV obtained by other methods [5]. However, the order of magnitude indicates that the concepts which lead to these values of 6.3-10.7 eV are worth further consideration.

Of particular interest is the indication that an increase in the charging voltage leads to an increase in the observed work function. This may be explained by the facts that the wires used in the explosions were not clean wires and that no attempt was made to get wires of pure iron. It has been observed by Smith [6] that adsorbed oxygen atoms on the surface of a metal may increase its work function by several electron volts due to the electron affinity of oxygen. If this fact is coupled with the knowledge that chemical reaction rates increase rapidly with temperature, as reviewed by Noyes [7], it seems plausible that increased oxygen activity at the surface of the wire might account for the values of the iron work function obtained by this method being high relative to those values obtained by other methods.

Unfortunately, the temperatures at which these work functions are measured are somewhat uncertain; but they are in the neighborhood of the melting point of iron, and an estimate of the temperatures will be derived from the data on ministriations.

There is a second possibility regarding these measurements which may be considered, and that is the possible emission of lower energy electrons than the conduction electrons in iron. The first two ionization potentials of iron are 7.83 and 16.16 eV.

It has been shown by Coffman [8] that the sudden ejection of an electron from a solid conductor should take approximately one-half the ionization potential. Therefore, the electronic work function for iron on this basis may be as low as 3.92 and as high as 8.08 eV.

Between the two explanations lies hope that the concepts which form the basis for the derivation of thermionic work functions from exploding wire data may be reasonable; and the possibility of getting work functions as a function of temperature may be of special value [9], because the usual determination of the work function involves a correction in the form of a calculation of the term $-T(d\Phi/dT)$. It can be seen from the following statements that this term may be evaluated from the data obtained from exploding wires.

POSTULATE OF DE BROGLIE WAVES

An explanation of the presence of ministriations lies in the concept of de Broglie waves to represent the current in the wire. The concept of de Broglie waves propagated at once from both ends of the wire may be obtained in answer to a question. When the switch is closed, from which direction does the electric field propagate into the wire? The answer must be from both ends. If an electron is near the negative terminal, it is forced toward the positive terminal by a displacement of charges actuated by closing the switch. If an electron is near the positive terminal, it is forced toward the positive terminal also; but it is proper to think of it being pulled toward the positive terminal because of displacement of charges there. Now the motion of charges begins near each

Table III. Ministriation Separation Distances and Current Densities

Figure number	Average ministriation separation distance $\bar{x} \cdot 10^6$, m	Drift velocity V, m/sec	Current density $J \cdot 10^{-12}$, A/m^2	Total current $I \cdot 10^{-4}$, A
1	3.9	93	1.3	7.7
2	4.1	89	1.3	7.7
3	4.2	87	1.2	7.1

terminal at once, and electrons move as they become aware of the field as though it were propagated from both terminals. Therefore, one can think of the resulting current as the sum of these motions.

Now an electron moving to the right is equivalent electrically to a positive charge moving to the left. Thus, the motion of the conduction electrons in a wire may be represented by the sum of negative charges moving toward the positive electrode beginning at the negative electrode and positive charges, or holes, moving toward the negative electrode beginning from the positive electrode.

Now the de Broglie wave which represents an electron moving to the right differs from the de Broglie wave which represents a hole moving to the left only in the direction of the velocity vector. The wavelength ascribed by de Broglie to such a particle is

$$\lambda = \frac{h}{mv} \tag{6}$$

which depends upon the mass rather than the sign of the charge. Therefore, these two waves, which add to make up the current in the wire, constitute a standing wave. The amplitude of the standing wave represents the density of conduction electrons at each point and has its nodes and antinodes and is assumed to be responsible for the additional striations recognized as ministriations.

MINISTRIATIONS AS EVIDENCE FOR STANDING WAVES

It is possible to make a check on the reasonableness of the assumption of de Broglie waves. Let the average ministriation separation distance from Table III be used to calculate the de Broglie wavelength. Then, Eq. (6) yields a value for the drift velocity of the electrons. From these results, the current density may be calculated from the relation

$$J = Nev \tag{7}$$

where N ($8.85 \times 10^{28}/\mathrm{m}^3$ for iron) represents the number of conduction electrons per unit volume and e is the charge on the electron in coulombs. If v is in m/sec, then J has units $\mathrm{A/m}^2$. Table III gives the resulting values of v, J, and the corresponding value of the applied electric field intensity for the explosions represented by the data in Figs. 1, 2, and 3. From these results and the cross section of the wire, total current is calculated. It is not clear whether these values represent the maximum current, but the order of magnitude of these currents is not unreasonable.

ELECTRICAL CONDUCTIVITY FROM EXPLODING WIRE DATA

A further check on the reasonableness of the assumption of de Broglie waves yields a new method for determining the electrical conductivity of a metal. The electrical conductivity of iron has been calculated from the data in Table III and is seen to be in good agreement with the value 10^7 mhos/m listed in textbooks. The results on electrical conductivity allow a further calculation of the skin depth [10] of the current in the wire which may be a useful quantity for providing a more precise Fourier analysis of the effects of sudden electrical discharges through conductors.

Table IV gives the values obtained from this data for the electrical conductivity of iron and the resulting skin depths for each of Figs. 1, 2, and 3. The skin depths have been calculated on the

Table IV. Electrical Conductivity and Skin Depth

Charging voltage V, V	Electrical conductivity $\sigma \cdot 10^{-7}$, mhos/m	Skin depth $\sigma \cdot 10^6$ $\delta = (\mu\sigma\omega/2)^{-1/2}$
1200	2.3	3.2
1600	1.6	3.7
2000	1.3	4.3

assumption of a fundamental Fourier frequency corresponding to the fundamental electromagnetic wave having a half wavelength equal to the length of the wire. It is assumed that the temperature of the iron corresponding to these data lies above the Curie temperature, which makes the speed of electromagnetic waves in the wire essentially the same as the speed of such waves in vacuum.

Thus, the data in Table IV are based on the following values: length of wire, 2.1 cm; speed of electromagnetic radiation, $3.0 \cdot 10^8$m/sec; angular frequency, $\omega = 2\pi f = 4.5 \cdot 10^{11}$/sec.

ESTIMATE OF TEMPERATURE CHANGE

While there is some uncertainty in the temperatures of the wires from which these data were collected, one can agree that the melting point of iron was surely a minimum temperature for the lowest applied voltage because molten droplets of iron constitute the striations. An increase in temperature for the higher voltages may be estimated from the conductivity data of Table III, for the conductivity declines with a voltage increase. Taking the coefficient for variation of conductivity with temperature as 0.005 mho/m-deg, one gets temperature changes of 38° and 61° on the Kelvin scale for the increases in temperature of the wire corresponding to increases in voltage from 1200 to 1600 V and 1600 to 2000 V, respectively.

Therefore, the change in temperature coupled with a minimum assured temperature allows the conservative assignment of temperatures to the three sets of data given in Figs. 1, 2, and 3. Assigned temperatures are 1808°, 1846°, and 1907°. Now it is seen that the temperature variation of electrical conductivity may be obtained from exploding wire data if the temperatures are measured independently.

It may be seen that the term $T(d\Phi/dT)$ has a value of the order of magnitude of several electron volts when these temperature changes are coupled with the variation in the electron work function with voltage reflected in the data of Table II. Perhaps this term is of considerable importance at high temperatures.

OTHER CHECKS

Corresponding to each set of data, one may obtain other checks. For example, it is found that for these temperatures radiation is completely negligible until the wire is melted, as melting requires on the order of 100 nsec. Also, by an extension of the Wiedemann–Franz–Lorentz law which relates the electrical conductivity σ to the thermal conductivity k by the relation

$$k = 2.3 \cdot 10^{-8} \sigma T \tag{8}$$

an estimate of the importance of thermal conductivity may be obtained. Such calculations show that skin effect is very significant in determining the final state of a wire and may explain the increase in temperature with voltage while some melting is doubtless still in progress. It is evident that in these explosions melting occurs so rapidly that there is little heat transfer by any means prior to melt. Finally, it should be noted that a check on the value of the constant in Eq. (8) exists in exploding wire data if experiments are designed to allow conduction to have a greater part. Thus, the conjecture on the extension of the Wiedemann–Franz–Lorentz law may be checked.

MICROSCOPIC VERSUS MACROSCOPIC VIEWS

Combining Eq. (6) with Eq. (7), one can see that the current density depends upon the ratio of charge to mass. Therefore, the charge density and the mass density of conduction electrons could be used in place of the particulate approach presented here. Thus, the conclusions from these data are not heavily dependent on the model; but the model serves simply as a convenience to introduce the concepts.

SUMMARY

New evidence for standing waves in electrically exploding wires has been presented together with its use in calculating current density, electrical conductivity, changes in temperature with applied voltage and electronic work functions of metals.

ACKNOWLEDGMENT

The author is indebted to several students who have worked to help construct the apparatus and take the data on the total of over 6000 interstriation distance measurements. Among these are Robert Rood, Constantine Antoniadis, Edward Parma, Michael Guin, and Ray Simons.

REFERENCES

1. H. Arnold and W.M. Conn, "About Distance in the Characteristic Pattern of Exploding Wires," in Exploding Wires, Vol. 2, W.G. Chace and H.K. Moore, eds., Plenum Press, New York (1961), pp. 77-86.
2. M.L. Coffman, "The First Picosecond in an Exploding Wire," in Exploding Wires, Vol. 3, W.G. Chace and H.K. Moore, eds., Plenum Press, New York (1964), pp. 89-102.
3. G.J. Woffinden, "Exploding Metal Films," in Exploding Wires, Vol. 3, W.G. Chace and H.K. Moore, eds., Plenum Press, New York (1964), pp. 193-210.
4. J. Nasiłowski, "Unduloids and Striated Disintegration of Wires," in Exploding Wires, Vol. 3, W.G. Chace and H.K. Moore, eds., Plenum Press, New York (1964), pp. 295-313.
5. R. Weast, ed., Handbook of Chemistry and Physics, 45th College Edition, The Chemical Rubber Company, Cleveland, Ohio (1964), p. E-44.
6. L.P. Smith, "Thermionic Emission," in Handbook of Physics. E.U. Condon and H. Odishaw, eds., McGraw-Hill Book Company, New York (1958), pp. 8-81.
7. R. M. Noyes, "Chemical Kinetics," in Handbook of Physics, E.U. Condon and H. Odishaw, eds., McGraw-Hill Book Company, New York (1958), pp. 5-141.
8. M.L. Coffman, "Franck–Condon Explanation of the Photoelectron Emission from Solids," Bull. of the Am. Phys. Soc., Series II, 12:372 (1966).
9. L.P. Smith, "Thermionic Emission," in Handbook of Physics, E.U. Condon and H. Odishaw, eds., McGraw-Hill Book Company, New York (1958), pp. 8-76.
10. A. Sommerfeld, Electrodynamics, Academic Press, New York (1964), p. 162.

DISCUSSION

Question by O.H. Zinke, University of Arkansas: I don't understand your de Broglie wave hypothesis and how it relates to the "amplitude of those electrons dumped out of the end of the wire," as you put it.

Answer by Author: Assume the wire extends left to right from a negative to a positive terminal. Then, as electrons move into the wire at the negative terminal, electrons will leave the wire at the positive terminal. These latter electrons may be thought of as holes entering the wire from the right, but these holes are assigned electron masses. Now, de Broglie waves depend on the mass but not on the charge of the particle they represent. Therefore, the wave moving to the right representing electrons entering the wire may be superposed on the wave moving to the left representing holes entering the wire. The result is a standing wave. At the antinodes the current effects add because electrons and holes have opposite charges. At nodes the current is zero from this superposition. Thus, according to this theory, the wire is heated at the antinodes and these antinodes are the source of ministriations.

Question by I.G. Halliday, Ion Physics Corp., Burlington, Mass.: Has the possibility of there being Taylor instabilities been ruled out as an explanation of the striated appearance of the wires, since we have a condition of a diverging deaccelerating shock running from a dense to light material, which is the classical condition for Taylor instabilities?

Answer by Author: Taylor instabilities are not ruled out as an explanation of the striated appearance of exploding wires, because this view is not claimed to be unique as an explanation. However, Taylor instabilities are liable to further microscopic explanation themselves.

Question by W. McDermott, University of California, Davis, Calif.: How do the e density fluctuations affect the atomic density as viewed in the experiment?

Answer by Author: Because the theory treats events in the wire which occur prior to any significant vaporization and expansion, the free electron density has been taken as the atomic density. In other words, the initial conditions for striations postulated by this theory are all set before vaporization or expansion begin. Therefore, striation positions are independent of vaporization or expansion phenomena when the applied electric field is represented by a square wave.

Question by C.P. Nash, University of California, Davis, Calif.: In our experiments, if free recondensation is allowed we see very small spherical particles. Can you account for this on the basis of striations? What assurance do you have that the material particles stick to the plate when they land?

Answer by Author: Striations consist of small spherical droplets which stream radially from the center of the wire. That the particles are spheres is a natural result of internal and surface tension forces. Their accumulation in planes perpendicular to the wire is explained by this theory. In some experiments, the glass plates were placed too close to the wire. The result was a melting of the glass, but no change in the appearance of striations resulted. In addition to this, any sliding of droplets as they cross the glass plate will be in the direction of their momentum, and the measured interstriation distances are not affected by their sliding to rest on the plate.

Question by T.J. Tucker, Sandia Corp.: In the de Broglie wavelength referred to, are you referring to the mass of the individual electron or the sum of the masses?

Answer by Author: In the particulate formulation, the charge-to-mass ratio of the electron enters the calculation. Considering many electrons, one still uses the same charge-to-mass ratio. In the paper, a paragraph is devoted to discussion of macroscopic versus microscopic view of the theory. Standing de Broglie waves may be formulated using Nm as the mass per unit volume. Then, the current density must be $J = N(Ne)\,v$ where $v = h/Nm$ and the charge is Ne to account for the fact that the de Broglie wavelength now accounts for the momentum per unit volume of conducting electrons instead of individual electrons. Therefore, one still gets $J = Neh/m$ for the current density.

Spectroscopic Investigation of the High-Density Plasma from Exploded Lithium Wires

B. Ya'akobi

*The Hebrew University of Jerusalem, Israel**

INTRODUCTION

Use of the exploding wire phenomenon has been shown in the past to constitute a simple and unique method for producing high-density metallic plasmas. Once the physical properties of the plasma produced under specified conditions are determined, some general phenomena characteristic of high-density plasmas can be investigated. Two examples are given here to demonstrate such a possibility. The first one employs the absorbed lines of the 2S-nP series of lithium for a new method of determining the time-averaged electron concentration as a function of the distance from the wire axis. The second example deals with the transition from quadratic to linear Stark broadening of the 2P–5D line of lithium.

EXPERIMENTAL PROCEDURE

The results of value to Section III below were obtained under the following experimental conditions: lithium wires 4 cm in length and 0.05 mm in diameter were exploded in air at atmospheric pressure, employing a capacitor of 0.36 μF capacity charged to 15 kV. The ringing frequency of the total circuit was 0.5 MHz. Under these conditions the current flows primarily through the wire itself, and no spectral lines from the ambient atmosphere were observed. Oxidation of the lithium wire was found to be negligible due to the

*Present address: Plasma Physics Laboratory, Princeton University, Princeton, N.J.

short time interval between exposing the wire to air and exploding it (typically 15 sec). The initial phase of melting and vaporizing the wire material and the pause were much shorter than the time period of the current. The amount of light emitted during these stages was found to be negligibly small as compared with that emitted later. Therefore, the spectrograms employed correspond to an average over the phase that follows the pause, when the wire is in the state of being an ionized gas. It was found that the intensity of the emitted optical radiation as well as the temperature and electron concentration decreased over a few time periods of the current, following a much faster rise. The diameter of the wire during this stage was found to be well defined and to increase very little beyond the main expansion during earlier phases.

Electrical, high-speed photographic and various spectrometric measurements were performed on the ionized gas produced under those conditions. The temperature and electron concentration were determined temporally but not spatially by various methods which gave consistent results [1]. The temperature was found to drop from 19,000°K to 3000°K, the electron concentration from 3×10^{18} cm^{-3} to 1.2×10^{17} cm^{-3} and the degree of ionization from 90 to 20 per cent. Local thermodynamic equilibrium was proved to exist during the corresponding time interval which started 1 μsec after the initiation of the explosion and lasted for 7 μsec.

SPATIALLY RESOLVED MEASUREMENT OF THE ELECTRON CONCENTRATION

Figure 1 displays the near ultraviolet spectrum of the radiation emitted by the wire. The spectrum consists of an intense continuum and the lines to the ground level which appear in absorption (except for the resonance line 2S–2P which does not appear in Fig. 1 and will not be dealt with here). The absorption lines are predominantly broadened due to the Stark effect and are therefore indicative of the existing electron concentration. The time-integrated spectrogram on the spectral plates is the optical image of the emitting region; it shows a strongly radiating axial region with diameter of about 8 mm, surrounded by a weakly radiating region with outer diameter of about 16 mm. The temporally resolved measurements of the electron concentration [1] correspond to an average over the axial region. Here, the complementary measurement is of interest: the time-integrated electron concen-

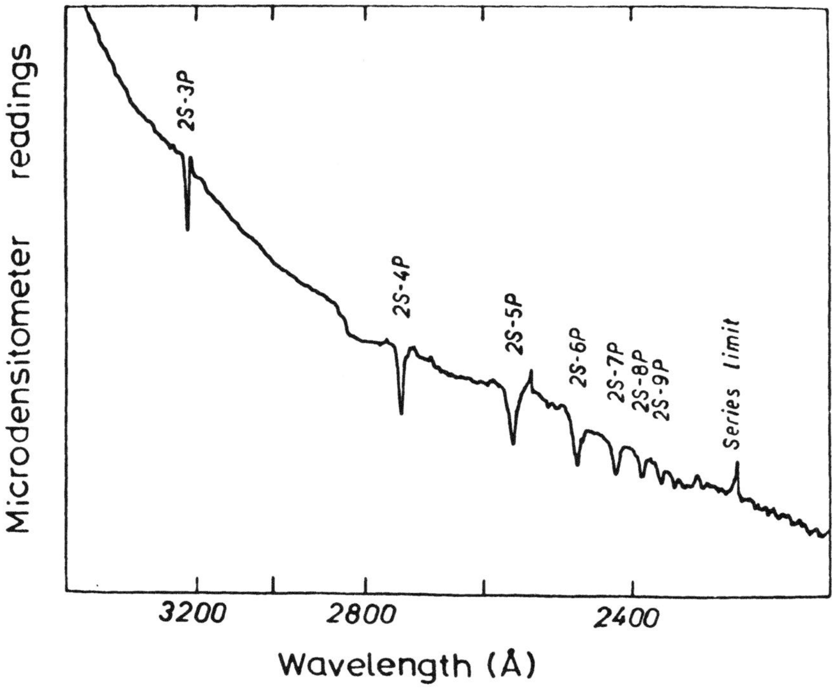

Fig. 1. Near ultraviolet spectrum emitted by exploded lithium wire when viewed along line passing through wire axis and perpendicular to it. Experimental conditions are given in Section II.

tration as a function of the distance from the axis. The results will correspond to an average over the time interval in which the light emission is significant.

The half-widths of most of the lines in Fig. 1 were calculated by Griem [2] as a function of the electron concentration for various temperatures (the dependence on temperature is very weak). They give progressively smaller values of electron concentration in the range 10^{15} to 10^{17} cm^{-3} when lines closer to the series limit are employed. The last few resolvable lines in Fig. 1 could not be formed near the axis where, because of the high electron concentration, their upper levels merge into a continuum. As the continuous radiation travels outward from the axis region and the electron concentration drops, progressively higher levels become detached from the continuum and higher series lines start to be absorbed. It remains now only to locate the absorption region of each line, where the local electron concentration is indicated by the line half-

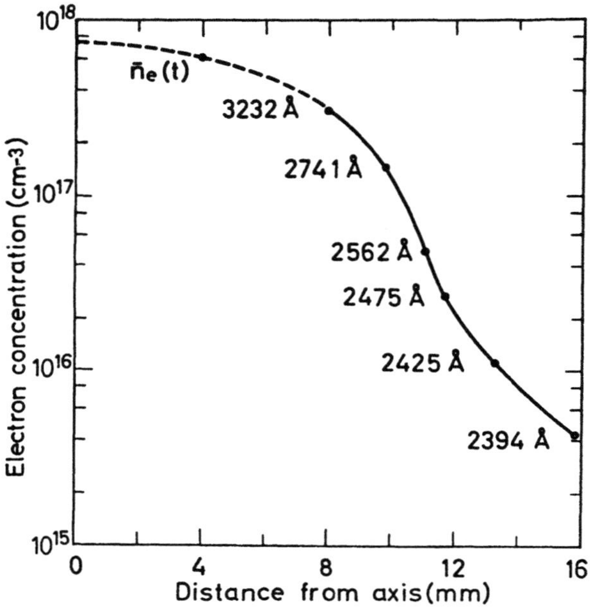

Fig. 2. Electron concentration as function of distance from axis of exploded lithium wire. Marked spectral lines were used to obtain corresponding values of electron concentration. $\bar{n}_e(t)$ is weighted average over time-resolved measurements.

width. The following simple procedure is adequate: the areas enclosed within the absorption contours in a graph of the optical density as a function of wavelength are proportional to the optical depths of the lines. The optical depths can be described as proportional to a product fr, where f is the absorption oscillator strength and is a characteristic length for absorption. The ratios of values of f for the lines under discussion were measured by Filippov [3]. Thus, by measuring the ratios of the areas enclosed within the absorption contours the desired ratios of values of r can be determined. The resulting curve of electron concentration as a function of the distance from the axis is shown by the solid curve in Fig. 2. The point marked $\bar{n}_e$ was obtained by averaging the time-resolved measurements of electron concentration with the light intensity in the near ultraviolet as a statistical weighting function. Since this measurement pertains to the whole axial luminous region of radius 4 mm the point marked $\bar{n}_e(t)$ was placed at half this radius. The two complementary methods of measurement are seen to be in agreement with each other.

FIRST-ORDER STARK BROADENING OF THE 2P–5D LINE OF LITHIUM

The results of value to this section were obtained under the following experimental conditions: lithium wire of the same dimensions as in Section II was exploded in air at atmospheric pressure, employing a capacitor of capacitance 0.1 μF charged to 25 kV. The ringing frequency of the total circuit was 1 MHz. The time-averaged electron concentration was determined for these experimental conditions by applying the Inglis–Teller formula to the 2P–nD series. It was alternatively determined by a detailed investigation of the Stark profile of the 2S–3P line [1]. Both methods yielded the value 2.5×10^{17} cm^{-3} for the time-averaged electron concentration within the axial region of the wire. The last resolvable line of the 2P–nD series under the present conditions is the 2P–5D line at 4132 Å. The energy levels with the principal quantum number $n = 5$ are broad enough to overlap. Thus, the upper state of the 2P–5D transition has the same degeneracy as the hydrogen atom has and

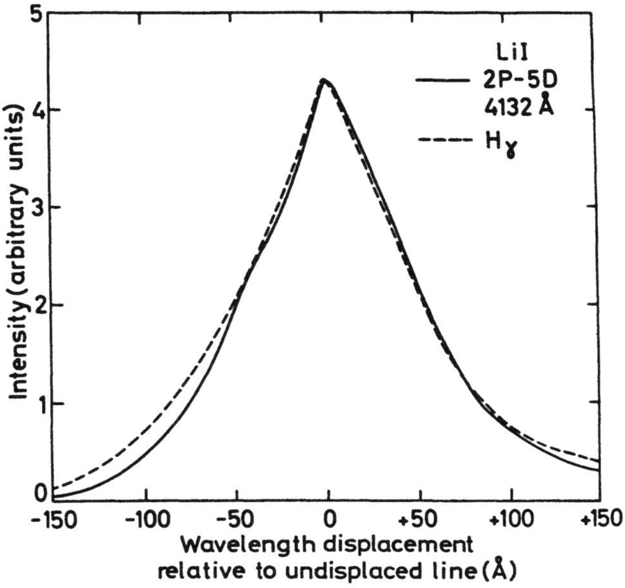

Fig. 3. Solid curve: experimental profile of 2P–5D line of lithium after calibration and substraction of underlying continuum. Dashed curve: profile of corresponding H_γ line of hydrogen calculated after Griem (1964) for $n_e = 2.5 \times 10^{17}$ cm^{-3} and $T = 10$ °K.

the line should therefore display a hydrogen-like broadening. Inspection of the line profiles indicates that this is the case: the 2P–3D and 2P–4D lines are asymmetric and shifted with respect to the unperturbed line as should be the case when quadratic Stark broadening prevails, whereas the 2P–5D line does not show any appreciable asymmetry or shift. A quantitative demonstration of this assertion is given by Fig. 3. Here the experimental profile of the 2P–5D line of lithium is shown to almost coincide with the theoretical profile of the corresponding transition in hydrogen (H_γ), calculated for the value of electron concentration which is known to exist when exploding the lithium wire, and correcting for the slight difference in wavelengths.

ACKNOWLEDGMENT

I am grateful to Dr. A. S. Kaufman for assistance in the preparation of this manuscript.

REFERENCES

1. B. Ya'akobi, A.S. Kaufman, and P. Avivi, in Proceedings of 7th (1965) Intl. Conf. on Ionization Phenomena in Gases, Vol. I, Graderinska Knjiga, Belgrade (1965), p. 799.
2. H.R. Griem, in Plasma Spectroscopy, Ch. 4, McGraw-Hill Book Co., New York (1964).
3. A.N. Filippov, Z. Phys. 69:526 (1931).

DISCUSSION

Question by: F.D. Bennett, Ballistic Research Laboratories, Aberdeen Proving Ground, Md. Since strong wire explosions produce an annular, or pipe-like, distribution of mass, why doesn't the measured electron density show this same effect? Compare X-ray data in Muller's paper in Exploding Wires, Vol. 1.

Answer by Author: Under the conditions employed here, it was found that there was no dip in the density near the axis. Furthermore, a dip in the total density does not necessarily involve a dip in the electron concentration.

Faraday Rotation Measurements of the Precursor Ionization from an Exploding Wire Discharge

Donald L. Jones*

Institute for Telecommunication Sciences and Aeronomy
Environmental Science Services Administration
Boulder, Colorado

Over the past years many workers have been concerned with the presence of weak precursor ionization well ahead of strongly ionized shock waves. Various mechanisms, such as radiation from the shock front, diffusion of electrons from the shocked region, and ultraviolet radiation from the explosion have been proposed to account for the observations. The precursor ionization from an exploding wire discharge has been measured by the Faraday rotation of a plane polarized microwave beam ($\lambda \simeq 1$ cm). While both the discharge and the subsequent approaching shock front have associated precursors, only the effects of the discharge will be presented. The source of the ionization is ultraviolet radiation from the exploding wire and from the hot plasma through which the discharge occurs. The quantity of ionization produced is a sensitive function of the pressure, the gas type, the capacitor energy, and the distance from the wire axis. Measurements have been obtained in helium, argon, and air at pressures ranging from 5 mm Hg to 300 mm Hg. By assigning proper values to the pressure and the voltage, electron densities in excess of 10^{13} cm^{-3} are obtained regularly in all three gases. With identical experimental conditions, however, the peak ionization in helium is an order of magnitude greater than in either air or argon.

INTRODUCTION

Precursor signals ahead of shock waves have been observed in many experiments over the past thirteen years [1–8]. In most cases the precursor associated with electrical discharges is ascribed principally to photoionization or diffusion of electrons from the approaching shock front [6–9]. The existence of an excited partially ionized medium ahead of the shock front can drastically

*Present address: General Electric Company, Philadelphia, Pennsylvania.

affect the hydrodynamics and the energy balance and relaxation times [10] as shown, for instance, in abnormally high ionization in shock fronts [11].

In the exploding wire discharge it is possible to isolate the precursor ionization related to the capacitor discharge from that connected with the approaching shock front. We will not be concerned with the shock-related ionization in this paper. Utilizing the Faraday rotation of a plane-polarized electromagnetic wave, propagating along the magnetic field, the electron density is measured continuously throughout the experiment with a microwave polarimeter [12, 13] operating near 35 GHz. The maximum precursor electron density attainable is observed to be well above cutoff ($N > 1.5 \times 10^{13}$ cm^{-3}). The minimum density detectable is $\sim 10^{11}$ cm^{-3}, thus a range of about 10^2 cm^{-3} is available for measurement. Ionization has been detected in helium, argon, and air at pressures ranging from 300 torr to below 5 torr. The quantity of ionization produced is a sensitive function of the pressure, gas type, capacitor energy, and distance from the wire axis. Shot-to-shot reproducibility is normally within a few percent for the same experimental conditions.

It is the purpose of this report to present microwave Faraday rotation measurements of the precursor plasma produced by an exploding wire discharge. In contrast to earlier work [14, 15], the data are analyzed using the full Appleton-Hartree magneto-ionic theory with collisional damping. The initial production of the plasma is associated with the first current maximum during which the wire explodes. However, the succeeding current maxima, arcing through the gap, are found to produce ionization densities that are an appreciable fraction of the peak value. The source of the ionization from both the wire explosion and the arcing is believed to be ultraviolet radiation from the discharge since: a) the time of the precursor ionization peak remains nearly constant throughout a wide range of pressures, b) the production peak times and amplitudes are independent of the magnetic field, c) the electron density decreases with distance from the wire axis, and d) the electron density decreases with increasing pressure.

The electron collision frequency is obtained from measurement of the attenuation of the microwave signal. The collision frequencies measured at the time of the ionization peak are an order of magnitude greater than those reported for thermal electrons by Pack and Phelps [16]. However, within a few microseconds the observed collision frequencies approach the reported values.

THEORY AND EQUATIONS

Measurement of electron densities in magneto-ionic media by the Faraday rotation of plane polarized electromagnetic waves has been achieved in the ionosphere [14] and in laboratory plasmas [15]. The rotation angle θ, in radians, is given by the formula [17]

$$\theta = \frac{e^3 2\pi}{m^2 c^2} \int_0^R \frac{BN}{\omega^2} \cos \alpha \, dr \tag{1}$$

in cgs units, where B is the magnetic field, N is the electron density, ω is the frequency, $\cos \alpha \, dr$ is the element of path length along the magnetic field, and e, m, and c are the electron change, the electron mass, and speed of light, respectively. For propagation through a uniform plasma, parallel to the magnetic field, Eq. (1) simplifies to

$$\theta = 5.36 \times 10^7 \frac{BNR}{\omega^2} \tag{2}$$

where θ is now in degrees. The derivation of Eq. (1) is based on the Appleton–Hartree equation [18] and assumes that the wave frequency is sufficiently greater than the electron collision frequency, the gyromagnetic frequency, and the plasma frequency, so that linear polarization is maintained. In the following discussion, Eqs. (1) and (2) will be referred to as the LZA (longitudinal zero absorption) approximation.

In many applications the assumptions on which Eq. (1) is based are violated. When the electron collision frequency becomes greater than about one-tenth the frequency of the sensing wave, Eq. (1) can be in error by an order of magnitude or more. Two separate effects combine to produce this error. The linearly polarized microwave beam entering the plasma is composed of a right-hand circular component, termed ordinary, and a left-hand circular component called the extraordinary wave. In a magnetoplasma, the extraordinary wave is absorbed more than the ordinary wave. This in turn leads to measurement errors when linear polarization is assumed. A more serious error is caused by collision effects on the refractive index of the ordinary and extraordinary waves. As the collision frequency nears the sensing frequency the difference between the two refractive indices approaches zero. Since the measured rotation angle, written in terms of the refractive indices, is

$$\theta = \frac{\omega R}{2c} (n_o - n_e) \tag{3}$$

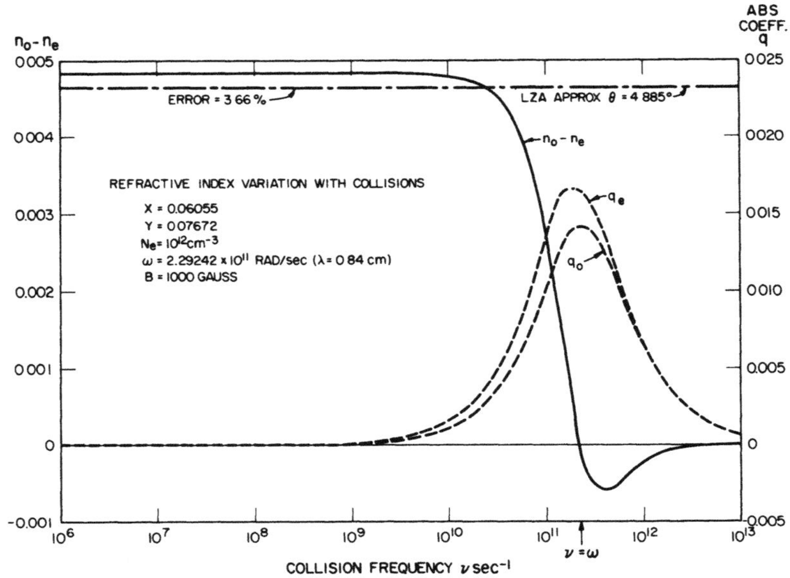

Fig. 1. Comparison of LZA approximation with full Appleton–Hartree equation for propagation along magnetic field. Solid curve is calculated from Eq. (A^1). The LZA curve does not take collisions into account. Dashed curves q_o and q_e are absorption coefficients calculated from Eq. (A^2).

large errors may result if Eq. (2) is applied in regions where it is not valid. An example of the refractive index differences calculated from the full Appleton-Hartree equation for longitudinal propagation

$$(n_{o,e} - iq_{o,e})^2 = -\frac{X}{1 \pm Y - iZ} \tag{4}$$

with $X = \omega_P^2/\omega^2$, $Y = \omega_H/\omega$, and $Z = \nu_{AH}/\omega$, is shown in Fig. 1. For this case, which corresponds to typical experimental conditions, the LZA approximation (broken line labeled LZA) follows the calculated curve (solid line labeled $n_o - n_e$) closely until the collision frequency approaches about 10% of the radian wave frequency ($Z < 0.1$). When $Z > 0.1$, the LZA solution becomes increasingly invalid. An interesting point is that the sign of the quantity $n_o - n_e$ changes at $\nu_{AH} \approx \omega$ and the formerly positive rotation now becomes negative. The dashed curves labeled q_o and q_e are the absorption coefficients calculated from (4).

It is apparent from Fig. 1 that unless the collision frequency is known to be sufficiently small, it is unsafe to apply the LZA approximation to Faraday rotation measurements. Experimentally, this means that, if damping of the wave is present, the data must be analyzed using the full Appleton-Hartree equation. When the damping can be measured, both the collision frequency and the electron density may be found. The procedure, which is given in detail in the appendix, is to first assume a frequency, a plasma slab thickness, and a magnetic field intensity. We designate the power in the arms of the polarimeter by P1 and P2, assuming the sign of the magnetic field such that rotation is toward arm 2. Now, using Eq (4), each pair of X and Z values determines a certain power level on each arm of the polarimeter. In this manner, sets of curves for constant Z and X may be drawn in the P1, P2 plane, as shown in Fig. 2. Note that when the power in each arm is unity, X and Z are both zero; this corresponds to the case of no plasma. The graph in Fig. 2 applies only to the first quadrant, i.e., $|\theta| < \pi/4$. Similar graphs can be drawn for larger rotation angles, keeping in mind that the power in the two arms becomes equal each time the wave is rotated through an angle of $\pi/2$ radians.

It is well known that the Appleton–Hartree equation is not strictly valid unless the electron-neutral collision frequency is independent of the electron velocity [19]. Sen and Wyller [20] have developed a generalized magneto-ionic theory which takes into account the velocity dependence of the collision frequency, and they

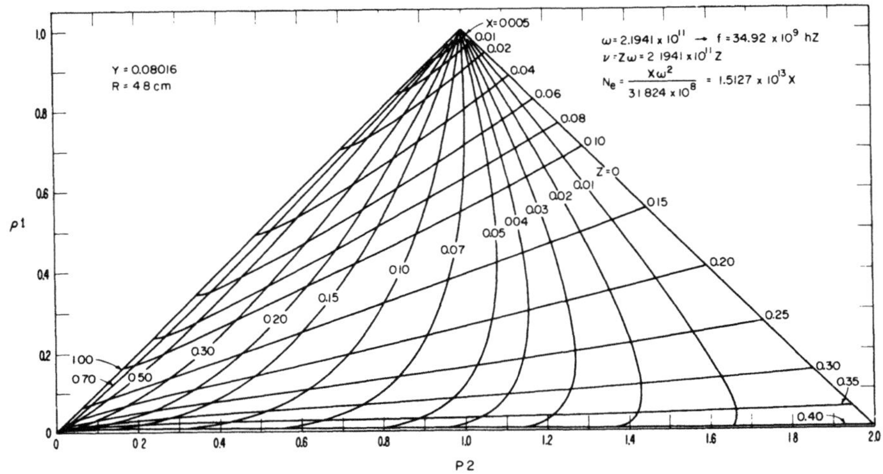

Fig. 2. Power measured on two arms of polarimeter shown as a function of $X = \omega_P^2/\omega^2$ and $Z = \nu_{AH}/\omega$. Magnetic field is such to cause rotation into arm 2. The rotation is limited to 45° or less.

have performed numerical computations for air ($\nu_m \propto v^2$). They find in general that the velocity dependence effects tend to decrease the birefringence of the medium, with the most marked deviation for transverse propagation. Comparison of the conditions for these experiments with those computed by Sen and Wyller indicates that the Appleton–Hartree equation is sufficiently accurate for the range of these experiments. Specifically, the propagation is longitudinal, and the gyromagnetic frequency, the plasma frequency, and the collision frequency are small ($Y < 0.1, X < 0.3, Z < 0.3$); all these conditions tend to minimize the difference between the Appleton–Hartree and the generalized magneto-ionic equations (see, for example, Sen and Wyller, Fig. 4).

Up to this point, only the longitudinal propagation of an infinite plane wave in an unbounded uniform plasma has been considered. In the experiment, however, the plasma is bounded by dielectric walls, the microwave beam width is of the order of a wavelength ($\sim$1 cm), and the radial plasma gradients exist.

The effects of dielectric walls on microwave plasma measurements have been studied by Jahn [21] and by Bachynski et al. [22]. In general, the presence of plasma between the walls alters the propagation conditions, and in certain circumstances, interference between the transmitted signal and reflected components can occur. In particular, when the collision frequency is small ($Z \lesssim 0.01$), the multiply reflected signal is not appreciably damped, and both the phase and the amplitude of the transmitted signal are affected. However, the phase shift at the interface does not enter into the Faraday rotation measurements, and the only source of error is the amplitude. For small Z, the LZA approximation is valid (see Fig. 1), and the electron density is found from Eq. (2). In this case, the absolute amplitude of the signal is unimportant because only the ratio of the amplitude on the two polarizations is used to obtain the rotation angle. It is only when Z is somewhat larger than 0.1 and X is 0.3 or larger that appreciable errors can occur.

The microwave beam width of approximately 1 cm limits the resolution and tends to smear out any inhomogeneities in the precursor plasma. Because of cylindrical symmetry the plasma is uniform parallel to the wire axis, but there is a radial electron density gradient (see Fig. 7). This gradient is sufficiently small so that with little error it may be assumed linear over a centimeter, and the Faraday rotation measurement over that cross section is a representative average value. The microwave signal amplitude of about 0.03 V/cm is much too small to perturb the plasma. Lastly, it is assumed that the plasma does not appreciably reflect the beam, an assumption that is observed to be true for attenuations of up to 6 db.

EXPERIMENTAL PROCEDURE

The exploding wire system is a simple RLC series circuit similar to the usual exploding wire circuitry [23] and will not be discussed here. A diagram of the device is shown in Fig. 3. The Plexiglas chamber is about 30 cm in diameter and the inside face-to-face width (R) can be varied, but in this work it is set at 4.8 cm. The wire length is also variable, but is normally 4 cm long. The energy bank is composed of two 20-kV capacitors capable of delivering 2000 J to the system at about 140 kHz. A magnetic field of up to 9000 G is provided by an electromagnet, between whose poles the chamber is inserted with the wire axis parallel to the magnetic field lines. A microwave Doppler system is used to determine the position–time relationship of the expanding shock–ionization front [11] simultaneously with the Faraday rotation measurements.

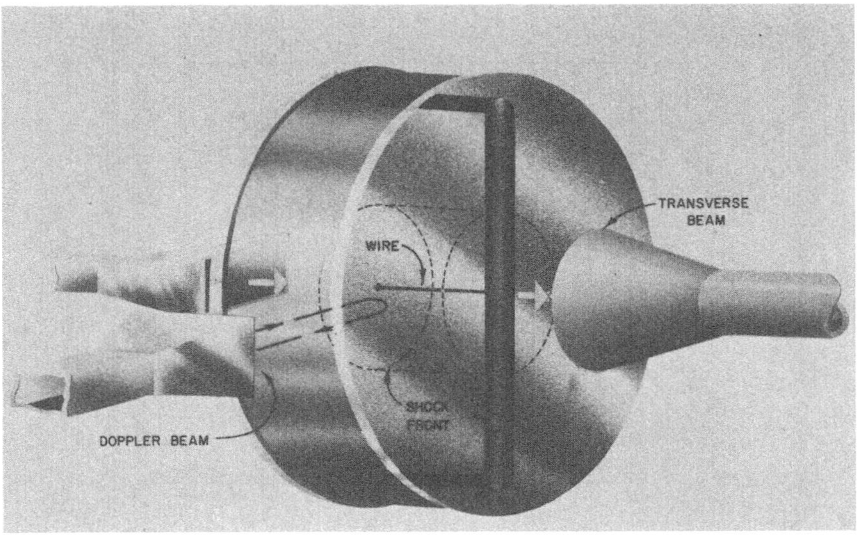

Fig. 3. Arrangement of exploding wire discharge. Transverse beam is parallel to and displaced radially outward from axis of wire. Doppler beam is normal to axis of wire. Magnetic field is parallel to axis of wire.

In operation, the transmitter arm of the plane-polarized transverse beam is placed along the axis of the magnet with the beam emerging from a hole in the center of the pole piece. The receiver arm with the conical horn lies along the axis on the opposite side of the magnet. The transverse beam is focused and aligned with Teflon lenses, resulting in a half-power beam width of about one wavelength.

EXPERIMENTAL RESULTS

Rotation (electron density = $1.88 \times 10^{11} \times \theta$ in degrees) measurements made at 12 cm from the axis, under the same experimental conditions, in helium, argon, and air display striking differences, as shown by Fig. 4. Helium is the most strongly ionized with a maximum electron density several times that of argon and more than an order of magnitude greater than air. The second peak at approximately 7 μsec on the helium curve occurs just at the second current maximum of the capacitor discharge. Thus, the spark in the gap is also an effective producer of photo-

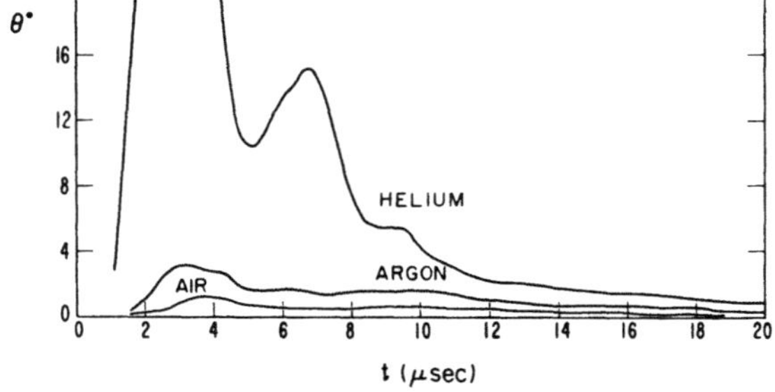

Fig. 4. Rotation angle graph for air, argon, and helium. Electron density is approximately 1.88×10^{11} $deg^{-1}cm^{-3}$. Second and third peaks result from ringing capacitor discharge. Microwave frequency is 34.92 GHz.

ionization. The lack of well-defined second and third ionization maxima in argon and air at this pressure is probably due to their larger photo-absorption cross sections. For instance, at 504 Å, argon and air both have photoionization cross sections approximately three times that of helium. At 1 torr, both argon and air are observed to display the second and third ionization peaks.

In the exploding wire system, the efficiency of transfer of energy from the capacitors to a shock wave is strongly dependent upon the total energy available from the capacitor bank (see [11], Fig. 10). Since in this experiment only 80 J/cm is available, the shock is initially very weak and damps out within a few centimeters. Assuming an unreasonably high 50 percent efficiency [11], the earliest possible arrival time of the shock wave at the radius 6 cm from the axis is 24 μsec [24], or 14 μsec after the completion of the measurements in Fig. 4. Thus, none of the ionization measured is from the shock wave.

An unknown quantity is the temperature of the electrons at the time they are produced by the photoionization, since the spectrum of the discharge has not been measured. However, at pressures of 1 torr or more the relaxation time is short [25], so that the electron will drop to the temperature of the background gas within a microsecond or so. For the gases used in this experiment, the electron-neutral collision frequencies at room temperature are about 2×10^8 sec^{-1} $torr^{-1}$ [16]. Since the magnetic field is 1 kG and the frequency is 35 GHz, linear polarization is maintained for even the peak ionization of 5×10^{12} cm^{-3} ($\omega_b/\omega = 0.08$, $\nu/\omega \approx 0.01$, $\omega_p/\omega < 0.5$).

Utilizing the same experimental conditions as for Fig. 4, the electron density at 10 μsec after the wire explosion is shown as a function of distance from the wire axis in Fig. 5. The two crosses which determine each experimental point represent separate measurements from different shots. The distance between the crosses gives a measure of the reproducibility from shot to shot. No point was obtained for helium 6 cm from the axis because the beam was cut off for the entire 20 μsec observation period.

The geometry of the 4-cm source wire at distances from 6 to 12 cm from the axis is between cylindrical and spherical, with the spherical approximation improving with distance from the axis. Therefore, if photoionization is the source of the electrons, the curves of Fig. 5 should show a decay lying between the inverse first and second powers of the distance from the wire axis. Further, an exponential decay due to photoabsorption must be superposed on the geometrical effect. The equation for this model is

$$\theta_a\, r_a^n\, e^{r_a \alpha} = \theta_b\, r_b^n\, e^{r_b \alpha} \tag{5}$$

where the subscripts a and b denote difference measurement points, n is the geometrical decay exponent (1 is for cylindrical and 2 is for spherical), α is the absorption coefficient $n_o\sigma$, n_o is the ambient gas density, and σ is the photoionization cross section. Application of Eq. (5) to helium and argon leads to cross sections of about 6×10^{-18} cm^2 for the cylindrical model and about 4×10^{-18} cm^2 for the spherical model in both gases. Qualitatively, these cross sections are of the proper order of magnitude [26] and we conclude that our photoionization model is consistent with the observations. Quantitative measurement of cross sections is not yet possible because a) n is varying from point to point, b) the lack of monochromaticity of the source with the result of a nonconstant α, c) at 10 μsec there may still be a contribution to the ionization from the arc, and d) losses due to recombination and diffusion are not considered.

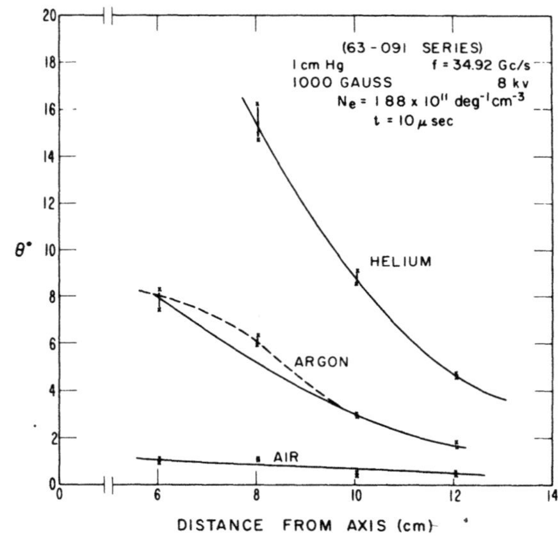

Fig. 5. Rotation angle as function of distance for air, argon, and helium at 10 μsec after switch-on. Electron density is $N = 1.88 \times 10^{11}$deg^{-4}cm^{-3}. Each cross represents measurement from separate shots.

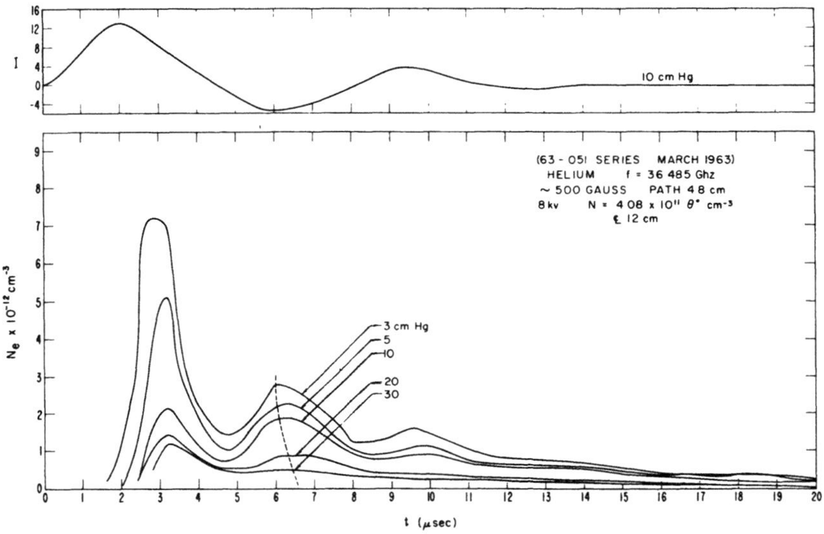

Fig. 6. Electron density graph for pressure range 3 to 30 cm Hg in helium. All shots made at 8 kV (77 J/cm) with 16.6 mil copper wire. Magnetic field strength is 500 G, transverse microwave frequency is 36.485 GHz. Path length is 4.8 cm.

In air, the level of ionization is so low under these conditions that the uncertainty from measurement errors precludes the estimation of cross sections. The relatively low level of ionization in air, however, recommends this gas for those applications in which precursor ionization is detrimental. Curves of precursor ionization in helium for pressures of 30 to 300 torr are shown in Fig. 6. The helium used was analyzed by a mass spectrometer. At 50 torr, no impurities were found, but at 5 torr nitrogen was barely detectable. Resolution of the mass spectrometer was 100 ppm. The current at 300 torr is highly damped with only a single peak, and the rotation-absorption data are also singly peaked. At 30 torr, the discharge is highly oscillatory and several peaks in the ionization are evident.

Collision frequency measurements are generally less precise than electron density measurements because the absorption, not just the ratio of amplitudes, must be correctly determined. Effects such as the previously mentioned dielectric boundaries can, in certain cases, cause large errors. Also, as transmission through the plasma approaches either 0 or 100%, the precision with which the collision frequency can be measured approaches zero.

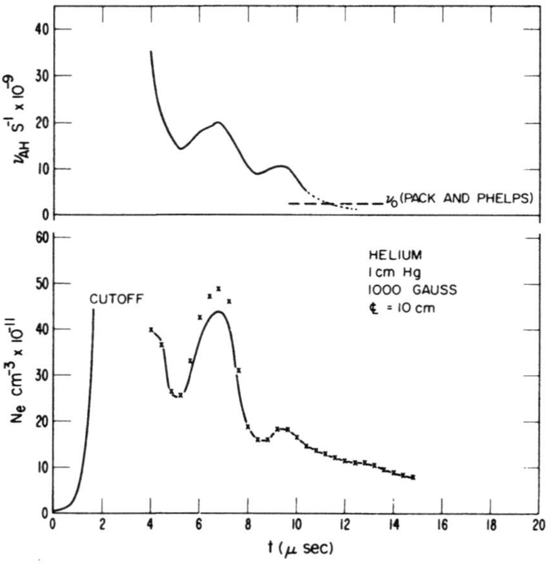

Fig. 7. Collision frequency for helium at 1 cm Hg shown in upper curve. When $\nu_{AH}/\nu_o \cong 1$, the electrons have returned to room temperature. Lower graph gives time evolution of electron density. Crosses represent LZA analysis of data; their displacement from solid curve is a measure of error in LZA approximation. Dashed portion of collision frequency curve indicates region where measurements become unreliable because of small attenuation.

The time evolution of the electron collision frequency for helium at 1 cm Hg is shown on the upper curve of Fig. 7. The two peaks at 6.8 and 9.2 μsec after switch-on correspond in time to the ionization peaks on the lower curve and to the current maxima of Fig. 6. The measured collision frequency ν_{AH} approaches the room temperature value [16] as the gas cools. As the collision frequency drops to the ambient value the precision of measurement deteriorates because the signal level has almost fully recovered and the mechanical errors of measurement have become large. The maximum measured value of ν_{AH} is 35×10^9 $\sec^{-1}$. This corresponds to a temperature of approximately 3 eV, if we can assume the average collision frequency of the gas, ν_m, is related to the measured value of $\nu_m/\nu_{AH} = 4/3$ [27] and if $\nu_m \propto T^{1/2}$.

Attenuation measurements leading to the collision frequency contain an ambiguity if the electron–ion collision frequency is appreciable. The collision parameter Z strictly refers to an effective collision number that causes damping of the wave. The damping can be caused either by electron-neutral (hard sphere) collisions or by electron–ion (long-range coulomb) collisions. Using the formulation of Spitzer [28] for the electron–ion collisions, it is easily shown that 300°K in helium at 10 torr, with 10^{12} electrons cm^{-3}, the coulomb collision frequency is approximately one-half the electron neutral frequency.

The mechanical errors of measurement of the electron density are estimated to be no more than 10%, except for signal amplitude near zero. Sources of error include scaling of the oscilloscope traces, which can be accomplished with a standard deviation of ± 0.04 degree of rotation; crystal detector calibration, which is responsible for errors of a few percent; and interpolations on the graph shown in Fig. 2, which also may be responsible for errors of a few percent. The errors of measurement of the frequency and magnetic field are negligible compared with the others.

CONCLUSIONS

We have shown that the precursor ionization produced by the exploding wire discharge can be of the order of 10^{13} electrons cm^{-3} or more at pressures of 10 torr and above in helium. At lower pressures, the level of precursor ionization generally exceeds 10^{13} cm^{-3} in all gases in our experimental device. We have also shown that the precursor plasma equilibrates rapidly, providing a convenient source of plasma at the temperature of the gas.

In agreement with other workers, we infer the source of the plasma to be ultraviolet radiation produced by the explosion of the wire. This result must be considered tentative until direct evidence of the spectrum of the discharge is available.

Lastly, we have provided a method, utilizing the Faraday rotation of a microwave, that allows measurement of the collision frequency in the plasma.

A part of this work was reported at the American Physical Society meeting at San Diego, California, in 1963 [7]. A short note discussing Figs. 4 and 5 is scheduled to appear in The Physics of Fluids.

APPENDIX

The procedure for developing graphs of the electron density and collision frequency as a function of the measured power on the two arms of a polarimeter, similar to Fig. 2, is given below. First solve Eq. (4) for the real and imaginary parts

$$n^2_{o,e} = \frac{+[(1 \pm Y)^2 - X(1 \pm Y) + Z^2] + \{[(1 \pm Y)^2 - X(1 \pm Y) + Z^2]^2 + (XZ)^2\}^{1/2}}{2[(1 \pm Y)^2 + Z^2]} \tag{A^1}$$

and

$$q^2_{o,e} = \frac{-[(1 \pm Y)^2 - X(1 \pm Y) + Z^2] + \{[(1 \pm Y)^2 - X(1 \pm Y) + Z^2]^2 + (XZ)^2\}^{1/2}}{2[(1 \pm Y)^2 + Z^2]} \tag{A^2}$$

For a given set of the X, Y, Z parameters, the refractive (n_o, n_e) and absorption (q_o, q_e) indices are calculated for the two characteristic waves. Since $q_o \neq q_e$, in general, the wave will be elliptically polarized. The amplitudes of the ordinary and extraordinary waves are

$$E_{ord} = E_o e^{-K_o R} = E_o e^{-\frac{\omega q_o}{c} R} \tag{A^3}$$

and

$$E_{ext} = E_o e^{-K_e R} = E_o e^{-\frac{\omega q_e}{c} R} \tag{A^4}$$

The semimajor axis of the ellipse is

$$E_v = E_{ord} + E_{ext} \tag{A^5}$$

while the semiminor axis is

$$E_l = E_{ord} - E_{ext} \tag{A^6}$$

Normally, the linear polarized wave that is received when no plasma is present is adjusted so that each arm of the polarimeter receives equal power, i.e., $\theta = \pi/4$.

The calculated rotation angle is then

$$\theta = \pi/4 - \left[\frac{\omega R}{2c}(n_o - n_e)\right] \tag{A^7}$$

and the relative power on the arms of the polarimeter is

$$P1 = (E_v \sin\theta)^2 + (E_l \cos\theta)^2 \tag{A^8}$$

and

$$P2 = (E_v \cos\theta)^2 + (E_l \sin\theta)^2 \tag{A^9}$$

For no damping, the measured rotation angle is

$$\theta_T = \tan^{-1}\left(\frac{P1}{P2}\right)^{1/2} \tag{A^{10}}$$

$$\theta = \pi/4 - \theta_T \tag{A^{11}}$$

which is used directly in Eq. (2) to find the electron density.

REFERENCES

1. R.G. Shreffler and R.H. Christian, J. Appl. Phys. 25:324 (1954); H.D. Weymann, Phys. Fluids 3:545 (1960); P. Gloersen, Phys. Fluids 3:857 (1960).
2. H.G. Voorhies and F.R. Scott, Phys. Fluids 2:576 (1954).
3. V. Josephson and R.W. Hales, Space Technology Laboratory Report, STL/TR – 60,0000 – 19313 (1960) (unpublished)..
4. D.W. Mahaffey, R.I. Schoen, and H.G. Ahlstrom, Proc. V Intl. Shock Tube Symp. (1965).
5. P.J. MacVicar-Whelan and P.R. Smy, Can. J. Phys. 45:2343 (1967).
6. E.A. McLean, A.C. Kolb, and H.R. Griem, Phys. Fluids 4:1055 (1961).
7. D.L. Jones, Phys. Fluids 5:1121 (1962); Phys. Fluids, Vol. 11 (to be published); Bull. Am. Phys. Soc. 9:308 (1964).
8. H. Groenig, Phys. Fluids 6:142 (1963).
9. L. Wetzel, Phys. Fluids 5:824 (1962).
10. A.C. Kolb and H.R. Griem, in Atomic and Molecular Processes, D. R. Bates, ed., Academic Press, Inc., New York (1962), p. 173ff.
11. D.L. Jones and R.M. Gallet, in Exploding Wires, Vol. 2, W.G. Chace and H.K. Moore, eds., Plenum Press, New York (1962), p. 127ff.
12. P.J. Allen and R.D. Tompkins, Proc. IRE 47:1231 (1959).
13. G.L. Ragan, in Microwave Transmission Circuits, MIT Rad. Lab. Series, Vol. 9, G.L. Ragan, ed., McGraw-Hill Book Co., New York (1948), p. 375.
14. C.G. Little and R. S. Lawrence, J. Res. NBS 640:335 (1960).
15. T. Consoli and M. Dagai, J. Nucl. Energy, Part C, 3:115 (1961).
16. J.L. Pack and A.V. Phelps, Phys. Rev. 121:798 (1961).
17. F.B. Daniels and S.J. Bauer (1959), J. Franklin Inst. 267:187.
18. J.A. Ratcliffe, The Magneto-Ionic Theory and Its Applications to the Ionosphere, Cambridge University Press, Cambridge, England (1959), p. 73.
19. L.G. Huxley, Phil. Mag. 23:210, 442 (1937); R. Jancel and T. Kahan, Nuovo Cimento 12:573 (1954); W. Pfister, The Physics of the Ionosphere, Cambridge University Press, Cambridge, England (1954), p. 394.
20. H.K. Sen and A.A. Wyller, J. Geo. Res. 65:3931 (1960).
21. R.G. Jahn, Phys. Fluids 5:681 (1962).
22. M.P. Bachynski, C.G. Cloutier, and K.A. Graf, RCA Victor Research Report 7-801-26 (1963) and M.P. Bachynski and C.G. Cloutier, Paper 9.5, 1963 PTGAP International Symposium, July 1963, Boulder, Colorado.
23. W.G. Chace, in Exploding Wires, Vol. 1, W.G. Chace and H.K. Moore, eds., Plenum Press, New York (1959), p. 15.
24. G.C. Vlases and D.L. Jones, Phys. Fluids 9:478 (1966).
25. E.W. McDaniel, Collision Phenomena in Ionized Gases, John Wiley & Co., New York (1964), p. 22.
26. R.W. Ditchburn and U. Opik, in Atomic and Molecular Processes, D.R. Bates, ed., Academic Press, Inc., New York (1962), p. 79ff.
27. W. Pfister, The Physics of the Ionosphere, Cambridge University Press, Cambridge, England (1954), p. 394.
28. L. Spitzer, Jr., Physics of Fully Ionized Gases, Interscience, New York (1956), pp. 78–80 (5–25).

DISCUSSION

Question by H.F. Swift, University of Dayton: Could a selective absorber be placed in front of your wire to guarantee that the early ionization is caused by radiation that can pass through the absorber?

Answer by Author: The spectral head for He is at ~ 500 Å. I believe any ultraviolet measurements would have to be in the vacuum ultraviolet.

Question by W.G. Chace, Air Force Cambridge Research Laboratories, Bedford, Mass.: You did not tell what wire you used. Does it make much difference what the wire is?

Answer by Author: The wire was commercially available copper. Copper and silver give essentially the same results; aluminum gave less precursor ionization. We did not try lithium wires. More important, however, is that the wire explode at the current peak. Thus, the size of the wire for a particular capacitor system should be properly chosen, as I pointed out at the Second Exploding Wire Conference.

Question by W. Lochte-Holtgreven, Kiel University, West Germany: Was the precursor radiation found at a very well-defined distance from the shock front?

Answer by Author: No. Within my ability to resolve the speed of the arrival ($\sim 10^{-8}$ sec), I could find no front. That is, the entire chamber was ionized very quickly.

Question by H. Knoepfel, Laboratori Gas Ionizzati, Rome, Italy: What is the speed of the precursor wave compared to that of the shock wave? I feel that your effect is not a precursor wave in the usual sense.

Answer by Author: I could not measure the speed of the front; it was greater than 10^8 cm/sec. These precursors, while not associated with the shock front, do modify the medium the shock traverses. Any thermodynamic studies depending on shock heating would have to take this precursor effect into account.

Time-Resolved Emission and Absorption Studies of Exploding Wire Spectra

Esther C. Cassidy and Stanley Abramowitz

High Voltage Laboratory
National Bureau of Standards
Washington, D. C.

Emission and absorption studies of the spectra produced by exploding wires in oxygen, nitrogen, hydrogen, or argon atmospheres are described. A drum camera and a rotating shutter are used for continuous and time-resolved recording of the explosion spectra. A high-intensity light source (a second exploding wire or Lyman flashtube), synchronized with the rotating shutter, is flashed through the explosion vapor for the absorption studies. Results from experiments performed under a variety of controlled conditions are presented. Effects of environment, pressure, and electrical energy on the spectrum are discussed, and conditions found conducive to production of selected features are indicated. The exploding wire is shown to be a rich source for study of the AlO molecule, and observation of a number of new bands (believed to be part of the $B^2\Pi_i - X^2\Sigma^+$ transition) is reported. Preliminary experiments with titanium wires exploded in oxygen suggest that the exploding wire may also be a desirable source for spectral studies of the TiO molecule. Some results from this work are included.

INTRODUCTION

Because of its convenience as a laboratory technique for generating extremely high temperatures (above 2000°C), the exploding wire is a promising source for spectroscopic studies of high-temperature atomic and molecular species. Nagaoka et al. [1] have conducted numerous studies of atomic species produced by exploding wires. Coheur, Delsemme, Lejeune, Malet, and Rosen [2–7], who produced and observed the spectra of TiO, AlO, FeO, CaO, and CuO in exploding wire experiments, are among those who have successfully applied the exploding wire method to studies of molecular spectra. The exploding wire has also been employed indirectly for spectral studies, as a flash source for photochemical reaction experiments [8, 9] and absorption studies [10]. More recently, finely divided metals obtained from exploding wires have been used as substrates for the study of the infrared spectra of adsorbed molecules [11].

Our purpose in the present work was to observe the spectra produced by wires exploded under selected controlled conditions, to search the spectra over the entire duration of light emission for features from previously unobserved species and, of necessity, to develop techniques of fast measurement which would permit time-resolved emission and absorption studies of the explosion spectrum. This latter task has involved the use of a second (synchronized) exploding wire as a source for absorption studies of the species in the main explosion mixture. In contrast to most earlier molecular studies by this method, where battery currents of about 100 A at 100 V were used, capacitor discharge currents as high as 150,000 A at voltages up to 20 kV were employed to explode the wire samples. In addition, because of the marked dependence of the emitted radiation upon the design and operating conditions of the exploding wire system, considerable care was taken to develop a versatile experimental system, with provisions for ready adjustment and variation of the amount and rate of electrical energy input, wire material, environmental gas and its pressure, and the interval of observation.

APPARATUS AND PROCEDURE

The capacitor discharge circuit (15 to 60 μF, 20 kV max., circuit inductance $\sim$ 0.16 μH, ringing frequency $\sim$ 50 kHz) has been described elsewhere [12]. The sample wires (0.14 mm in diameter, 9.5 cm in length) were enclosed in sealed cylindrical (7.6 cm ID, 9.5 cm in length) vessels, equipped with demountable quartz windows. Although other materials were tried, glass was found to be the most suitable vessel material. Bakelite vessels were often ignited by the explosion, particularly when the wire was exploded in an oxygen medium. The walls of acrylic plastic vessels were also frequently blackened and burned by the explosion vapor. Polytetrafluoroethylene was unsuitable because it introduced impurities to the mixture (C_2 and CN bands predominated and confused the spectrum). Glass, on the other hand, showed none of these effects. Cleansing of the vessel was also simplified; the entire vessel could be soaked and washed in hydrochloric acid. Explosion of the vessel was the only difficulty encountered with a glass vessel. With the pressures (up to 1 atm) and energies used here, a wall thickness of 4 mm was required. Even then, vessel explosions occurred when titanium wires were exploded at higher voltages (above 15 kV) and pressures (above 0.5 atm), and it should be expected that this hazard will be increased with faster discharge circuits.

The electrodes which clamp the wire also require careful selection when spectral studies are to be conducted. For clarity in the results, the electrodes should be made from the same material as the wire. It was found, however, that oxygen-free, high-conductivity copper was suitable for this purpose; it introduced only a few strong Cu I lines to the spectrum and suffered minimal melting and erosion effects from the discharge.

Figure 1 shows the explosion chamber with a wire clamped in place. Rubber O-rings were placed between the end flanges of the vessel and the electrodes to effect a vacuum-tight seal. Bakelite ring clamps placed over the electrodes were drawn together by nuts at the ends of four threaded rods, to compress the seals. Before each experiment, the chamber was evacuated to approximately 4×10^{-4} torr, and then charged with argon, oxygen, hydrogen, or nitrogen to a controlled pressure between 0.01 and 1 atm. However, since the molecular features of the spectra were found to be more intense and more enduring at reduced pressures, most of the experiments were conducted with the gas pressure fixed at about 76 torr.

A spectrograph (f/6.3, focal length = 0.75 m) with interchangeable plane gratings (a 600 lines/mm grating blazed at 5000 Å with dispersion 20 Å/mm in the first order and a 2160 lines/mm grating blazed at 3000 Å with dispersion 5 Å/mm in the first order) was employed for the spectroscopic observations. Quartz lenses were used to focus the exploding wire radiation on the bilateral straight entrance slit. Best results were obtained with an entrance slit height of 3 mm and the slit width adjusted to 50 μ. In experiments with the rotating disc (described below), slit widths down to 5 μ were used.

During the explosion, the spectrum at the focal plane of the spectrograph was photographed with a high-speed drum camera. The speed (600 rps max.) of the drum was adjusted to give maximum time resolution of the entire duration of detectable light emission over the 31.9-cm length of 70-mm film. (With the present system, at maximum drum speed, the time history of the spectrum was re-

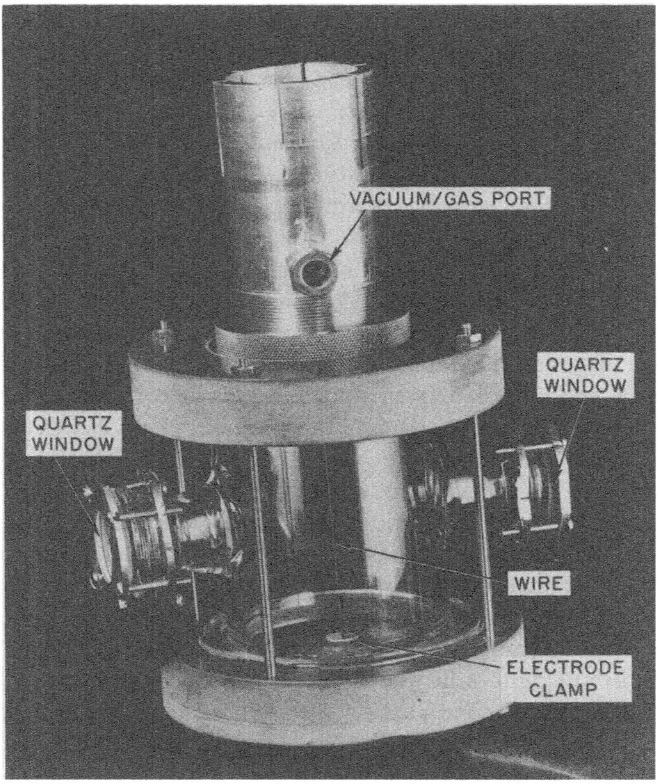

Fig. 1. Explosion chamber with wire installed.

solved to 15 μsec.) These continuous photographs, which showed that the explosion spectrum varied markedly with time, were searched for intervals which appeared more conducive to study of the features from selected species in the explosion mixture. Then, since the drum camera limited our observations to the visible region and reduced the effective dispersion of our optical system, more detailed information on the structure of the spectrum and broader wavelength coverage were obtained from time-resolved photographic plates, also taken at the focal plane of the spectrograph. Time resolution was achieved by placing a rotating shutter disc, similar to that described by Bartky and Bass [13], before the entrance slit. The disc was driven by a reversible synchronous motor (1800 rpm, $^1/_{50}$ hp). Slots, as shown in Fig. 2, were machined in the disc for high-speed shuttering of the spectrograph and for generating timing pulses. The latter were obtained as light from a small 6-V bulb passed through a slot in the rotating disc to a photomultiplier. The photomultiplier's response was inverted, amplified, and delayed by the technique indicated in the figure. This signal was then further amplified and passed through a thyratron to a pulse transformer. The output of the latter produced a spark which ionized the spark gap switch in the main exploding wire circuit. It was thus possible to synchronize the shutter slot's arrival at the entrance slit of the spectrograph with a selected interval of the explosion to within ±50 μsec. With the present disc design, the shutter slot (two degrees) allowed light to pass into the spectrograph for any selected 180-μsec interval. Portions of typical drum camera records are given in Figs. 3 and 4. The AlO (in 3) and Fe (in 4) plate spectra were included for reference purposes. A typical time-resolved plate from a wire explosion (under conditions identical to those of the drum record), taken by use of the rotating shutter 1 msec after initiation of the discharge, is given at the bottom of Fig. 4.

For absorption studies, radiation from an intense pulsed light source was focused on the explosion mixture as shown in Fig. 2.

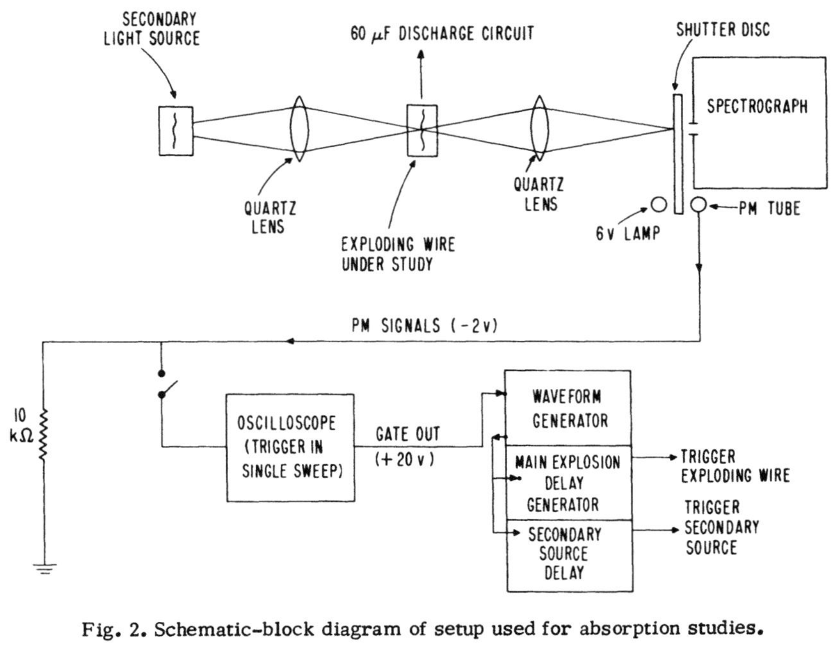

Fig. 2. Schematic-block diagram of setup used for absorption studies.

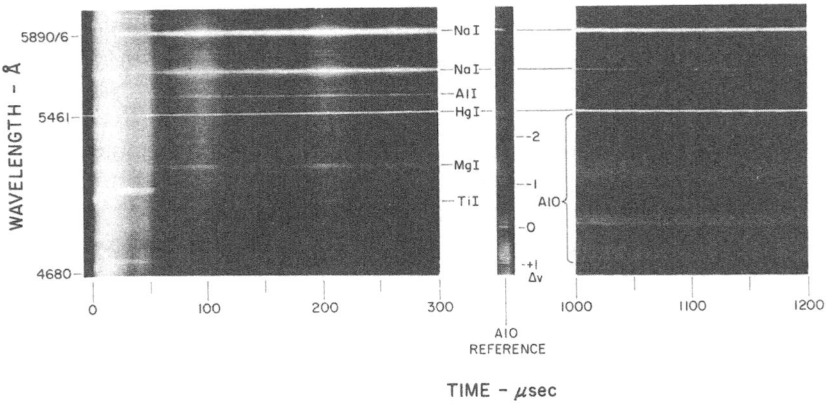

Fig. 3. Drum camera record from aluminum wire exploded in nitrogen with trace of oxygen (60 μF at 14 kV).

A photographic plate of the spectrum, with features from the species in the system appearing in absorption against the strong background continuum of the secondary source, was thus obtained at the focal plane of the spectrograph. The secondary flash source was synchronized with the selected interval and with the arrival of the shutter slot at the entrance slit of the spectrograph by use of an additional delay generator. In some experiments, an exploding titanium wire (0.14 mm in diam., 6.2 cm in length), enclosed in a sealed glass vessel with a quartz window was used as the secondary source. The vessel was charged with oxygen (pressure ≈ 0.5 atm). In other experiments, a Lyman flashtube [14] was used as the secondary source. In general, the continuum radiation emitted by the latter was less intense than that from the titanium wire. However, because of the simplicity of its operation (no reinstallation of wires was required), the flashtube was used for most of the experiments. Its use was also desirable because very few atomic lines and no bands appeared in its spectrum; confusion of spectral features from the lamp with those from the explosion mixture under study was therefore less likely. However, maximum intensity was required for observations at wavelengths shorter than 2500 Å, because the explosion vapor scattered or absorbed most of the radiation in this region. In fact, several band sequences (at shorter wavelengths) of the ultraviolet systems of AlO could be observed only during the first 400 μsec of the main explosion, when the initial background continuum radiation from the wire under study supplemented that from the secondary source. In such cases, the titanium wire source was used because of its high intensity.

RESULTS FROM EMISSION STUDIES

Upon initiation of the discharge, the drum records showed that the explosions emitted intense continuum radiation, with broadened, very intense line features from constituents of the wire material. After the initial continuum, the spectral distributions produced by aluminum or titanium wires exploded in vacuum, argon, or nitrogen were characteristic of the wire material (and its impurities). Numerous lines from atomic and ionic species were quite distinct, particularly during the interval between 600 and 1100 μsec (after the intensity of the background continuum has subsided). Al I, Al II, Ti I, Ti II, Na I, Mg I, Fe I, Ca I, Ca II, Cu I, Zn I, and Mo I were among the many species produced. Most of the lines were

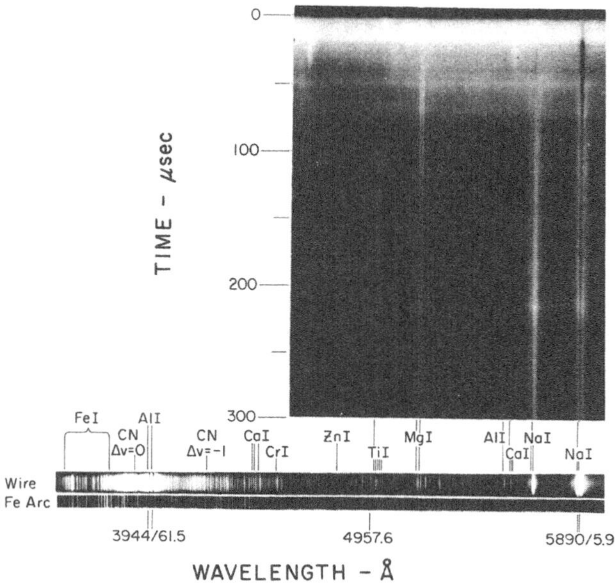

Fig. 4. Drum camera (top) and time-resolved plate (bottom) of spectrum from exploding aluminum wire in nitrogen (60 μF at 14 kV).

observed in emission, with the exception of the Al I and Na I doublets at 3944 and 5890 Å, respectively, which were at times self-reversed. Pulsations in the spectral intensity [15] were observed for about 400 μsec. These pulsations cannot be attributed to the oscillating discharge current because they persisted after the discharge current had ceased. (Simultaneous measurements of the discharge current and of the voltage across the wire sample indicated that the electrical discharge was completed in less than 90 μsec [12].) The intensity of the line features increased and decreased periodically in coincidence with the pulsations in the continuum intensity. Moreover, the more intense lines appeared broadened during the intervals of greater intensity, thus suggesting that the pressure of the radiating vapor is higher at these times and that the pulsations are caused by reflecting pressure or shock waves. Further study of this phenomenon would be most interesting and perhaps useful for correlation with hydrodynamic studies of the wire explosion.

Band features were observed with wires exploded in vacuum, argon, or nitrogen in only two instances: once when AlO bands due to air leaked into the explosion chamber (see Fig. 3) and again when C_2 and CN bands were seen due to the use of a polytetrafluoroethylene (rather than glass) vessel.

Titanium wires exploded in oxygen produced the most intense radiation (5 and 10 times more intense than that from aluminum wires). The spectral distribution was found to depend upon oxygen pressure. Figure 5 shows spectra produced at three different pressures. At very low pressures (e.g., 0.75 torr) a complex line spectrum, similar to those produced in argon, nitrogen, and hydrogen, was produced. At higher pressures, the continuum intensity was found to be increased, and weaker lines could not be distinguished. At 76 torr continuum radiation was predominant for several milliseconds, and weak absorption features, which were more easily distinguished on the drum camera records, of the $C\,^3\Pi - X^3\Pi$ and $^1\phi - {}^1\Delta$ transitions of TiO were observed.

With aluminum wires exploded in oxygen and hydrogen, chemical reaction occurred almost immediately (in less than 50 μsec), producing AlO and AlH bands, respectively. The latter appeared only in absorption, and were most distinct between t = 50 and 500 μsec. With oxygen, AlO bands dominated the spectrum for several milliseconds. It was interesting to find that the Δv = 0 and +1 sequences of the Blue-Green System could be observed in emission for as long as 12 msec after initiation of the discharge. When no second-

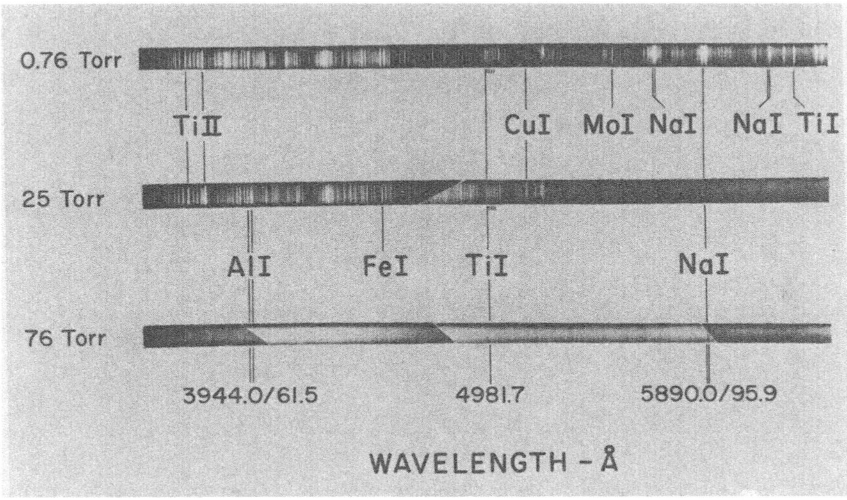

Fig. 5. Time-resolved spectra ($t \approx 150$ to 330 μsec) from titanium wires in oxygen at various pressures (60 μF at 14 kV).

ary source was used, the interval between 200 and 400 μsec was found most suitable for observation of the AlO bands in emission, while the interval between 40 and 100 μsec was most favorable for absorption studies.

Plates (a) and (b) in Fig. 6 show portions of the Blue-Green and Ultraviolet Systems, as observed at different times during the explosion. At $t \approx 1$ msec (a), only the most intense of the atomic lines remained, and many bands in the $\Delta v = +3$ through $+5$ sequences of the Blue-Green System, which could not be distinguished at $t \approx 300$ μsec (b), were observed in emission. However, the $\Delta v = +2$ through -1 sequences of the $A\,^2\Sigma^+ - X\,^2\Sigma^+$ transition (which are less distinct at 1 msec) are sharply defined in (b), and the $B\,^2\Pi - X\,^2\Sigma^+$ transition (which was not observed at 1 msec) is seen in absorption.

Plate (c) in Fig. 6 shows the spectrum at $t \approx 300$ μsec when a higher energy was employed for explosion of the wire. It was hoped that greater energy input would increase the intensity of the background continuum, and thereby permit observation of the band structure in the ultraviolet region. It was found, however, that greater energy had several effects: (1) it increased the intensity of the continuum, (2) it produced and intensified more lines from atomic and ionic species, and (3) it prolonged the total period of radiation. Our results suggested, therefore, that higher energies are desirable if the explosion is to be utilized as an intense light

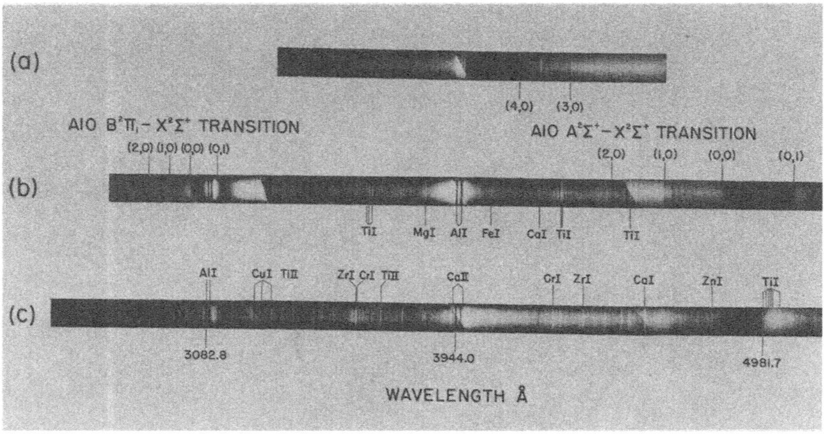

Fig. 6. Time-resolved spectra from aluminum wires in oxygen (pressure, 76 torr) at different intervals and energies: (a) $t \approx 950$ to 1130 μsec; 60 μF at 14 kV; (b) $t \approx 200$ to 380 μsec, 60 μF at 14 kV; and (c) $t \approx 200$ to 380 μsec, 60 μF at 18 kV).

source or for study of atomic and/or ionic species. On the other hand, for studies of molecular species, more energy only seems to obscure the structure of the molecular features by producing more intense continuum and line radiation.

ABSORPTION STUDIES

The fact that the molecules AlH and TiO were observed only in absorption suggested that they were produced only in the ground or some other long-lived electronic state. It was thought, therefore, that their spectral features might be observed more effectively by using a secondary continuum source for absorption studies of the explosion products. Typical results from exploratory absorption experiments (with the 600 lines/mm grating), in which aluminum wires were exploded in hydrogen, nitrogen, and oxygen (pre-explosion pressure = 76 torr), are given in Fig. 7. In these initial experiments, an exploding titanium wire was used as a secondary source. All of the plates were taken about 1 msec after initiation of the main discharge. In spite of the emission lines from the titanium wire, band features from the AlH $A^1\Pi - X^1\Sigma$ and the AlO $A^2\Sigma^+ - X^2\Sigma^+$ transitions and line features from atomic species (e.g., Al I, Na I, Cu I, etc.) in the explosion mixtures are clearly evident in absorption. The 2160 lines/mm grating was then installed in the spectrograph for further studies at higher resolution of the spectrum of AlO.

THE SPECTRUM OF AlO

The visible bands of AlO have been studied extensively. Tyte and Nicholls [16] have recently collected and correlated the results of the most significant studies in the first section of their Identification Atlas Series.

The Blue-Green System as observed in the present work was distinct and well developed. However, examination of the spectral plates indicated that significant additions to the extensive observations of Shimauchi [17] and Becart and Mahieu [18] were not likely. The details of the spectra were, therefore, not carefully measured, and attention was directed to the ultraviolet region.

In 1941, Coheur and Rosen [3] reported observation of three systems of bands which they attributed to AlO: (1) a complex group of bands between 3322 and 2879 Å; (2) a group of three broad bands between 2879 and 2686 Å; and (3) a weak, complicated system be-

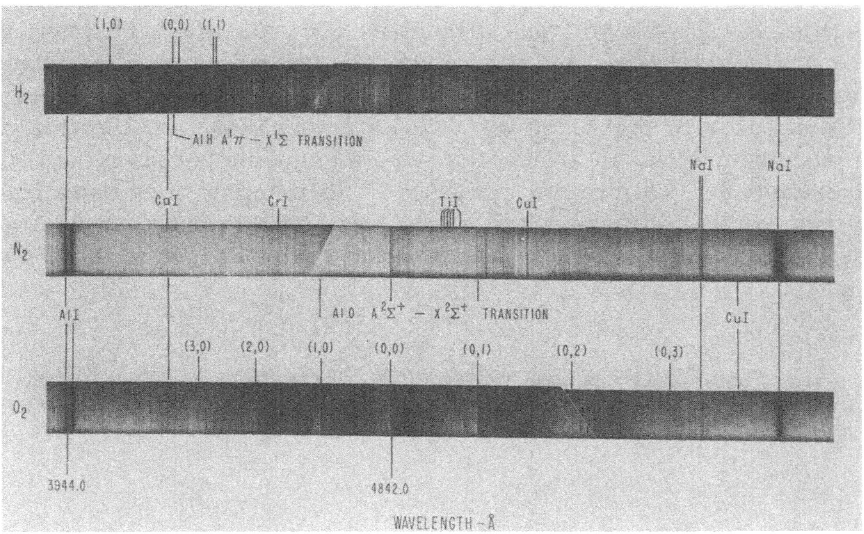

Fig. 7. Time-resolved absorption spectra from aluminum wires exploded in various gases; pre-explosion pressure, 76 torr.

tween 2686 and 2390 Å. As in the present study, they worked with an exploding wire source and with rather low dispersion instruments. In 1959, Goodlet and Innes [19] studied a few of the more intense bands of the system near 3000 Å with higher resolution, using a hollow-cathode source. They also observed a fragmentary system near 2500 Å. Finally, the recent work of Krishnamachari et al. [20], which appeared after completion of the present work, reports analysis of several bands (attributed to a $^2\Sigma^+ - X\,^2\Sigma^+$ transition) in the region 2300 to 2800 Å, as observed in a dc aluminum arc in oxygen at low pressures. In all of these ultraviolet studies, the bands were observed in emission.

In the present work, a rather extensive, complex series of bands was observed in absorption in the region from 2200 to 3400 Å. Comparison of our results with the fragmentary measurements of Coheur and Rosen [3] showed that there was good agreement between the wavelengths of the more intense bands. However, our observations showed many more bands, which also fitted into the schemes predicted by Coheur and Rosen for two of the ultraviolet systems (the system between 3322 and 2879 Å and the system between 2686 and 2390 Å). A complete listing of the observed bands is given elsewhere [21].

A portion of the spectrum which we have, in accordance with Coheur and Rosen, attributed to the $B\,^2\Pi - X\,^2\Sigma$ transition is shown in Fig. 8. This plate was taken about 500 μsec after initiation of the main discharge, by use of the shutter wheel and the Lyman flashtube. The oxygen pressure before the explosion and the energy stored prior to discharge were 0.03 atm and 2430 J, respectively. A spectrum from an iron arc in air was superimposed on the AlO spectrum for calibration purposes. In printing of the spectral plates, several different exposures were used to bring out details of the various sequences. The spectra were enlarged to the scale

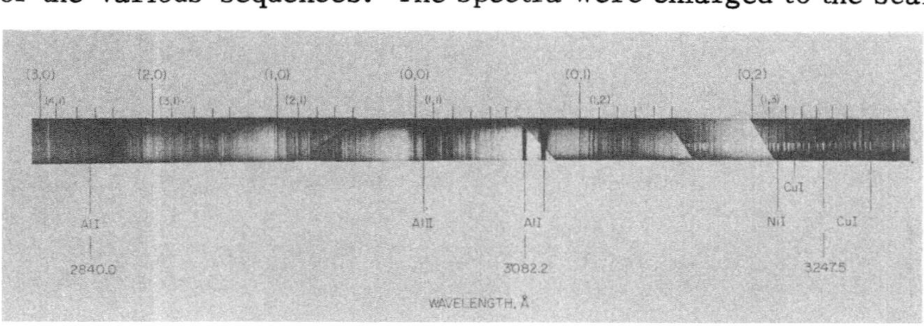

Fig. 8. Time-resolved absorption spectrum of $B\,^2\Pi_i - X\,^2\Sigma^+$ transition of AlO.

of the Vatican Tables of Iron Line Spectra [21], in order to permit direct comparison of the observed spectra with the reference Fe I lines. The wavelength measurements are believed to be accurate to ±0.5 Å. Unfortunately, detailed analysis of the spectrum was not possible. Higher resolution instruments than those available in the present investigation are required to distinguish the overlapping features more clearly and to establish the symmetry of the upper state in the system near 2500 Å.

The shorter wavelength portions of the spectra were quite difficult to observe, probably because the intensity of the continuum radiation from the secondary source was either scattered or absorbed by the exploding wire vapor. It was, therefore, possible to observe features below about 2800 Å for only 400 μsec after initiation of the explosion-discharge. Observations after $t \approx 400$ μsec showed no background continuum radiation and thus no absorption features from atomic or molecular species. Prior to 400 μsec, however, when the initial continuum radiation from the exploding wire itself supplemented that from the secondary source, it was possible to distinguish sharp Al I lines and numerous bands. The quality of the spectra was, however, reduced by the relatively low intensity of the continuum radiation and by emission lines from atomic species (Al I, Cu I, etc.) which are very intense in the early stages of the explosion. The number of lines from impurities was also increased, because, as mentioned earlier, it was necessary to use an exploding titanium wire as the secondary flash source. Ti I lines from the titanium wire were, therefore, superimposed on the spectrum. It was not possible to distinguish any features below 2100 Å, because no radiation below this wavelength was passed by the present system.

CONCLUSIONS

In conclusion, it is perhaps well to note some general observations from our work:

(1) Time resolution is in most cases essential for the observation of spectral features which occur later in the explosion. Integrated photographic plates are overexposed by the initial continuum, so that only the most intense later features can be distinguished.

(2) Greater energy has several effects: it increases the intensity of the continuum, it produces and intensifies more lines from atomic and ionic species, and it prolongs the total period of radiation. Our results suggest that higher energies are desirable if the explosion is to be utilized as an intense light source or for study of atomic and/or ionic species. On the other hand, for studies of molecular species, more energy seems only to obscure the structure of the molecular features by producing more intense continuum and line radiation.

(3) In explosions where no chemical reaction occurs or where the extent of reaction is small, atomic lines from the wire, the electrodes, and their impurities predominate the spectrum for the full duration of radiation (several milliseconds).

(4) In cases where reaction does occur, bands from the molecular species produced may be observed in less than 50 μsec, and only the most intense line features endure.

(5) Aluminum wires exploded in oxygen are shown to be rich sources for spectroscopic study of the AlO molecule. Absorption studies, conducted with an intense secondary source, were particularly informative. It is anticipated that these techniques should be equally suitable for study of other high-temperature species.

Finally, though the resolution of the present AlO spectra is not adequate to permit vibrational and rotational analyses, it is hoped that the techniques employed will prove useful in future studies of high-temperature species.

REFERENCES

1. H. Nagaoka, D. Nukiyama, and T. Futagami, Proc. Imp. Acad. (Japan) 3:208–212, 258–264, 319–333, 392–418, 499–502 (1927).
2. F.P. Coheur, Bull. Soc. Roy. Sci. Liege 12:98 (1943).
3. F.P. Coheur and B. Rosen, Bull. Soc. Roy. Sci. Liege 10:405 (1941).
4. A. Delsemme and B. Rosen, Bull. Soc. Roy. Sci. Liege 14:70 (1945).
5. J.M. Lejeune, Bull. Soc. Roy. Sci. Liege 14:318 (1945).
6. J.M. Lejeune and B. Rosen, Bull. Soc. Roy. Sci. Liege 14:81 (1945).
7. L. Malet and B. Rosen, Bull. Soc. Roy. Sci. Liege 14:377 (1945).
8. G.K. Oster and R.A. Marcus, J. Chem. Phys. 27:189 (1957).
9. R.A. Marcus, in Exploding Wires, Vol. I, W.G. Chace and H.K. Moore, eds., Plenum Press, New York (1959), p. 307.
10. V.A. Loginov, Opt. Spectry. (USSR) 6:67 (1959).
11. C.P. Nash and R.P. De Sieno, J. Phys. Chem. 69:2139 (1965).
12. E.C. Cassidy, S.W. Zimmerman, and K.K. Neumann, Rev. Sci. Instr. 37:210 (1966).
13. I.R. Bartky and A.M. Bass, Appl. Opt. 4:1354 (1965).
14. J.E.G. Wheaton, Appl. Opt. 3:1247 (1964).

15. E.C. Cassidy and K.K. Neumann, in Proc. 7th International Congress on High-Speed Photography, Zurich, 1965, O. Helwich, ed., Verlag Dr. Othmar Helwich, Darmstadt, West Germany (1967), p. 178.
16. D.C. Tyte and R.W. Nicholls, Identification Atlas of Molecular Spectra, 1: The AlO $A\,^2\Sigma - X\,^2\Sigma$ Blue-Green System, University of Western Ontario, Canada (1964).
17. M. Shimanchi, Science of Light (Japan) 7:101 (1958).
18. M. Becart and J.M. Mahieu, Compt. Rend. 256:5533 (1963).
19. V.W. Goodlett and K.K. Innes, Nature 183:243 (1959).
20. S.L.N.G. Krishnamachari, N.A. Narasimham, and M. Singh, Can. J. Phys. 4:2513 (1966).
21. E.C. Cassidy, S. Abramowitz, and C.W. Beckett, Natl. Bur. Std. Monograph No. 109, in press.
22. A. Gatterer, Grating Spectrum of Iron, Specola Vaticana, Vatican City (1951).

DISCUSSION

Question by L.P. Harris, General Electric Research and Development Center: Is it likely that some of the impurities observed come from the glass container rather than the wire?

Answer by Author: I do not believe that the impurity features in the spectra were caused by the glass container, because these features appeared with approximately the same intensity when the wire was exploded in vessels of other materials (Bakelite and Teflon). A few features, such as C_2 and CN bands, were found to be caused by use of a Teflon vessel.

Question by D. Jones, U.S. Dept. of Commerce: Do you have any intent to do vacuum ultraviolet below 500 Å?

Answer by Author: Not at the present time.

Comment by A. Guenther, Air Force Weapons Laboratory: Utilizing the system described by Trolen et al. during the Third Conference on the Exploding Wire Phenomenon we have observed copious amounts of hard ultraviolet radiation emitted at wavelengths well below 100 Å. The radiative characteristics were of a strong underlying continuum with numerous discrete spectral lines both in emission and absorption. The experimental conditions were as follows: maximum voltage $\sim$320 kV, initial current rise $\sim 10^{13}$ A/sec, maximum current $\sim$70,000 A, wire materials $\sim$Be, Al, Ni, and Cu, vacuum environment $\sim 10^{-7}$ torr, wire diameter $\sim$1 to 5 mil, and wire length $\sim$3.6 mil. The spectrograph utilized was a 3-m grazing incidence vacuum ultraviolet instrument employing sensitized SWR emulsions. In addition, expansion velocities in excess of 10^7 cm/sec were recorded using image converter cameras. And, finally, x-rays were produced in sufficient quantities to expose appropriate emulsions. However, it is not known yet whether this high energy radiation is emitted by the very hot wire or is due to Brehmstrahlung from this anode.

Question by R. L. Musselman, University of California, Davis, Calif.: At what pressures were the oxide webs obtained? How high a vacuum were you able to obtain in your latest chamber?

Answer by Author: Oxide webs were produced by explosions in oxygen (pre-explosion pressures: 0.1 to 1 atm) and by explosions in a sealed vessel containing air at 1 atm. Our best vacuum was about 4×10^{-4} torr.

Question by B. Siegel, Aerospace Corp.: You noted the existence of AlO species for periods of many milliseconds. This seems a very long time for Al to react with O_2 in the aftermath of a wire explosion and suggests that the AlO was formed very early in the explosion and has a long lifetime. Do you believe this is true? Our experimentation with methane pyrolysis in wire explosions indicates that reactions are complete in less than 50 μsec, and methane pyrolysis should be slower than gaseous Al reactions with oxygen.

Answer by Author: Our experiments showed that AlO was formed early in the explosion (in less than 50 μsec). As for the lifetime of the molecule, we made no attempt to measure lifetime in the present work.

Comment by B. Ya'akobi, The Hebrew University of Jerusalem, Israel: I wish to remark on the question concerning the long time of molecular lines. These lines are probably emitted near the surface of the expanding wire. There the temperature can for most of the explosion be conducive to emission of molecular lines and since the slit is parallel to the wire this area is continuously observed by the spectrograph.

Kerr-Cell Time-Resolved Spectra of an Exploding Wire

Robert P. DeSieno* and Charles P. Nash

Department of Chemistry
University of California
Davis, California

During the first five μsec after the restrike in a "pause"-type explosion of a copper wire, the spectrum is an intense continuum pierced by many absorption lines attributable to transitions of the copper atom. A very few emission lines of the Cu^+ ion also appear. The absorption lines are unshifted and fairly symmetrical. Fifteen μsec later the ion lines have disappeared, the continuum intensity is markedly diminshed, and the atom lines now appear as unsymmetrically broadened, shifted emissions, some of which retain an absorption "notch" at the location of the unperturbed line. By assuming the pure absorption spectrum originates from a source which is a locally thermally equilibrated line-absorbing layer overlying a continuum emitter (the Schuster–Schwarzchild approximation), we have used the equivalent widths of absorption lines originating from different atomic states to estimate a skin temperature of 1 eV. If it is assumed that the absorption lines are broadened by a resonance collisional mechanism, the density of the skin may be estimated. We find a density which is too large by comparison with photographs of the wire at the time of restrike and, hence, we conclude, as have others on different evidence, that at this stage the exploding wire is "hollow."

INTRODUCTION

Spectroscopic studies of the exploding wire date from Anderson's [1] classic work in 1920. With few exceptions, however, the dozens of papers which deal with this aspect of the phenomenon report only time-integrated results and, hence, yield little if any information about even the average behavior of the wire material. Most noteworthy of the early studies are the papers by Smith [2], Anderson and Smith [3], and Futagami [4]. By using rotating mirror spectrographs, these workers found gross behavior which

*Present address: Rohm and Haas Corporate Research Laboratories, Springhouse, Pennsylvania.

is in good qualitative agreement with the results to be reported here. The introductory section of a very recent paper by Jäger and Lochte-Holtgreven [5] alludes to spectra obtained in their laboratory which also change with time in a manner similar to those shown here. As yet, however, no detailed analysis of an exploding wire spectrum aimed at a quantitative deduction of the state of the wire has come to our attention, with the exception of the work of Bartels and Bortfelt [6], and Mayfield [7], who confined their attention to the continuum. Our analysis will be based entirely on the absorption lines which overlie the continuum.

EXPERIMENTAL

The condenser bank and oscilloscope used in these experiments have been described previously [8]. In the present experiments, the 28.4-μF bank, whose ringing frequency is 30 kHz, was charged to 14.5 kV. Spectra were obtained with a Bausch and Lomb Model 11, 1.5-meter Stigmatic Grating Spectrograph, using a fixed 60-μ slit. This instrument, if used only in the first order, covers the wavelength region 3700 to 7000 Å with a dispersion of 15 Å/mm.

Films used were Kodak 103-0 and Kodak Tri-X, processed according to the manufacturer's specifications [9]. Film darkening was read with an Applied Research Laboratories No. 5400 Film Comparator Densitometer fitted with a 10-μ-wide slit. The densitometer was modified to scan and to record, the scan rate being 0.37 Å/sec. Wavelengths were established by placing mercury arc spectra above and below the wire spectrum, thus providing wavelength calibrations which we estimate to be valid to ± 0.2 Å. Film darkening versus relative incident energy was determined for the two films at several wavelengths by photographing the spectrum of a mercury arc through an 8-step (1.5:1/step) rotating sector driven at 200 rps for a total exposure time of 5 sec. We have assumed the wavelength spectral sensitivity curves for our emulsions to be the same as those cited by the manufacturer [9]. The relative intensities of lines in our mercury arc calibration spectra, corrected according to the manufacturer's response curves, compare well with those given in the literature [10].

Time-resolved spectra were obtained by using as a shutter an Electro-Optics Model KSC-50 Camera System fitted with a 5-μsec pulse network which is tapped to provide a monitoring signal to determine the time at which the cell fires. A two-channel, variable-delay, pulse chassis was used to trigger both the condenser bank and the Kerr cell system. Thus, the cell could be made to open at any time during the course of the explosion, and the instant of firing could be determined from a dual-trace oscillogram showing both the voltage across the wire and the output of the monitor tap on the same time base. Two Kerr cells were used: an acetonitrile cell, fitted with Glan prisms, which transmits uniformly without extinction for wavelengths above 3200 Å, and a nitrobenzene cell fitted with Polaroid HN22 windows, whose transmission is limited to 4400 $< \lambda <$ 6000 Å. By using various combinations of films and Kerr cells, high-quality spectra were obtained over the wavelength interval 4000 to 6000 Å. A 1-μsec pulse network was tested and rejected for lack of sufficient intensity in the film plane to yield satisfactory spectra. A pair of quartz lenses were used in the optical path to cast a diffuse image (2 to 3 mm wide) of a line source in the exploding wire position on the spectrograph slit. By probing with a point source, it was established that the spectra were averaging over a 1-cm-high $\times$ 0.5-cm-wide portion of the exploding wire, centered on the wire axis, at the midpoint of its length. In all the present experiments, the wire axis and the spectrograph slit were parallel to one another.

Survey spectra of exploding copper, silver, and gold wires were taken with qualitatively similar results for all. For the detailed studies, we selected 7.5-cm lengths of #34 gage (0.16-mm diameter) copper wires. This material provides a spectrum with a satisfactory number of lines in the wavelength interval of interest. More importantly, however, the spectrum has been fully analyzed and oscillator strengths have been tabulated for virtually all the reasonably intense lines.

RESULTS

The oscillogram for the explosion of the copper wires in air shows a typical "pause phenomenon" discharge. Attempts to obtain spectra at times up to the termination of the pause were unsuccessful. Spectra were obtained, however, at 5-μsec contiguous intervals for the 40 μsec immediately following the pause. In neither the time-resolved nor the time-integrated spectra were oxygen or nitrogen lines observed. This result is in agreement

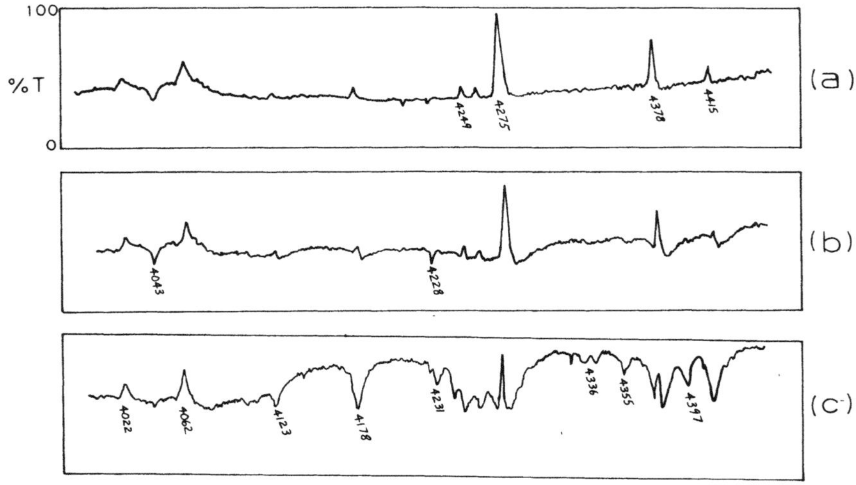

Fig. 1. Densitometer traces of exploding wire spectra.
(a) < t > = 2.5 μsec.
(b) < t > = 7.5 μsec.
(c) < t > = 17.5 μsec.

with that of Reithel et al. [11], who similarly failed to observe air excitation in a pause experiment.

Figure 1 shows densitometer traces of portions of the spectra of wires recorded at mean times 2.5, 7.5, and 17.5 μsec after reignition of the discharge. Figure 1a shows a structureless continuum interrupted by many absorption lines, all characteristic of the copper atom. In this spectrum, only 3 emission lines are observed: the 4043, 4228, and 4555 Å lines of the Cu^+ ion. Five μsec later (Fig. 1b) the continuum has receded somewhat and the absorption lines show emission edges displaced to the red. The lines at 4178 and 4415 Å are about half-inverted. In Fig. 1c, the entire aspect of the spectrum has changed. The continuum is nearly gone and essentially all the lines in the copper atom spectrum appear as extraordinarily broad emissions. In every case, the absorption lines appear in their "proper" positions, whereas many of the emission lines in Fig. 1c are shifted to longer wavelengths. Particularly noteworthy is the fact that in Fig. 1c the continuum intensity is relatively very low and also the ion lines have disappeared. In the frames subsequent to those shown here, the emission lines are observed to dim and to narrow until the last frame appears to be a fairly conventional arc spectrum of copper.

DISCUSSION

The spectroscopic results here shown are consistent with a model of a "pausing" exploding wire in which the restrike occurs well within the cylindrical boundary surface which defines the (by now greatly expanded) "wire." This model, first proposed by Anderson [1] and reiterated by many subsequent investigators, was clearly confirmed experimentally by Müller [12]. Müller's results may be used to explain, in a very qualitative way, why the continuum intensity (core temperature) declines although the system still is absorbing energy from the condenser. Several of his figures show an expansion of the spark channel with time at a rate which nearly parallels our continuum decay. The entire question of the origin and spectral distribution of the continuum, and its time dependence, is too complex for the present experiments to provide even a partial answer. Underway are spectroscopic studies in which both space and time resolution are achieved, and which hopefully will allow some insight into processes within the core of the wire plasma.

The present results do, however, admit an interpretation which yields information about the properties of the wire boundary layer, since Fig. 1a is presumably its absorption spectrum. We assume all the line reversals in Fig. 1a arise in a (thin) layer which is in local thermodynamic equilibrium. The model of an absorbing layer overlying a continuum emitter is known, in astrophysical applications, as the Schuster–Schwarzchild (S-S) model [13]. Our analysis does not, however, depend on the details of this model in any essential way.

When the spectra of Fig. 1a are converted to plots of intensity versus wavelength for the absorption lines, many of these plots may be represented quite adequately by a Lorentz-shaped line centered on the proper wavelength, as determined from the tabulated copper spectrum [14]. Several other plots approximate Lorentz profiles slightly distorted on the long wavelength side. These lines are also, however, unshifted. For all the lines used in our analysis, the distortion from Lorentz shape is fairly small, and the line breadths are large. In particular, we have selected lines whose breadths found here are much larger than the natural breadths reported by Allen [15]. Their breadths are also large by comparison with numerous of the emission lines observed when a wire is exploded in vacuo and, hence, we assume we may safely neglect instrumental effects.

Five of the absorption lines originate from resonance states of the copper atom; i.e., states which allow a direct optical transition to the ground state. We assume that the widths of these five lines may be described by the width of the resonance level which is, in turn, derivable from the theory of resonance collisional broadening. The several theories of this effect, as discussed, for example, by Breene [16], all yield resonance level widths (and, by inference, line widths for lines which originate in this level) which are within a factor of two of

$$\Delta\nu_{1/2} \approx \frac{e^2}{m}\left(\frac{f_{ir}}{\omega_0}\right)N_1 \qquad (1)$$

Here $\Delta\nu_{1/2}$ is the (whole) half-width of the absorption line in frequency units, e and m are the charge and mass of the electron, f_{ir} is the absorption oscillator strength for the ground state–resonance level transition, and ω_0 is the angular frequency of the resonance line (not the measured line). The quantity N_1, the number density of ground state atoms in the absorbing system, is the only unknown in Eq. (1). In Table I are listed the quantities pertinent to an estimate, based on the present model, of the ground state atom density in the absorbing layer. Oscillator strengths were obtained from the tables of Corliss and Bozman [17] and Corliss [18]. The values of N_1, which show considerable scatter, indicate a mean density of ground state atoms $\sim 1 \times 10^{21}$ cm^{-3}.

We next estimate the temperature of the absorbing layer and from it, and the results in Table I, the total density in this layer. The equivalent width of an absorption line W_λ, is defined as the width of a completely black rectangle which removes from the continuum the same amount of energy as does the actual absorption line. For intense absorption lines, all theories of line-formation

Table I. Ground State Copper Atom Density in the Absorbing Layer at $\langle t \rangle = 2.5$ μsec

λ, Å	$\Delta\nu_{\frac{1}{2}}$, sec^{-1}	f_{ir}	ω_0, sec^{-1}	N_1, cm^{-3}
4378	4.7×10^{11}	1.5×10^{-2}	7.6×10^{15}	9.4×10^{20}
4259	2×10^{11}	1.5×10^{-2}	7.6×10^{-15}	4.0×10^{20}
5218	4.75×10^{12}	0.32	5.8×10^{15}	3.3×10^{20}
4415	2×10^{11}	3×10^{-3}	7.7×10^{15}	2×10^{21}
4249	2×10^{11}	3×10^{-3}	7.7×10^{15}	2×10^{21}

Table II. Relative Populations of Atomic States at $\langle t \rangle = 2.5\ \mu$sec

E_1, cm^{-1}	λ, Å	W_λ, Å	f	Γ, sec^{-1}	rel N'
13245	5700	2	2×10^{-3}	1.5×10^{12}	1.0
13245	5782	6	7×10^{-3}	4.5×10^{12}	0.83
40909	4651	8	2×10^{-1}	12×10^{12}	3.0×10^{-2}
39019	4275	4	1×10^{-1}	6×10^{12}	3.5×10^{-2}

yield the result [13, 19] that W_λ/λ is proportional to a quantity $(N'f\Gamma)^{1/2}$, where f is the oscillator strength of the transition, and for a Lorentz line $\Gamma = 2\pi\Delta\nu_{1/2}$. The quantity N' depends for its precise identification on the theory which is being used. For our purposes it is sufficient to realize that it is proportional to the number of atoms acting to produce the line and, hence, to the number of atoms in the absorbing layer which are present in the quantum state from which the transition begins. In Table II we list data from which follow relative values of N' for atoms in states which are roughly 1.6 eV and 5.0 eV above the ground state of the copper atom. Table II also illustrates the dependence of W_λ/λ on $(f\Gamma)^{1/2}$ for states whose N' values are (virtually) equal. The energy levels and term symbols for the copper atom were obtained from Moore's compilation [20] based on the assignments of Shenstone [14]. When we now assume thermal equilibrium in the absorbing layer, we deduce from Boltzmann's equation, and the relative values of N' from Table II, a temperature in the absorbing layer of 10,000 plus or minus 1000°K. The estimated temperature now is used to calculate the electronic partition function of the copper atom, for which we find, using the first 12 energy levels cited by Moore [20], the value 4.7 at 1 eV. When the partition function and the estimated ground-state density found from resonance broadening theory are substituted into Boltzmann's equation, we find a total density in the absorbing layer $\sim 2 \times 10^{21}$ cm^{-3}. The initial density of the wire is 8.6×10^{22} cm^{-3} and, hence, if uniform radial expansion had occurred, we would predict about a 10-fold increase, or a diameter ~ 2 mm at the time of restrike. This prediction contrasts sharply with the observation in this, and many other laboratories, that at the time of restrike the cylinder diameter is not measured in millimeters, but rather in centimeters.

The seeming failure of resonance broadening theory to predict the "correct" density in these experiments by several powers of ten is in excellent accord with a similar result noted very recently by Ya'akobi [21]. When the breadth of the lithium resonance line (in emission) observed from an exploding lithium wire was compared to the breadth predicted by resonance broadening theory, using a density inferred from the photographic dimensions of the wire, Ya'akobi noted a discrepancy of a factor $\sim 10^3$, the line being too broad. Ya'akobi's conclusion, namely that resonance broadening theory completely fails, had been reached earlier by several workers cited in his paper. We suggest, however, that at least part of the apparent inapplicability of the theory to his experiment, and to ours, may be rationalized if the peripheral density of the exploding wire is, in fact, high. Experimental support for this proposal may be found in the x-ray pictures published by Müller [12], and in a less convincing way in the vapor deposits obtained by Conn [22] and by Moran [23]. Each of these papers concludes that the wire vapor cylinder is hollow, and our spectroscopic results may also be used to support this conclusion.

We conclude with a discussion of the extent to which the spectrum supports our assumption that all the absorption lines originate in a layer which may be described by unique values of temperature and density. To this end we shall examine the central intensities of the absorption lines. The absorption coefficient at line center is obtained experimentally from the Beer–Lambert law

$$I = I_o e^{-\tilde{N} \alpha d} \tag{2}$$

when $\tilde{N}$ is the number density of absorbers contained in a path of length d in a medium having absorption coefficient α. I and I_o are the emergent and incident intensities of radiation. Provided that the thickness of the absorbing layer is constant, relative values of $\tilde{N}$ for the several atomic states of interest may be found, we assume, from the Boltzmann equation with $T \cong 1$ eV. Thus, the spectra can be made to yield relative values of α at the centers of the absorption lines in the copper spectrum.

Theoretical values (relative) for α at line center, given the experimental half-widths, may be found from the theory of combined Doppler and collisional broadening for Lorentz lines. The absorption coefficient at the center of such a line is given by [24]

$$\alpha = \frac{Pf}{\Delta\nu_D} \frac{a}{\pi} \int_{-\infty}^{\infty} \frac{e^{-y^2}}{a^2 + y^2} dy = \frac{Pf}{\Delta\nu_D} [\exp a^2][1 - erf(a)] \tag{3}$$

Table III. Absorption Coefficients at Line Center

Line, Å	α_{rel} (experimental)	α_{rel} (theoretical)
4275	1	1
4378	0.6	0.4
4586	0.3	0.4
5218	0.3	0.2
5700	0.05	0.06
5782	0.04	0.07

Here P is a group of natural constants, $\Delta\nu_D$ is the Doppler width of the line (at $T = 1$ eV), and the quantity a is defined, in our notation, by

$$a \equiv \frac{\Delta\nu_{1/2}}{2\Delta\nu_D} \tag{4}$$

For the present lines, $a \sim 50$. When a is very large, $[1 - erf(a)]$ may be expanded as [25]

$$[1 - erf(a)] \sim \frac{e^{-a^2}}{a} [1 - \frac{1}{a^2} + \cdots] \tag{5}$$

and under these circumstances Eq. (3) assumes the simple form

$$\alpha = \frac{Pf}{\Delta\nu_D} \cdot \frac{1}{a} \left[1 - \frac{1}{a^2} \cdots\right] \approx \frac{Pf}{\Delta\nu_D a} \tag{6}$$

Experimental half breadths have been used in Eq. (6), together with oscillator strengths from the literature, to obtain theoretical relative values for absorption coefficients at the centers of several lines. Theoretical and experimental absorption coefficients are listed in Table III. The agreement is well within the accuracy claimed for both our experimental intensity data and the values of the oscillator strengths employed. This kind of agreement provides gratifying support for our postulate of a thermally equilibrated absorbing layer.

ACKNOWLEDGMENT

This work was supported by the U.S. Atomic Energy Commission through the Lawrence Radiation Laboratory, Livermore, California.

REFERENCES

1. J. A. Anderson, Astrophys. J. 51:37 (1920).
2. S. Smith, Astrophys. J. 61:1 (1925).
3. J. A. Anderson and S. Smith, Astrophys. J. 64:295 (1926).
4. T. Futagami, Sci. Papers Inst. Phys. Chem. Research (Tokyo) 31:1 (1937).
5. H. Jäger and W. Lochte-Holtgreven, Z. Physik. 198:351 (1967).
6. H. Bartels and J. Bortfeldt, in Exploding Wires, Vol. 3, W.G. Chace and H.K. Moore, eds., Plenum Press, New York (1964), p. 9.
7. E.B. Mayfield, in Exploding Wires, Vol. 1, W.G. Chace and H.K. Moore, eds., Plenum Press, New York (1959), p. 147.
8. C.P. Nash and C.W. Olsen, in Exploding Wires, Vol. 2, W.G. Chace and H.K. Moore, eds., Plenum Press, New York (1959), p. 5.
9. Kodak Publication P-9, Eastman Kodak Co., Rochester, New York (1967).
10. American Institute of Physics Handbook, 2nd Edition, McGraw-Hill Book Company, New York (1963), pp. 7–125.
11. R.J. Reithel, J.H. Blackburn, G.E. Seay, and S. Skolnik, in Exploding Wires, Vol. 1, W.G. Chace and H.K. Moore, eds., Plenum Press, New York (1959), p. 19.
12. W. Müller, in Exploding Wires, Vol. 1, W.G. Chace and H.K. Moore, eds., Plenum Press, New York (1959), p. 186.
13. L.H. Aller, Astrophysics, 2nd Edition, The Ronald Press Co., New York (1963).
14. A.G. Shenstone, Phil. Trans. A241:297 (1948).
15. C.W. Allen, Phys. Rev. 39:42 (1932); Phys. Rev. 39:55 (1932).
16. R.G. Breene, Jr., The Shift and Shape of Spectral Lines, Pergamon Press, New York (1961), Chapter 5.
17. C.H. Corliss and W.R. Bozman, Nat. Bur. Stds. Mon. 53 (1962).
18. C.H. Corliss, J. Res. NBS 66A:497 (1962).
19. A detailed treatment for the S-S model is given by V.A. Ambartsumyan, in Theoretical Astrophysics, Pergamon Press, New York (1958), p. 155.
20. C.E. Moore, Nat. Bureau Stds. Circular 467, Vol. II (1952).
21. B. Ya'akobi, J. Quant. Spect. Radiat. Transfer 6:909 (1966).
22. W.M. Conn, Z. Angew. Phys. 7:539 (1955).
23. K.E. Moran, in Exploding Wires, Vol. 3, W.G. Chace and H.K. Moore, eds., Plenum Press, New York (1964), p. 285.
24. S.S. Penner, Quantitative Molecular Spectroscopy and Gas Emissivities, Addison-Wesley Publishing Co., Inc., Reading, Mass. (1959), p. 46.
25. E. Jahnke and F. Emde, Tables of Junctions, 4th Edition, Dover Publications, New York (1945), p. 24.

DISCUSSION

Question by O.H. Zinke, University of Arkansas: How did you estimate the radius of the wire, i.e., what technique do you use?

Answer by Author: From optical pictures. Very recently we have taken space- and time-resolved spectra which give diameters which agree well with the photographic results.

Comment by B. Ya'akobi, Princeton Plasma Physics Laboratory, Princeton, N.J.: The average density in the wire during the time of measurement is 3 to 4 orders of magnitude smaller than the measured density. This means that all but a very small fraction of wire material is concentrated in an extremely thin shell inside which there would be not enough material to produce a strong continuum. This difficulty could be connected with some reported difficulties concerning resonance broadening (see, for example, Reck et al. 1965).

Question by W. Lochte-Holtgreven, Kiel University, West Germany: In case the absorption lines are broadened by resonance, other pressure effects should be small. But you may have apparent broadening from thick layers?

Answer by Author: I think if the optical thickness of the absorbing layer were very large the spectra would show saturation effects which we do not observe. Our intensities at line center vary over a factor of nearly one hundred.

Question by H. Jäger, University of Kiel, West Germany: Did you ever consider the radial distribution of the particle densities?

Answer by Author: The present experiments do not allow us to say anything except about the thin layer at the surface. Space- and time-resolved spectra should give insight into the densities elsewhere.

Question by M. Levine, Air Force Cambridge Research Laboratory, Bedford, Mass.: How thick would the "shell" need to be to account for your absorption?

Answer by Author: The absorption at line center falls off like $\exp(-ND\alpha)$ so both N and d are required. If densities are ~ 10 then the shell thickness would be about a micron.

Question by F.D. Bennett, Ballistic Research Laboratories, Aberdeen Proving Ground, Md.: Do you observe the dense, annular vapor distribution, and do you find any Doppler broadening on account of its radial motion?

Answer by Author: Yes, we infer a dense perimeter on the basis of widths of lines from resonance lenses. Radial motion would produce a Doppler shift rather than a broadening, and our lines are not shifted. Pictures show, however, that the contact surface to a first approximation stands still over the time of our experiment.

Interaction of High-Powered Focused Laser Beam with Exploding Wire

Victor E. Scherrer, Philip W. Davis,
Matthew J. Hillsman, and Dick Manser

NASA Electronics Research Center
Cambridge, Massachusetts

A combination of the high electric field intensity attainable with a focused laser beam and the high electric current capability of an electrical pulsed power system is an efficient technique for heating small diameter wires. Measurements have been made on the combined electrical–laser heating of 25-μ-diameter, 6-mm-long bare copper wires. Electrical heating was provided by a 0.05-μF, 20-kV capacitor, discharging through the wire with a frequency of 3 MHz. A ruby laser, generating a single 10-nsec pulse with a peak power of 40 MW, was used for optical heating. With the laser beam focused in a 1000-μ-diameter spot on the wire, a power density of 5.1×10^9 W/cm^2, corresponding to an average electric field of 1.4×10^6 V/cm, was measured. The focused laser beam produces a high degree of ionization in the wire material during the "discharge pause."

INTRODUCTION

One of the serious difficulties in attaining high temperatures in exploding wires is the well-known "discharge pause" phenomena, during which a relatively small current flows through the wire, allowing it to expand and cool. Reducing this pause significantly, or eliminating it, without the generation of a peripheral arc, would make it possible to deposit more energy in the wire and thereby attain higher temperatures. With the elimination of the pause, it may be possible to attain conditions where the magnetic pressure at the surface of the wire exceeds the thermal pressure from within the wire and thereby impede the hydrodynamic expansion of wire material. The resulting high temperature and density produced in the wire would make possible the investigation of properties of materials in temperature and density regions not previously attained by exploding wires.

Previous efforts [1, 2] to attain magnetic containment were not successful due to the development of a peripheral arc around the wire. In the NRL experiments [1] thin quartz tubes were used in an unsuccessful effort to eliminate the peripheral arc. Another technique [3] used at NRL to eliminate the "pause" in vacuum wire explosions involved the use of imploding cylinders. This method was successful, and electron temperatures of 1.8 keV were attained. Bennett [4] found that the pause was of no concern when the electrical circuitry was carefully matched to the load so that all of the energy was delivered in the first pulse. He observed increased energy density in the wire.

In this paper we describe a method of eliminating the "discharge pause" by heating the wire material with a high-powered laser. This optical heating takes place at a time when current cut-off would normally occur. The high power densities attainable from lasers (at least 10^{12} W/cm^2) are very efficient in producing ionization and would be capable of eliminating the pause in wire explosions if the laser energy could be uniformly deposited in the wire.

Two advantages of the dual electrical–laser heating of small wires are: (1) efficient heating, and (2) efficient ionization due to the high electric field gradient associated with the focused laser beam. For example, a 100-MW, Q-switched laser, when focused to a 100-μ-diameter spot, produces an electric field gradient of 2×10^7 V/cm. Studies of laser heating processes are required since, as in the case of electrical heating, it is poorly understood at this time. The laser may be heating the wire material during the solid, liquid, vapor, and plasma states. In this paper, we describe an experiment for making such measurements and report the results of these preliminary studies.

When compared with the results of other plasma experiments, exploding wires have not yielded very high temperatures. Lochte-Holtgreven [5] reported temperatures as high as 70,000°K (6.6 eV). Bennett [6] measured input energies corresponding to 12 eV/particle. Zinke et al. [7], using a time-of-flight mass spectrograph deduced temperatures of 10^{6}°K (94 eV). Vitkovitsky [3] measured electron temperatures of 1.8 keV for imploding metal cylinders in vacuum, but it is not certain that thermal equilibrium existed in this experiment.

Similar temperatures have been obtained for the laser heating of materials. Linlor [8] and Haught and Polk [9] heated various small metal targets in vacuum. Linlor used a 40-nsec, 5.4-MW focused laser beam. He measured temperatures by ion time-of-flight and deduced ion temperatures of 1 keV. He determined that 94 to 99% of the laser light energy was absorbed in the target.

The mechanisms by which laser light is absorbed varies significantly for solid, gas, and plasma targets and with the density of the gas or plasma. An opaque solid absorbs light very efficiently. The thickness of the sample that can be uniformly heated depends upon the skin depth of the material at the laser frequency, and the heat diffusion rate into the sample. Solid spheres with diameters up to 20 μ have been uniformly heated [10].

For gaseous targets (i.e., an exploding wire during the discharge pause) it is necessary for the laser light to ionize the gas before significant energy absorption takes place. For a gas at high density (10^{19} atoms/cm^3), the mechanism is E field breakdown. After breakdown, at high densities, the principal energy absorption mechanism is thought to be the inverse bremsstrahlung effect [10], which is very efficient.

The electric field due to a focused laser beam is given by [10] $E = 1.1 \times 10^4 \sqrt{W}/r_o$ V/cm, where E is the average intensity of the electric field, W is the total laser power in MW, and r_o is the radius of the laser focus in centimeters.

Air at atmospheric pressure breaks down and becomes a good absorber for a laser power density of 5×10^{10} W/cm^2 [11]. The corresponding E field is 4.4×10^6 V/cm.

The variation in the electric field intensity required to ionize a gas resembles a typical Paschen curve. That is, starting with good vacuum, as the gas pressure is increased, the electric field required for breakdown decreases, attains a minimum value, and increases again. For argon, the minimum breakdown field occurs at a particle density of 5×10^{21} atoms/cm^3 and a power density of 5×10^8 W/cm^2.

DESCRIPTION OF THE EXPERIMENT

A schematic diagram of the experiment is shown in Fig. 1. Electrical heating of the wire (25-μ-diameter, 6-mm-long bare copper) occurs when the capacitor (0.05 μF) is connected across the wire by a triggered air gap switch. The capacitor stores 10 J of energy at a potential of 20 kV and rings at a frequency of 3 MHz through the exploding wire circuit.

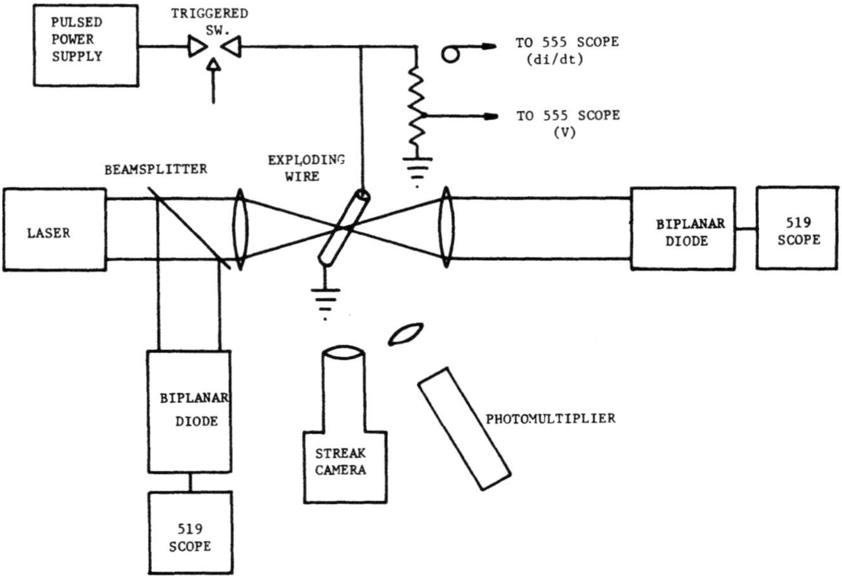

Fig. 1. Schematic diagram of experimental setup.

The exploding wire is heated by a laser beam focused to a diameter of 1000μ. The laser is an Optics Technology Model 130, emitting a single 10-nsec pulse with a wavelength of 6943 Å and a peak power of 40 MW (0.4 J energy). The corresponding electrical field strength in the focal volume for a 1000-μ-diameter focus is 1.4×10^6 V/cm. Trigger circuits are arranged so that the laser can be fired at any preset time relative to the beginning of the electrical discharge.

In the electrical heating circuit, the current through the wire and voltage across the wire are monitored. During laser heating, the incident light and the light transmitted through the exploding wire are monitored with 1-nsec time resolution, using biplanar photodiodes and fast oscilloscopes. A Beckman and Whitley Model 200 streak camera having a streak resolution of 20 nsec is used to observe motion of heated wire material. The light output of the heated wire material is observed with 1.5-nsec time resolution using an Amperex Model 1020 photomultiplier detector with a quartz window. A close-up view of the experiment is shown in Fig. 2. The disc capacitor, switch, and exploding wire are mounted on an optical bench in the foreground.

EXPERIMENTAL RESULTS

The experimental results obtained to date are primarily for the purpose of testing the concept of optically heating exploding wires to eliminate the discharge pause and, therefore, must be regarded as preliminary. Experiments were performed to measure the absorption of laser light during the gaseous phase of the wire explosion (during the discharge pause) and during the ionized phase of the explosion, after the electrical discharge was completed.

Figure 3 shows the characteristics of the exploding wire without laser heating. The wire was bare copper 25 μ in diameter, and 6 mm long. The capacitor was charged to 4 kV (0.4 J energy). Trace "A" shows the output of the photomultiplier operating with high optical sensitivity (filter ND 3.0). Trace "B" shows the output of the photomultiplier detector operating in the linear region (filter ND 6.0). Trace "C" shows the current through the wire. In this figure, the discharge pause is evident on traces "A" and "C."

Figure 4 shows the effects of laser heating timed to occur after the electrical heating was completed. Trace "A" shows the emitted

Fig. 2. General view of experimental setup.

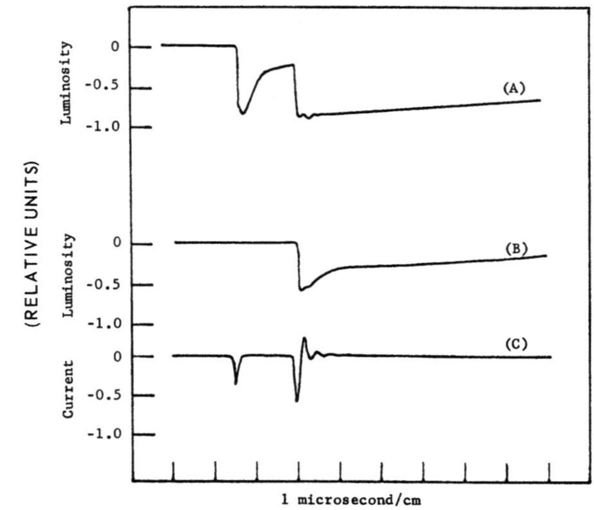

Fig. 3. Exploding wire characteristics. (A) Luminosity, PM output, ND = 3.0. (B) Luminosity, PM output, ND = 6.0. (C) Current.

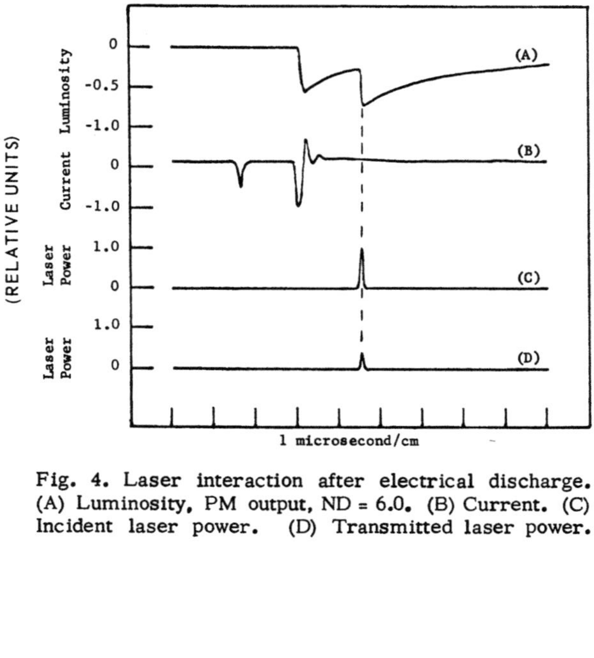

Fig. 4. Laser interaction after electrical discharge. (A) Luminosity, PM output, ND = 6.0. (B) Current. (C) Incident laser power. (D) Transmitted laser power.

light from the wire, resulting from combined electrical and laser heating. Trace "B" is the current through the wire. Trace "C" shows the signals from the photodiode monitoring light incident upon the wire. Trace "D" is the transmitted laser power.

The laser pulse had a peak power of 40 MW. The pulse width at half-maximum was 10 nsec. The diameter of the laser focus in the plane of the wire was 1000 μ, and the optical power density at that plane was 5.1×10^9 W/cm^2.

Figure 5 shows the results of laser heating during the discharge pause. Trace "A" shows the output of the photomultiplier detector. Trace "B" shows the electrical current through the wire. Trace "C" shows the incident laser light pulse. Trace "D" is the transmitted laser power. It is evident that the wire material becomes ionized during the pause and the luminosity due to laser heating is comparable to that of electrical heating. The expansion velocity due to laser heating is much greater than that due to electrical heating in all of the above examples.

CONCLUSIONS

Indications are that further experiments would give more quantitative information for cases where the total length of the

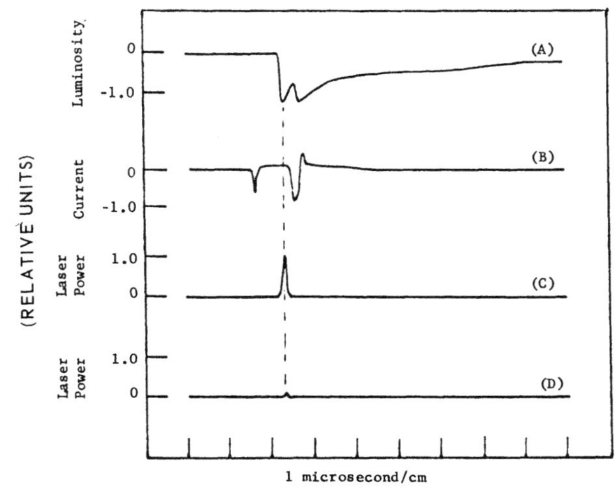

Fig. 5. Laser interaction during discharge pause. (A) Luminosity, PM output, ND = 6.0. (B) Current. (C) Incident laser power. (D) Transmitted laser power.

wire material is optically heated by using a cylindrical lens to focus the optical beam on the wire. The possibility of attaining a high-density "pinch" of the wire material in vacuum remains for future investigation. Heating of small wires using a combination of electrical and optical heating has produced promising preliminary results that indicate this to be a useful technique.

REFERENCES

1. V.E. Scherrer, in Exploding Wires, Vol. 1, W.G. Chace and H.K. Moore, eds., Plenum Press, New York (1959), p. 118.
2. J. Katzenstein, in Exploding Wires, Vol. 1, W.G. Chace and H.K. Moore, eds., Plenum Press, New York (1959), p. 135.
3. I. Vitkovitsky, Physics of Fluids 7:612 (1964).
4. F. Bennett, BRL Report No. 1339, October 1966.
5. W. Lochte-Holtgreven, Reports on Progress in Physics, Vol. XXI, The Physical Society, London, England, pp. 358-363.
6. F.D. Bennett, Physics of Fluids 8:1106 (1965).
7. O.H. Zinke, R.B. Owen, and C.K. Manka, in Exploding Wires, Vol. 3, W.G. Chace and H.K. Moore, eds., Plenum Press, New York (1964), p. 103.
8. W.I. Linlor, Appl. Phys. Letters 4:169 (1964).
9. A.F. Haught and D.H. Polk, Physics of Fluids 9:2047 (1966).
10. Yu. P. Raizer, Soviety Physics—Uspekhi 8:650-2 (1966).
11. P. Nelson, P. Veyrie, M. Berry, and Y. Durand, Phys. Letters 13:226 (1964).

DISCUSSION

Comment by A.H. Guenther, Air Force Weapons Laboratory, Kirtland AFB, N.M. It is important to realize that the ionization, that is, the electron concentration produced by the laser is probably 100 to 1000 times that of the exploding wire proper.

Answer by Author: Yes, this is important.

Question by J.C. Bryner, North American Rockwell Corp.: Exploding wires normally produce temperature of less than 10 °K. Have you any idea how hot your laser-enhanced plasma is?

Answer by Author: No. We soon hope to make temperature measurements.

Question by H. Knoepfel, Laboratori Gas Ionizzati, Rome, Italy: The one-sided illumination of the exploding wire by the laser will make any interpretation of your results difficult. Do you intend to use two or more lasers in future experiments?

Answer by Author: No. We are investigating a single lens optical system with which we hope to irradiate the entire wire surface uniformly.

Question by D. Jones, U.S. Department of Commerce: I found that by varying the wire size I could eliminate the "pause" in my device. Possibly you could vary the electrical characteristics of your circuit and attain the same result. Have you considered doing this?

Answer by Author: This may be true, but we prefer our method, which may work for a wide range of conditions instead of just one.

Question by W. Lochte-Holtgreven, Kiel University, West Germany: Did you make a spectroscopic investigation? As you observe light, this seems necessary.

Answer by Author: No, but we hope to initiate such observations in the near future.

Question by B. Ya'akobi, Princeton Plasma Physics Laboratory, Princeton, N.J.: This experiment may be considered essentially a laser-produced plasma experiment in which the wire provides preionization which permits using relatively low-power lasers. The fact that the intensity of light emitted from a small region of the wire upon the laser irradiation is about equal to that emitted by the whole wire without this irradiation is indicative of higher temperatures due to the laser.

Answer by Author: We agree. It is our hope that, in the future, the high temperature capability of laser heating can be combined with the high magnetic fields possible at the surface of fine wire to give a more interesting source.

High-Temperature Plasmas Produced by Exploding Wires*

David P. Ross† and O.H. Zinke

Department of Physics
University of Arkansas
Fayetteville, Arkansas

The time-of-flight technique for the analysis of transient plasmas discussed at the previous Exploding Wire Conference by Zinke, Owen, and Manka has been perfected to include a flow velocity, and the "containment time" is now interpreted as a "plasma release time." Plasmas have been generated from gold wires which have temperatures of 2×10^6 °K, densities of 10^{18} cm^{-3}, and flow velocities of about 3×10^6 cm/sec. The critically damped discharge current is "slow" by Bennett's criterion, and analyses of the current, voltage, and light data indicate heating by a collapsing pinch effect at the current maximum of a plasma produced through gaseous breakdown in vapor generated by Bennett's expansion wave. The plasma is apparently stabilized from electromagnetic ejection by a liquid core. Flow velocities do not seem to be electromagnetic in origin. Wedemeyer's objections to the time-of-flight technique will be discussed.

INTRODUCTION

The time-of-flight technique for the analysis of transient plasmas produced in vacuum reported earlier by Zinke, Owen, and Manka [2] has been perfected by Manka, Crawford, and Zinke [1]. The analysis technique now allows for a net flow velocity of the plasma, and the data procurement technique provides a built-in check upon the assumptions. The technique has been found to apply well to the plasmas produced by a coaxial plasma gun. This paper is an extension of the perfected technique to the plasma produced by the explosion of a wire in vacuum. Analysis of the positive ion flux, the total light output of the plasma, and the current–voltage

*Work supported by the Southwestern Electric Power Company and the National Science Foundation.
†Present address: Department of Physics, University of Houston, Houston, Texas.

data shows that the data support the following behavior of the wire explosion in vacuum: The initial heating of the wire material causes the wire to successively melt and become a superheated liquid in the same manner as those explosions characterized by Bennett [3] as "fast." However, since in some explosions there is not enough energy added to the wire to vaporize it, the explosions may also be characterized as "slow." As the current in the critically damped discharge circuit rises to maximum, the vaporization wave described by Bennett [4] starts inward from the surface of the wire. Since the explosion takes place in vacuum, the wave does not travel inward for more than half a microsecond before the vapor so produced breaks down electrically and shunts the current from the liquid core to the plasma now formed around the surface of the liquid. This mode of conduction has been observed by Edelson and Korneff [5]. The plasma so formed is then adiabatically compressed in a z-pinch, generating high temperatures in a small fraction of the wire mass. The liquid core stabilizes the plasma against ejection from the circuit and against disintegration as a result of plasma instabilities. The pinched plasma is ejected from the explosion vicinity by the "transplosion" of the superheated liquid core in the manner described by Chace [6] and observed by Edelson and Korneff [5].

The plasma expands into the drift tube and the positive ion flux is detected by biased ion pickup plates. Positive ion temperatures in the ranges of 1×10^{5}°K to 2×10^{6}°K for gold, 3×10^{4}°K to 6×10^{5}°K for copper, and 2×10^{4}°K to 3×10^{5}°K for aluminum and flow velocities of around 3×10^{6} cm/sec for all metals are measured. Plasmas are released from the circuit configuration around 10 μsec after the initial current rise and seem to be contained (by evidence of total light emission) for times of the order of 1 μsec. Densities as high as 10^{18} cm^{-3} are estimated from time-of-flight data and from radius estimates based on photographs of total light using E.G. & G. "XR" film, which show that the plasma radius under conditions of most intense emission is about twice the original wire radius.

APPARATUS AND MEASUREMENTS

Gold, copper, and aluminum wires 1 cm long and 0.062 cm (#22 A.W.G.) and 0.045 cm (#25 A.W.G.) in diameter were used.

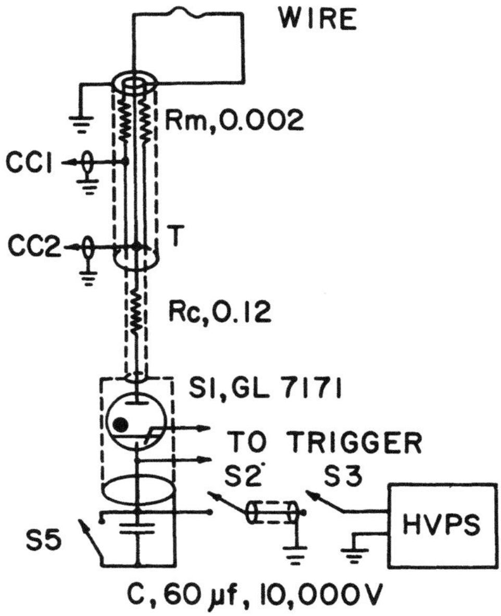

Fig. 1. Wire explosion circuit. Coaxial connectors CC1 and CC2 provide current and voltage signals respectively. Resistor R_m is the current shunt resistor, R_c the critical damping resistor. Grounded tube T provides a shield for the shunt and the reference ground for the system. Capacitor C is charged through switches S2 and S3 by the high voltage power supply HVPS. S5 is provided to short capacitor.

They were exploded in the circuit shown in Fig. 1. A 60-μF energy storage capacitor charged to voltages in the range of from 4 to 8 kV was used. The circuit has an inductance of 0.3 μH and is externally critically damped with the resistor R_c (0.12 ohm). The external critical damping resistor was added to facilitate the resistance analysis by preventing the current from going through zero. The time from initial current rise to current maximum is 6 μsec and the peak current is typically of the order of 30 kA for an initial charging voltage of 6 kV.

The drift tube is shown in Fig. 2. Note that there are pickup rings at three distances from the explosion. The flux incident on all three plates is monitored simultaneously. Since but two of the ion signals are necessary to determine the parameters of the plasma, the third pickup plate provides a check upon the validity

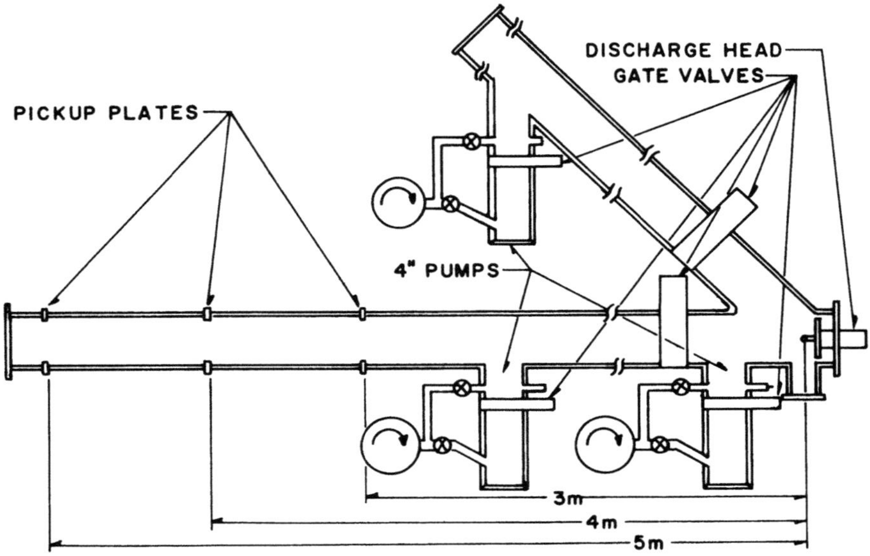

Fig. 2. Drift tube. Ionized debris from the wire exploded in discharge head is allowed to expand into tube and is detected at pickup plates. The upper drift tube was not used.

of the assumptions, since for data to be acceptable, it is demanded that the theoretically generated flux fit all three flux curves.

The current in the discharge circuit is measured with a Park type shunt [7]. The risetime of the circuit is long enough so that the skin effect should not affect the measurement. The resistive portion of the voltage across the wire is obtained by electronically subtracting a dI/dt signal from a small pickup loop placed near the discharge circuit from the measured voltage drop across the wire. It can be easily shown that the voltage measured is given by

$$V = I\left(R + \frac{1}{2}\frac{dL}{dt}\right)$$

where L is the inductance of the circuit, R the resistance of the wire, and I the current. In our case, the resistance of the circuit is at all times much greater than the time derivative of the inductance so that the resistance of the wire may be measured unambiguously in this fashion. The subtraction circuit is calibrated with an unexploded wire.

Some representative data are shown in Figs. 3 and 4. Figure 3 is a plot of the ion flux observed at three pickup plates at 3, 4, and 5 m from the explosion, respectively. The wire was a 0.045-cm-diameter gold wire, exploded with an initial capacitor voltage of 8 kV. The parameters corresponding to the open figures are $T = 2.2 \times 10^{6}$°K, $V = 2.5 \times 10^{6}$ cm/sec, and $H = 11.44\ \mu$sec, where T is the temperature of the positive ions of the plasma, V is the net flow velocity of the plasma, and H is the plasma release time, measured from the time of initial current rise. Figure 4 is a plot of the current and voltage for the same wire explosion as the flux in Fig. 3. Also plotted in Fig. 4 is the output of an E.G. & G. SD-100 photodiode which monitored the total light output of the plasma. Note that there is no photodiode response until the maximum of the voltage trace.

RESULTS AND CALCULATIONS

Table I presents the results of the calculations performed on the analyzable positive ion flux. It is to be noted that the only

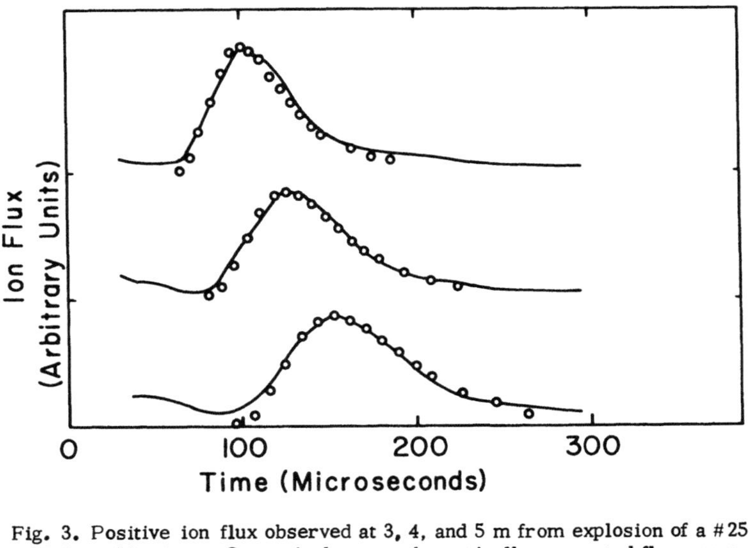

Fig. 3. Positive ion flux observed at 3, 4, and 5 m from explosion of a #25 A.W.G. gold wire. Open circles are theoretically generated flux corresponding to parameters $T = 2.2 \times 10^{6}$°K, $V = 2.5$ cm/μsec, and $H = 11.4\ \mu$sec.

Table I. Parameters of Plasmas to Which the Time-of-Flight Technique Was Found to Be Applicable*

D, A.W.G.	V_o, kV	T, 10^{5}°K	V, cm/μsec	N, 10^{16}	H, μsec	$H - T_p$, μsec
			GOLD			
25	4.0	2.0	3.0	0.31	18.4	8.0
25	4.0	1.2	3.3	0.23	16.0	5.6
25	4.0	19.4	2.2	6.8	24.2	13.4
25	4.5	2.2	3.6	0.29	14.7	5.2
25	5.0	4.6	3.5	1.1	13.4	3.9
25	5.0	1.8	3.3	0.29	18.0	8.0
25	5.9	7.7	3.3	1.5	–	–
25	5.5	6.6	3.8	2.2	12.04	1.04
25	5.5	1.0	2.9	0.27	13.8	4.5
25	6.0	5.2	3.1	0.20	21.3	13.1
25	6.0	21.5	2.5	1.5	13.4	5.4
25	6.0	19.2	2.5	0.96	19.3	10.5
22	8.0	11.0	3.3	1.91	19.5	11.5
22	8.0	21.0	3.0	3.0	22.3	11.8
22	8.0	6.8	3.2	0.91	23.4	13.2
			COPPER			
25	6.0	0.30	3.2	0.17	14.7	0.82
25	6.0	1.96	3.3	2.9	26.1	9.85
25	6.0	2.7	2.5	9.1	22.3	11.4
25	8.0	1.8	3.3	2.5	13.7	2.95
25	10.0	1.3	3.4	4.6	13.2	4.95
25	4.0	0.7	3.4	0.32	18.5	9.0
25	4.5	6.0	1.6	10.0	23.5	9.38
			ALUMINUM			
22	4.0	0.21	3.3	0.31	21.3	9.8
22	6.0	0.22	0.12	1.4	24.5	–
22	6.0	0.50	3.5	0.58	19.2	–
25	5.0	2.8	3.1	25.0	15.8	5.7
25	5.0	0.84	3.2	17.0	28.4	5.8
25	7.0	0.59	3.8	1.5	15.5	8.5
25	8.0	1.09	4.0	2.4	19.4	5.6
25	8.0	0.54	4.1	1.2	13.8	5.8

*The table is divided into three sections, one for each metal used. The parameters listed are the wire size (D), the initial voltage on the 60-μF energy storage capacitor (V_o), the measured positive ion temperature (T), the measured flow velocity of the plasma (V), the measured number of particles in the plasma (N), the measured plasma release time (H), and the difference between the plasma release time and the time of the voltage maximum ($H - T_p$).

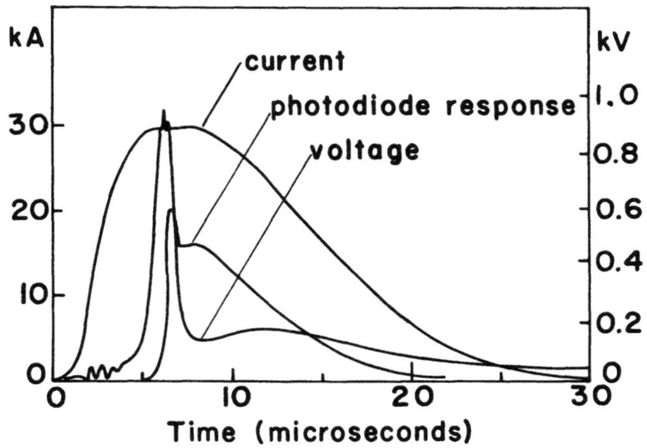

Fig. 4. Current, voltage, and total light output data for same wire explosion as for flux data of Fig. 3.

systematic variation of the measured parameters that is evident is the increase of the positive ion temperature with increasing ion mass. Also note that the flow velocities did not depend greatly upon the discharge current or the type of wire used. The data presented here represent but a small fraction of the data procured. Most ion flux data are rendered unanalyzable by the appearance of more than one peak in the data. This is attributed to the formation of more than one plasma in the wire explosion.

Calculations performed with the current–voltage data give a great deal of insight into the mechanisms of plasma production and heating. Figure 5 is a plot of the resistance of the wire calculated from the current–voltage data of Fig. 4. The sequence of events during the wire explosion may be deduced to be as follows: The wire material is heated until it reaches the melting point and melts at time t_1. The time of the melt may be found by comparing the total energy deposited with that required for melting the wire. Time t_2 is the time at which the wire material has reached the boiling point of the liquid wire material. Time t_3 is the time at which the vaporization wave described by Bennett and his coworkers [4] starts to travel inwards from the surface of the wire. At time t_4 the vapor so produced breaks down electrically and produces a plasma which is adiabatically compressed during the time interval from t_4 to t_5. The presence of the plasma is taken to be indicated by the intense light emission between t_4 and t_5. The plasma is subsequently

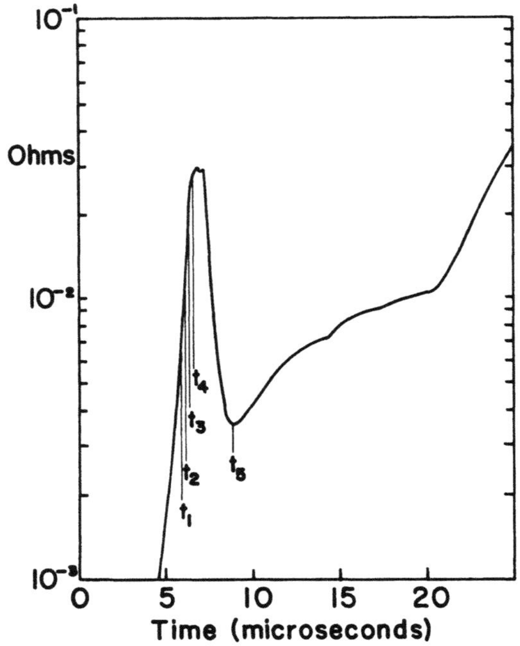

Fig. 5. Resistance calculated from current–voltage data of Fig. 4. t_1 is time of melt, t_2 time at which wire reaches boiling point, t_3 time at which vaporization wave starts inward, and t_4 time of initiation of peripheral arc. Plasma is heated and accelerated between t_4 and t_5.

ejected from the region of the explosion in the manner already described. Note that in the work of Korneff and Edelson [5] there is about a one-microsecond lag between the initial light emission and the ejection of the plasma. We find a similar lag here.

The vaporization wave may be detected by a comparison of the measured resistance of the wire with the resistance of the wire calculated from the assumption of linear resistance rise with specific energy. Following Bennett, Kahl, and Wedemeyer [4], one may derive an expression for the radius of the conducting liquid.

$$r(t) = \left[\frac{\rho_o L}{\pi R(t)}\right]^{1/2} \left[1 + \frac{2\beta}{\rho_m \rho_o L^2}\int_{t_o}^{t} V^2(t)\,dt\right]^{1/4}$$

where ρ_o is the resistivity of the liquid at the melting temperature,

L is the length of the wire, $R(t)$ is the measured resistance, ρ_m is the mass density of the liquid, β is the specific energy coefficient of resistivity, $V(t)$ is the voltage across the wire, and t_o is the time at which the melting temperature was reached. This expression is derived for the case of explosion at the current maximum, which is the case we are considering. The wire radius in all cases decreased to a value not smaller than $^2\!/_3$ of the original wire radius before breakdown occurred. Calculated wave velocities were about 100 m/sec for gold and about 50 m/sec for copper and aluminum. All these are within the ranges given by Bennett [8].

The plasma is assumed to generate in the vapor along the surface of the liquid core. From the temperature of the core, the thermal velocity of the gold atoms in the vapor may be estimated to be of the order of 8×10^4 cm/sec. In the microsecond or so during which the vapor was released from the core, the vapor could be estimated to have expanded to a radius of about 0.1 cm, or about four times the wire radius. The time required for the discharge to spread through the volume of the vapor would be very difficult to estimate, but from the data of others [5] and the time required for the detected light to rise, the discharge probably spreads throughout the volume of the vapor in about 0.5 μsec. From time-integrated photographs of the wire explosion made with E.G. & G. "XR" film, a film capable of recording subjects whose brightness ranges over seven decades, we may estimate that the original plasma diameter is from 3 to 5 times the original wire diameter, while the final value may be estimated to be about twice the original wire diameter.

Using an external radius of 0.04 cm for the plasma, the magnetic pressure at the surface of the plasma is about 1000 atm. The density of ions is estimated to be 4×10^{18} cm^{-3}. The final equilibrium temperature is calculated by the ideal gas law to be 1.5×16^{6}°K, which is in good agreement with experiment. The change in the volume that is necessary to heat the plasma to the measured temperatures can be calculated as a check on the model, with the assumption of an adiabatic compression which started in the vapor at an initial temperature of 5000°K, the temperature of the core at the time of plasma formation. For a quasi-static, adiabatic compression, the volume ratio is [9]

$$\frac{V_i}{V_f} = \left[\frac{T_i}{T_f}\right]^{1/1-\gamma}$$

where V_i and V_f are the initial and final volumes respectively, and T_i and T_f the initial and final temperatures. The quantity γ (the ratio of specific heats) is given by

$$\gamma = \frac{2 + \delta}{\delta}$$

where δ is the number of degrees of freedom appropriate to the compression. The case at hand is a cylindrical compression, so δ is 2. These values yield $V_i = 1.28\ \text{cm}^3$, corresponding to an initial radius of 0.62 cm as compared with 0.1 cm estimated by the rate of expansion of vapor.

The volume estimated above may be too large, since the temperature multiplication ratio in the dynamic pinch may be higher than that given by the ideal gas law. The compression of the ions may not be quasi-static and the ratio predicted by the ideal gas law may be too low by an order of magnitude if the plasma–vacuum interface moves rapidly [9]. Another check on the temperature can be performed through the photographically measured radius, the measured resistance, and application of the Spitzer–Härm conductivity formula. From this it was determined that the electron temperature was 2.2×10^{6}°K.

To perform another check on the mechanism proposed, we investigate the mass dependence of the temperature predicted by a magnetic compression theory. Jackson [10] gives a scaling law for the velocity of the plasma–vacuum interface in a dynamic compression

$$V \propto \frac{E_o{}^{1/2}}{\sqrt{\rho}}$$

where E_o is the applied electric field, and ρ the initial mass density of the plasma. Since the ions are heated by acquiring the velocity of the interface (in a nonquasi-static compression), the scaling law for the ion temperature is just

$$T_+ \propto 1/k \frac{m^{1/2} E_o}{\sqrt{n_+}}$$

where m is the mass of an ion and n_+ is the number density of ions. Now, if one assumes that the electric field and the number density of ions are the same for different ions (and these may be rather poor assumptions), one can formulate a scaling law for the temperature of the positive ions for metals with different atomic masses m_1 and m_2

$$\frac{T_1}{T_2} \propto \sqrt{\frac{m_1}{m_2}}$$

For the three metals chosen—gold, copper, and aluminum—the temperatures of the positive ions should be in the ratio $T_{Au} : T_{Cu} : T_{Al} = 2.7 : 1.8 : 1$. The experimental ratio, obtained from an average over all wires exploded, is $T_{Au} : T_{Cu} : T_{Al} = 9.3 : 2.6 : 1$. One can only note that the trend evident is one of higher temperature for higher atomic mass, as the scaling law predicts. The scatter in the data and the differences in initial conditions make the mere observation of the trend gratifying.

It is expected that the electrons of the plasma will have reached the same temperature as the ions in the time of containment of the plasma, since ion–electron relaxation times predicted from the Fokker–Planck equation for a plasma of the temperature and density of the one produced are of the order of a microsecond. The Planck radiation law would then predict that the plasma would radiate at the rate of 10^{11} W. Fortunately, the Planck radiation law is not applicable to a high-temperature plasma. Rose and Clark [11] discuss the radiation from a hot plasma and point out that the radiation is due to electron cyclotron radiation and bremsstrahlung. Calculation of the rate of radiation from these two mechanisms yields radiation rates of the order of kilowatts. Since at the time of plasma generation and heating, the external circuit is supplying energy at the rate of 10^7 W, there is more than enough energy available to supply radiation without cooling the plasma. Indeed, the heated and accelerated plasma has a total energy of about one joule, as compared with ten or more joules supplied by the external circuit during heating.

The observed flow velocity of the plasma does not vary more than a few percent from 3 cm/μsec over a wide range of initial conditions. If the flow velocity is electromagnetically produced by $J \times B$ forces, one would expect the flow velocity to vary with the integral of the current squared, which is calculated to vary by more than a factor of 100 over the range of initial conditions used. A possible clue to the origin of the flow velocity is afforded by the photographic data of Edelson and Korneff [5], who observe an opaque core ex-

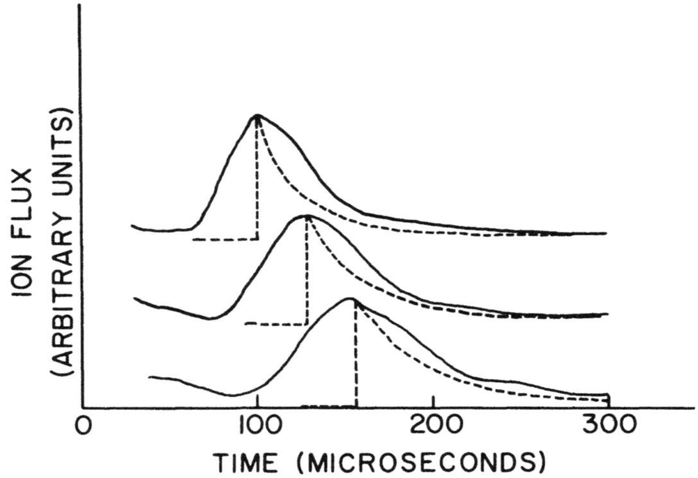

Fig. 6. Same flux data as in Fig. 3, with flux predicted by Wedemeyer superimposed. Temperature necessary for fit is 5.2×10^6 °K.

panding within the plasma sheath. The velocity of expansion of this core is about 1 cm/μsec which is within a factor of three of the flow velocities observed here. The specific energy deposited in the core is roughly constant at 3×10^3 J/g. One would expect the core expansion velocity to be dependent upon the specific energy of the core, and perhaps not upon the density of the wire material, in much the same way as the velocity of a sound wave in a monatomic gas is independent of the mass of the atoms of the gas.

There has been some criticism of the assumption of free-molecular flow for the ions in the drift tube, notably by Wedemeyer [12]. Wedemeyer has found an approximate solution of the Euler equations for the expansion of a gas into vacuum. It is his contention that the approximate solution is a better approximation to the data than the free-molecular-flow solution and that lower temperatures will result. Figure 6 is a plot of the data presented in Fig. 3 with the flux predicated by Wedemeyer's theory superimposed. For this fit, it was assumed that the plasma started to expand at the time of maximum light output. The temperature of the unexpanded gas necessary for this fit is 5.2×10^6°K.

CONCLUSIONS

It would seem that the exploding wire in vacuum may be, under the proper conditions, a source of high-temperature, high-density plasma. Here, the adiabatic compression of the linear z-pinch is the mechanism for the generation of high temperatures. The advantage of the use of an exploding wire as the source of the plasma is that the dense liquid core stabilizes the plasma against the development of instabilities and premature ejection from the circuit and delays plasma formation until the time of current maximum.

REFERENCES

1. C.K. Manka, J.R. Crawford, and O.H. Zinke, Phys. Fluids 10:767 (1967).
2. O.H. Zinke, R.B. Owen, and C.K. Manka in Exploding Wires, Vol. 3, W.G. Chace and H.K. Moore, eds., Plenum Press, New York (1964) p. 103.
3. F.D. Bennett, "High Temperature Exploding Wires," BRL Report 1339, U.S. Army Ballistic Research Laboratories, Aberdeen Proving Ground, Md. (1966).
4. F.D. Bennett, G.D. Kahl, and E.H. Wedemeyer in Exploding Wires, Vol. 3, W.G. Chace and H.K. Moore, eds., Plenum Press, New York (1964) p. 65.
5. H.D. Edelson and T. Korneff in Exploding Wires, Vol. 3, W.G. Chace and H.K. Moore, eds., Plenum Press, New York (1964) p. 267.
6. W.G. Chace, Proc. 4th Int. Conf. on Ioniz. Phen. in Gases, IV, 1191, Uppsala (1959).
7. J.H. Park, J. Research Natl. Bur. Standards 39:191 (1947).
8. F.D. Bennett, Phys. Fluids 8:1425 (1965).
9. D.J. Rose and J. Clark, Jr., Plasmas and Controlled Fusion, M.I.T. Press, Massachusetts Institute of Technology, and John Wiley & Sons, Inc., New York (1961) pp. 137-139.
10. J.D. Jackson, Classical Electrodynamics, John Wiley & Sons, Inc., New York (1962) pp. 320-326.
11. D.J. Rose and J. Clark, Jr., Plasmas and Controlled Fusion, M.I.T. Press, Massachusetts Institute of Technology, and John Wiley & Sons, Inc., New York (1961) pp. 233-251.
12. E.H. Wedemeyer, "Asymptotic Solutions for the Expansion of Gas into Vacuum," BRL Report No. 1278, U.S. Army Ballistic Research Laboratories, Aberdeen Proving Ground, Md. (1965).

DISCUSSION

Comment by F.D. Bennett, Ballistics Research Laboratories, Aberdeen Proving Ground, Md.: Wedemeyer's continuum model of the expansion into vacuum both eliminates the need for a "containment time" and leads to lower estimates, by a factor of 10, of the ion temperatures. It seems clear that at early times the wire must expand according to a continuum model; it is not clear at what stage one may change to a free molecule flow picture, but certainly this cannot occur until particle densities have fallen by more than four orders of magnitude.

Answer by Author: Wedemeyer's estimate of lower ion temperatures is based upon the earlier results reported in *Exploding Wires*, Vol. 3, by Zinke et al. It has been shown here that Wedemeyer's theory yields temperatures higher than those measured by free molecule flow by a factor of about two in cases where the time of plasma formation may be easily identified. We feel that these results demonstrate the equivalence of the two approaches. Indeed, Strang and Tarnove, through a numerical integration of the Euler equations, have shown that the density in the continuum expansion is essentially the same as that predicted by free molecule flow.

Question by H.F. Swift, University of Dayton, Ohio.: Have you considered the acceleration of your exploding wire plasma?

Answer by Author: Calculations of the impulse delivered by the $J \times B$ force yield flow velocities lower by a factor of at least 10 than those measured. The constancy of the flow velocity over a wide range of discharge condition is also at odds with a $J \times B$ acceleration.

Influence of Adsorbed Gases on the Emission of X-Rays from Exploding Tungsten Wires in Vacuum

S. K. Händel and B. Stenerhag

Institute of Physics
University of Uppsala, Sweden

Undegassed and degassed tungsten wires, initially coated with a thin impurity layer consisting of tungsten oxide (WO_3) and graphite, have been exploded in vacuum, 2×10^{-6} torr. The 0.1-μF capacitor was discharged at an initial voltage of 20 kV. The loop inductance of the short-circuited system was 150 nH. Correlated measurements of x-ray emission from the "anode side" and dI/dt show that the emission of x-rays starts prior to the shunting discharge. The hard x-ray burst emitted is extremely short (half amplitude width of about 25 nsec) and reproducible both in time and intensity, provided the initial conditions are kept constant. Explosions of undegassed and degassed wires give maximum x-ray photon energies of 23 keV and 11.5 keV, respectively. The dI/dt signals obtained from both kinds of explosions are identical; i.e., the gases initially adsorbed by the wire surface affect the emission of x-rays but not the main discharge current. On the other hand, the main discharge current but not the hard x-ray emission is influenced by the impurity layer. Finally, the x-ray output versus the pressure of the surrounding medium has been investigated. It is found that the x-ray burst is reproducible inside the pressure interval 10^{-6} torr to 1×10^{-3} torr.

INTRODUCTION

Since 1959, low-inductance flash x-ray systems have been developed in our institute. A modified type of such a system has been used for studies of wire explosions in air as well as in high vacuum. One of the main aims of the exploding wire investigations was to study the x-ray emission from tungsten wires. Recently, the authors have reported that hard x-rays are emitted from the "anode side" of an exploding tungsten wire in vacuum [1].

A remark made by Bennett et al. [2] concerning the existence of a bright cylindrical glow during an early stage of a wire

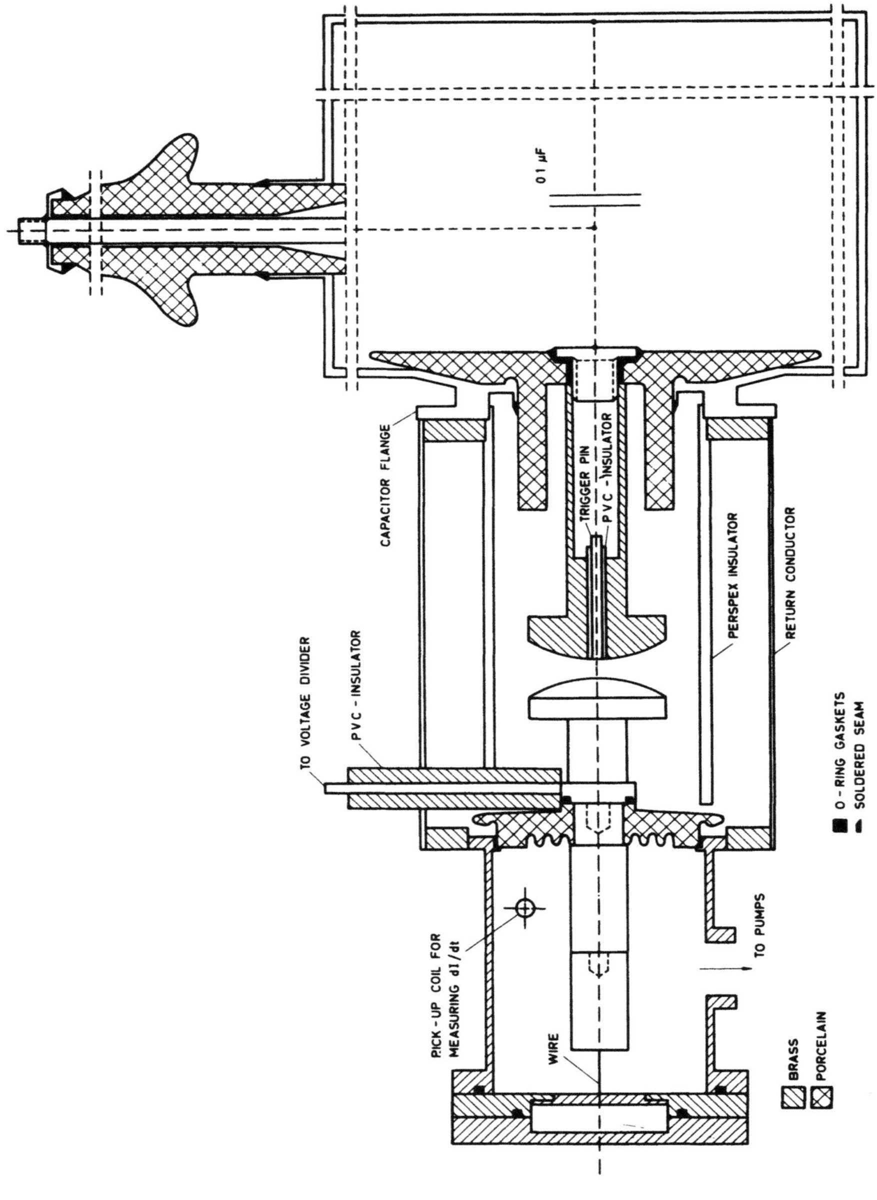

Fig. 1. Cross-sectional view of exploding wire apparatus.

explosion in vacuum, reported by Kvartskhava et al. [3], led the authors to make the present investigation. Bennett et al. inferred that the glow was due to a discharge in released gases initially adsorbed by the wire. According to this suggestion, the authors were interested to find out whether such released gases influenced the x-ray emission from the exploding wire or not. Therefore, correlated x-ray and electrical measurements of undegassed as well as degassed tungsten wires were carried out. In addition, this work presents some preliminary results concerning the relation between the x-ray emission from undegassed wires and the pressure of the surrounding medium.

APPARATUS

The exploding wire apparatus consisted essentially of an energy storage capacitor, an air spark gap, and a brass tube in which the wire was axially suspended. The electrical circuit was coaxially arranged. A cross-sectional side view of the apparatus used is shown in Fig. 1.

The energy storage capacitor (ASEA Type CTU, 100 kV, 0.1 μF) was provided with two high-tension bushings, one for connection to the high-voltage supply, and another for the discharge circuit. The capacitor was also provided with a flange on its casing on which the return conductor of the coaxial circuit was fastened. The tube in which the wire was suspended had an outside diameter of 100 mm and a wall thickness of 3 mm. It was provided at both ends with 10-mm-thick flanges, giving it a total length of 120 mm. The central conductor of the tube was divided into two parts. The first was a fixed bolt which was mounted through a central hole of a porcelain insulator soldered to the tube flange. The second part, wireholder, was threaded on the fixed bolt. The two conductor parts had the same diameter, 25 mm. Wireholders of various lengths were used, thus permitting wires of various lengths to be suspended.

An air spark gap was arranged between the capacitor and the tube. The central conductor of the tube carried, axially, one of the practically hemispherical electrodes of this gap. The other, equally formed but provided with a central triggering pin, was connected to the high-tension side of the storage capacitor. Two flanges, one on the tube and another on the capacitor casing, served as bearing surfaces for the coaxial return conductor, made of a 1-mm perforated aluminum sheet fastened with steel strips.

The discharge vessel used had two radially placed inspection channels, marked with A and B in Fig. 2. These were 4 mm in diameter and hermetically sealed by a 50-μ aluminum window. The distance between their central lines was 13 mm. Through these channels the x-radiation from the anode surface (A) as well as from the wire itself (B) was detected by a plastic scintillator, NE 102, coupled to a photomultiplier, Philips 56 AVP. When hard

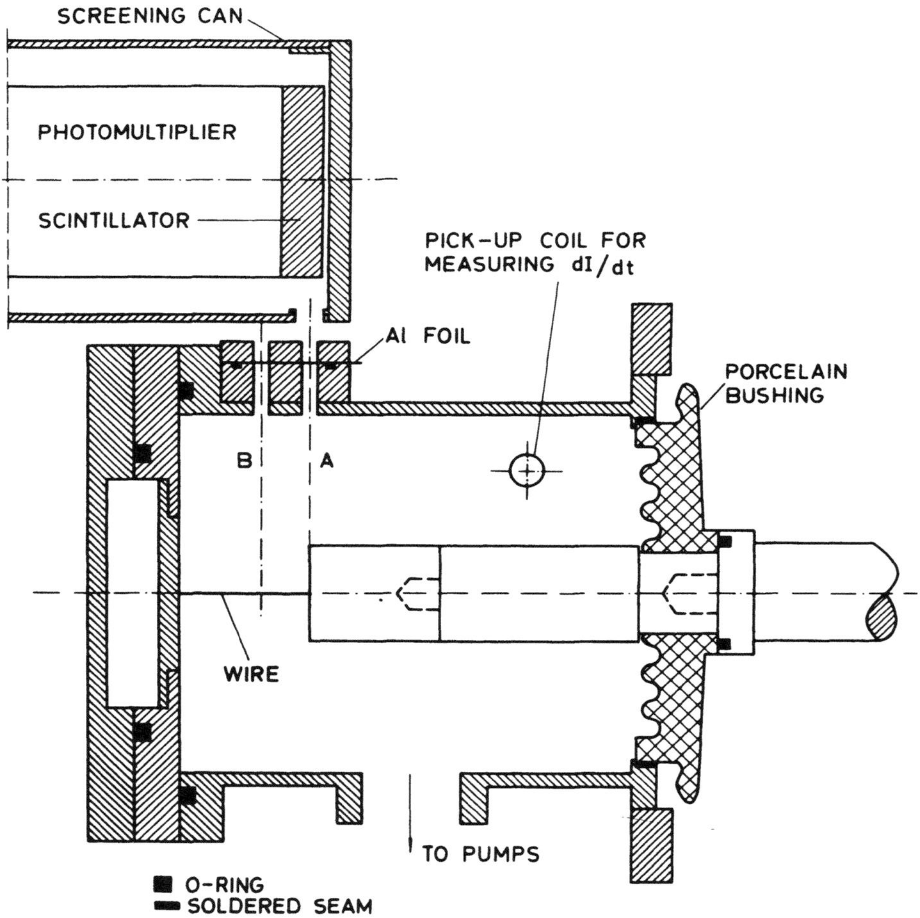

Fig. 2. Cross-sectional view of discharge vessel and x-ray detector. Scintillator is here arranged for detection of x-radiation transmitted through channel A.

Fig. 3. Total view of experimental arrangement.

x-radiation was investigated, filters of various thicknesses were placed between the aluminum window and the scintillator.

The time rate of change of current (dI/dt) was measured by means of a magnetic probe radially inserted in the discharge vessel. The photomultiplier and dI/dt signals obtained during a wire explosion were fed to a dual beam oscilloscope, type Tektronix 555, provided with a Robot camera.

Undegassed as well as degassed tungsten wires, 0.07 mm in diameter and 33 mm in length, were exploded in vacuum, 2×10^{-6} torr, at 20-kV initial capacitor voltage; i.e., the stored energy was about 20 J. According to the wire manufacturer, the tungsten wires used were covered with a thin impurity layer consisting of tungsten oxide (WO_3) and graphite. The degassing of the wires was carried out in such a manner that the wire was suspended between the electrodes of the vessel which then was evacuated to about 10^{-6} torr. Thereafter, a 6-V ac voltage was applied to the wire for an hour and a half, so that the wire was heated to a yellow-white color. The degassing procedure was continuously controlled by the pressure gauge. During this process also the impurity layer disappeared.

EXPERIMENTAL RESULTS

The dI/dt and photomultiplier traces obtained during the wire explosions were reproducible when the initial conditions were identical. Therefore, a comparison between the two discharge characteristics representing undegassed and degassed wires could be made.

The discharge characteristics — dI/dt, current, total voltage across the wire, and, finally, x-ray output — versus time obtained from explosions of undegassed (unbroken lines) and degassed (broken lines) wires are shown in Fig. 4. The current was obtained by numerical integration of the dI/dt trace. The x-ray output was detected behind a 1-mm aluminum filter. The total voltage V_w ($V_w = V_R + V_L$, where V_R is the resistive and V_L the inductive voltage drop, respectively) was calculated according to the expression

$$V_w = V_o - \frac{1}{C} \int_o^t I \, dt - L_e \frac{dI}{dt} - R_e I \tag{1}$$

where V_o is the initial capacitor voltage, C the capacitance of the capacitor, L_e the external inductance, and R_e the external resistance of the discharge circuit. The values of R_e and L_e were determined according to the formulas presented in an earlier paper [4].

During the first phase, i.e., from $t = 0$ to $t = t_1$ (cf. Fig. 4), the dI/dt traces overlap each other. This means that the adsorbed gases and the impurity layer do not influence the main discharge current during this phase. A rough estimation of the wire temperature based on the formula for homogeneous heating showed that the wire temperature at $t = t_1$ is approximately the same as the melting point of the wire material. Thus, the wire temperature seems to be high enough to release adsorbed gas molecules from the metal surface. In addition, some vaporization can also be presumed at the instant $t = t_1$.

At $t = t_1$, the time derivative dI/dt changes rapidly. The electric field between the electrodes is high, about 7×10^5 V/m, at this instant. Furthermore, the wire is surrounded by a vapor cloud of varying density (radially seen). Some thermionic emission from the wire surface also exists here. This emission is estimated to be about 10 to 20 A.

In the case of undegassed wires the initially adsorbed gases probably have formed a layer outside the metal vapor cloud. This statement is verified by the fact that the gases are released at a

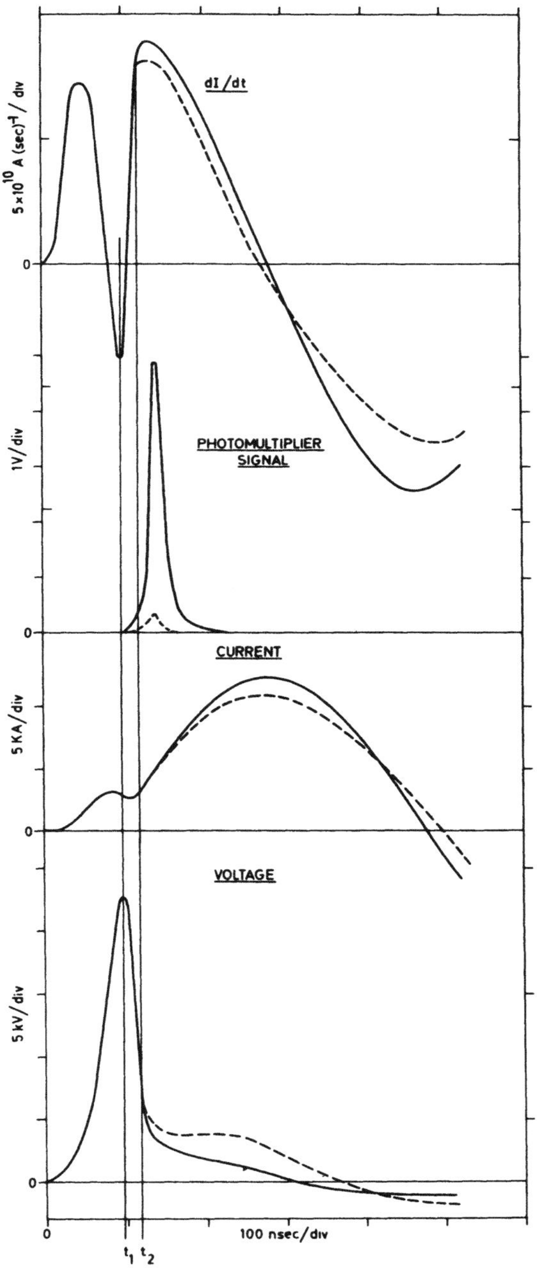

Fig. 4. Curves showing discharge characteristics — dI/dt, current, voltage, and x-ray output (photomultiplier signal) — versus time of undegassed (unbroken lines) and degassed (broken lines) tungsten wires exploded in vacuum, 10^{-6} torr. Length and diameter of the wire, 33 mm and 0.07 mm, respectively.

temperature lower than that of the melting point. In addition, their mobility is higher than that of the metal vapor.

The dI/dt traces obtained from both degassed and undegassed wires also overlap each other during the interval $(t_1;t_2)$. Also here, the impurity layer and adsorbed gases do not influence the main discharge current. Accordingly, the dI/dt change at $t = t_1$ seems to be caused by the ionization of and breakdown in the metal vapor outside the wire.

During the last stage of the explosion (i.e., from $t = t_2$ and onwards) the dI/dt, current, and voltage curves show that the explosions of undegassed and degassed wires do not develop identically.

In order to investigate whether or not the discharge characteristics were influenced by the impurity layer, the following experiments were carried out. A wire was degassed in vacuum; i.e., adsorbed gases and impurity layer were removed. The wire was then exposed to air of atmospheric pressure for about 10 min so that gases again could be adsorbed by the purified wire. This was then exploded in vacuum. In this case, the adsorbed gases but not the impurity layer were present. The dI/dt signal obtained during this explosion was identical with the dI/dt signal obtained from an explosion of a wire in which both the impurity layer and the adsorbed gases were removed. The corresponding photomultiplier pulse (x-ray output) was identical with that representing an explosion of a wire having both the impurity layer and the adsorbed gases present. These experiments reveal the fact that the adsorbed gases affect the x-ray emission but not the main discharge current. In addition, the impurity layer affects the current (only during the final stage of the explosion) but not the x-ray emission.

X-RAY OUTPUT

The photomultiplier pulses shown in Fig. 4 indicate that x-radiation is produced during a short time after $t = t_1$. However, the time correlation experimentally obtained between the dI/dt and photomultiplier pulses is incorrect. In order to establish the real time correlation between them, one must take into account the delay in the magnetic probe, the delay in the scintillation crystal, the transit time of the electrons in the photomultiplier tube, the cable transit times, and, finally, the displacement between the two scope sweeps.

The delay in both the probe and scintillator was about 3 nsec, thus eliminating each other. The cables had the same length; i.e., the transit times were equal. The displacement between the scope sweeps was corrected when plotting the curves. According to the tube manufacturer, the transit time of the electrons in the photomultiplier tube was 36×10^{-9} sec. Thus, the photomultiplier signal was displaced about 40×10^{-9} sec with respect to the dI/dt signal. If these corrections are made, it is found that the x-ray burst starts about 40×10^{-9} sec before $t = t_1$ and that the x-ray peak corresponds to $t = t_1$; i.e., to the dI/dt dip (cf. Fig. 4).

The voltage curve in Fig. 4 shows that its peak value coincides with the dI/dt dip. However, a careful calculation based on differentiation of Eq. (1) indicates that V_w must have its peak value within an interval $(t_1; t_1 + \Delta t)$, where Δt is less than 4 nsec. Thus, the dI/dt dip, voltage peak, and x-ray output peak can be considered to coincide at $t = t_1$.

The determination of the x-ray energies was made by comparing the photomultiplier signals obtained from separate discharges. The x-radiation was filtered by aluminum filters of various thicknesses.

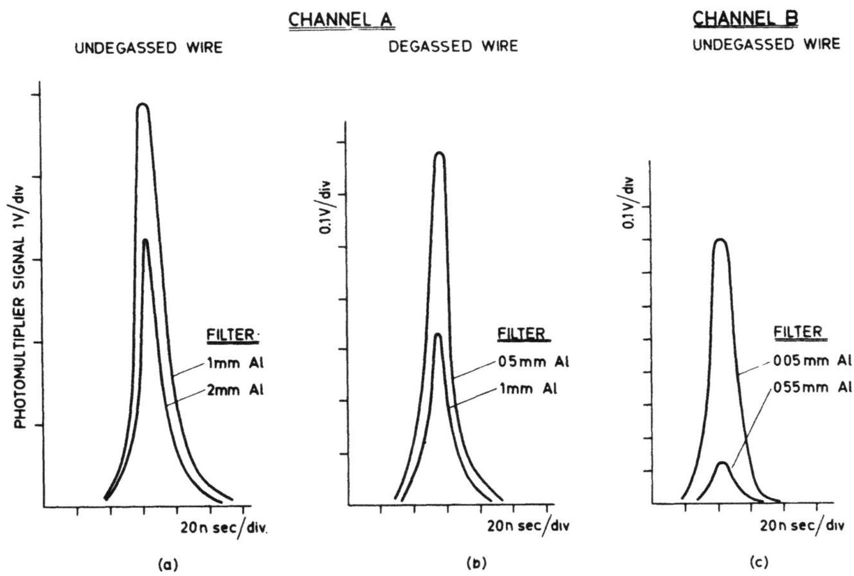

Fig. 5. Photomultiplier signals indicating hard x-ray output from undegassed and degassed tungsten wires exploded in vacuum.

The photon energy was obtained according to the extinction law

$$J_x = J_o e^{-\mu x} \tag{2}$$

where J_x and J_o are the remaining and initial intensity, respectively, μ the mass absorption coefficient, and x the filter thickness traversed. The extinction law is valid for monochromatic radiation only, but can be used for heterogeneous radiation if the radiation is homogenized by filters [5]. The intensities measured behind additional filters were used for calculation of the photon energy.

The photomultiplier pulses obtained behind various filter thicknesses are shown in Fig. 5. They represent explosions of degassed as well as undegassed wires. In Fig. 5a and b the pulses are detected at channel A, while in Fig. 5c the pulses are detected at channel B. The photon energies obtained were (i) undegassed wires about 23 keV (Fig. 5a), (ii) degassed wires about 11.5 keV (Fig. 5b), and (iii) undegassed wires (part of the wire) about 17 keV (Fig. 5c). The calculations were based on the relative peak intensities of the pulses. As a comparison, the voltage curve shown in Fig. 4 indicates a peak value of about 23 kV. Thus, the voltage peak value almost coincides with the maximum x-ray energy of 23 keV.

X-RAY OUTPUT VERSUS PRESSURE

In Fig. 4, it could be seen that the x-ray emission from undegassed exploding wires had higher penetration power than that from degassed wires. Therefore, it was of interest to investigate how the pressure of the surrounding medium influenced the penetration power as well as the reproducibility of the x-ray burst. The pressure was varied between 10^{-6} torr and atmospheric pressure. Some preliminary results from this investigation are shown in Fig. 6. It was found that, inside the interval 10^{-6} torr to 1×10^{-3} torr, the dI/dt and photomultiplier signals were identical from explosion to explosion. The x-ray burst could here be detected behind a 5-mm aluminum filter at channel A (cf. Fig. 2). However, at a pressure of 1×10^{-2} torr, it is observed that the reproducibility of the x-ray pulses is lost, but the x-radiation produced showed high penetration power down to 1×10^{-1} torr. At pressures equal to or higher than 1 torr, no x-radiation could be detected.

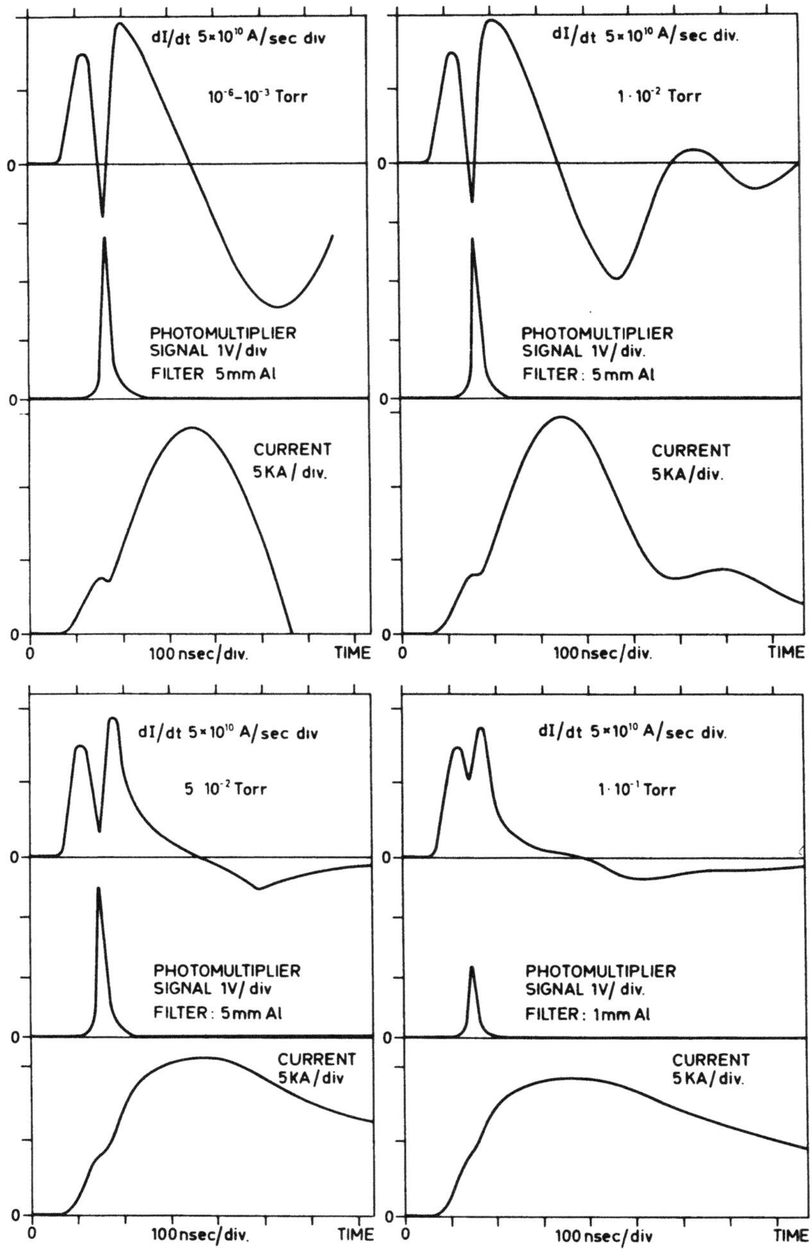

Fig. 6. *dI/dt*, current, and x-ray output traces representing undegassed tungsten wires (33 mm in length and 0.07 mm in diameter) exploded at various pressures.

EXPLODING WIRE-PULSED X-RAYS IN HIGH-SPEED PHOTOGRAPHY

It is well known that conventional flash x-ray tubes emit x-ray bursts of poor reproducibility; e.g., the x-ray output varies both in time and duration from flash to flash. Mostly, the x-ray pulse emitted shows multiple character which causes much trouble when studying high-speed events. However, wires exploded in vacuum seem to be an excellent source of short duration and reproducible x-ray bursts. It is the opinion of the authors that in the future exploding wires can be used for studies of such high-speed phenomena where visible high-speed photography can not be used; e.g., when the object is opaque or when the event studied is itself an intense source of light. However, much more work has to be done in this field in order to increase the penetration power and the x-ray intensity.

ACKNOWLEDGMENT

The authors' thanks are due to Prof. P. Ohlin who has followed this investigation of exploding wires with great interest and who has given much invaluable help. In addition, we wish to express our thanks to the members of the flash x-ray group at this institute for many valuable discussions about wire explosions.

REFERENCES

1. S. K. Händel, B. Stenerhag, and I. Holmström, Nature 209:1227 (1966).
2. F. D. Bennett, H. S. Burden, and D. D. Shear, Phys. Fluids 5:102 (1962).
3. I. F. Kvartskhava, V. V. Bondarenko, R. D. Meladze, and K. V. Suladze, Soviet Phys., JETP 4:637 (1957).
4. B. Stenerhag, S. K. Händel, and I. Holmström, Z. Physik 198:172 (1967).
5. W. Schaaffs, in Encyclopedia of Physics, Vol. XXX, S. Flügge, ed., Springer-Verlag, Berlin (1957), p. 9.

DISCUSSION

Question by J. C. Bryner, North American Rockwell Corp.: Did you make any intensity measurements of your x-rays?

Answer by Author: No.

Question by H. Swift, University of Dayton, Dayton, Ohio: Have you attempted x-ray photography to date?

Answer by Author: No.

Current Distribution for Wire Exploded in Vacuo*

D.C. Chern and T. Korneff

Temple University
Philadelphia, Pennsylvania

This investigation explores the current sheath surrounding an exploding copper wire in vacuum. Rogovski coils are used to map the current sheath. The results are plotted as a function of time and radius. Soon after the transplosion of the wire, the current leaves the wire in an initially well-defined cylinder, which diffuses rapidly with time. Initial radial expansion is of the order of 10^5 cm/sec. A "reverse current effect" is also observed.

INTRODUCTION

The behavior of exploding wire phenomena under reduced pressure has been studied by several authors [1-6]. The most distinguishing feature is a sheath of plasma which appears around the wire at a very early stage of the explosion. Together with the ionization of the ambient atmosphere, these sheaths develop into a cylinder of highly conductive vapor. This effect, which is presumably responsible for the disappearance of the "dwell," shunts a large part of the total discharge current [3, 4].

The spatial distribution of the current at various stages of the explosion is studied by the Rogovski coil method. The investigation is concerned with low pressure ($\sim 1\ \mu$) and thin wire (0.4 mm diameter). The situations under higher pressures were also examined for comparison.

APPARATUS

As shown in Fig. 1, a copper wire (length, 4 cm; radius, 0.02 cm) is exploded inside a cylindrical vacuum chamber. This chamber,

*Work done under Contract No. AF 19628-67-C-0250, Air Force Cambridge Research Laboratories, Office of Aerospace Research, USAF.

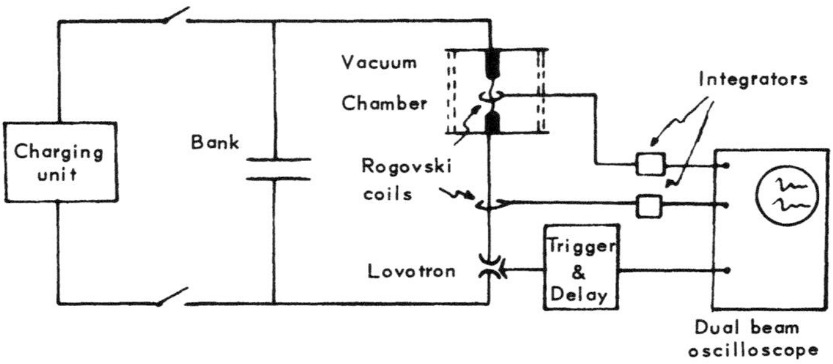

Fig. 1

consisting of a glass wall and metal ends, measures 29 cm in length and 18 cm in radius.

The energy is supplied by a capacitor bank of 44.7 μF, charged to 4 kV. The discharge is triggered by a lovotron switch, and the charging unit is disconnected before triggering. The ringing frequency is 1.85×10^4 cps.

Two Rogovski coils are used in each measurement. One encircles the discharging power line and measures the total current. The other, situated inside the vacuum chamber, measures the current passing through a particular region under investigation. Their integrated outputs, which are proportional to the currents passing through the planes circumscribed by the coils, are photographed on a dual-beam oscilloscope.

A single coil–integrator combination is used to measure the total current and is retained throughout the experiment. It serves as an indicator for the beginning of the discharge, transplosion and zero points, and as a standard of calibration.

The calibration of this "standard" coil is as follows: for a capacitor bank discharge through a shunt, or short circuit, the first peak of current is given by

$$I_{T/4} = V_o C \left(\frac{2\pi}{T}\right)\left(\frac{A_3}{A_1}\right)^{1/4}$$

where T is the period of oscillation, C is the capacitance of the bank, V_o is the initial voltage, and A_3/A_1 is the ratio of two successive peaks of the same polarity. V_o and C are known, and T and A_3/A_1 can be found from an oscillograph of the damped oscillation produced by the integrated output of the "standard" Rogovski coil.

For our case: $C = 3 \times 14.9\ \mu F$, $T = 52\ \mu sec$, $V_o = 4000$ V, and $A_3/A_1 = 1.1/1.3$, so that $I_{T/4} = 2.05 \times 10^4$ A. The oscilloscope calibration factor becomes

$$\text{Calibration factor} = \frac{I_{T/4}}{\text{Amplitude on scope (cm)} \times \text{scope setting } (V/\text{cm})}$$

$$= 6.4 \times 10^4 \text{ A/V}$$

The calibrations of the integrated outputs of other Rogovski coils can be made simply by comparing the gains of two coil–integrator combinations in one discharge. A notch, which occurs at 4 to 5 μsec after the beginning of the discharge, is always observed in the total current wave form of an explosion. It is assumed that the notch indicates the beginning of the explosion, or transplosion.

If the second Rogovski coil is centered on the wire, the notch and the wave form before the notch are always exactly the same shape as the total current wave form. A convenient way to calibrate the coil is to compare the height of the notches in the two wave forms. Once calibrated, the coil may be moved to another position without changing its calibration factor.

As shown in Fig. 2, a Rogovski coil is a conductive wire wound in the form of a toroid with a back strap winding inside it. The idea of the Rogovski coil is no more than an approximation of the integral form of Ampere's law:

$$\oint H \cdot dl = \iint J' \cdot dA = I$$

differentiated against time,

$$\text{Output emf} = \frac{d}{dt}\oint H \cdot dl = \frac{dI}{dt}$$

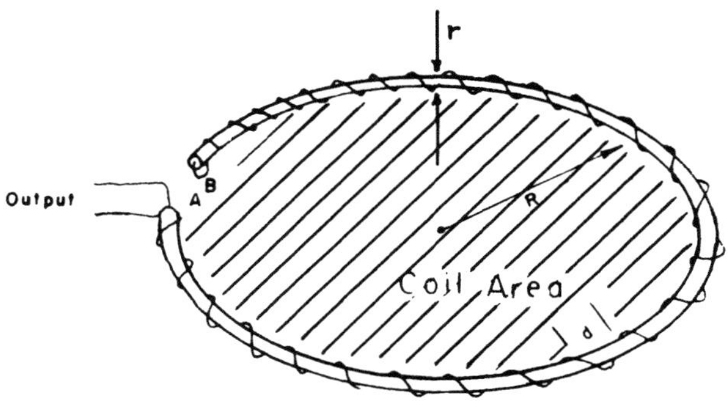

Fig. 2

where I indicates the current passing through the coil area. In order to have a valid approximation of the contour integration, the inequality $d \ll r \ll R$ must be satisfied, d and r should be kept uniform throughout the coil, and the gap between A and B should be made small. It is easily seen that the voltage output of the coil is proportional to the square of r and inversely proportional to d.

The purpose of the back strap winding is to obviate the finite pitch of the toroid. A thin wire should be used as the back strap in order to avoid eddy current effects. As an alternate method, a double-layer structure may be used. In this structure, the back strap is replaced by another toroid wound upon the original one in the same sense so that the induced emf's do not cancel, but opposite inclinations to obviate the finite pitch of the first toroid. It has the advantage that the gain rises by a factor of two, and is theoretically more desirable because of the greater symmetry. However, technically, it is more difficult to maintain the uniformity of r and d.

To test the efficiency of a Rogovski coil, a damped sine wave current from a shunted discharging circuit may be used. A perfect Rogovski coil, when current passes outside the coil area, should pick up no signal. However, some signal is always detected when the coil is brought near enough to one of the discharge circuit con-

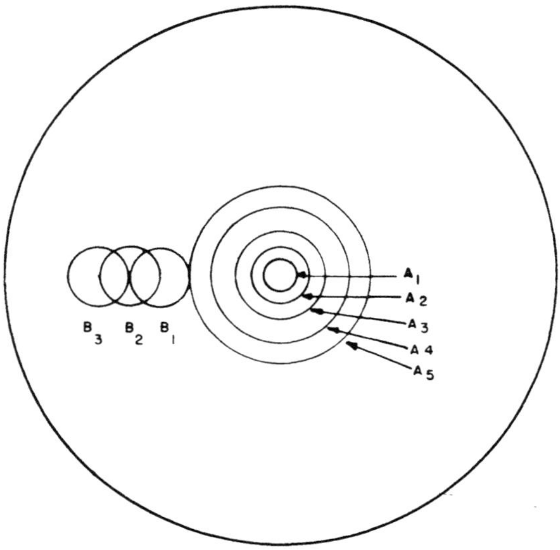

Fig. 3

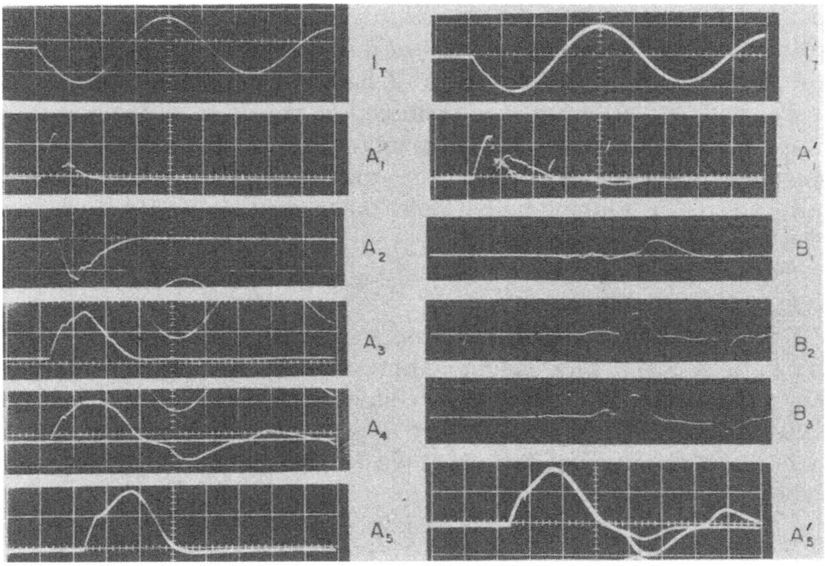

Fig. 4

ductors. The efficiency of the Rogovski coil is thus defined as

$$\text{Efficiency} = 1 - \frac{A_m}{A}$$

where A stands for the amplitude of the signal when the current passes through the coil area, and A_m stands for the amplitude of the largest signal that can be measured when the current passes outside the coil area.

To test the response of a Rogovski coil–integrator combination, an exploding wire in air, containing a "dwell," may be used. The sharp drop and rise at the two ends of the "dwell" provides a test for the combination. Rogovski coils with poor efficiency or response have been rejected.

DATA

Current versus time oscillograms are obtained from the second Rogovski coil for various diameters and for different positions.

In Fig. 3, the plane of the paper shows the cross section of the vacuum chamber with the wire sitting at its center. The outer circle represents the glass wall of the chamber; the other circles show the positions and diameters of the Rogovski coils. For A_1, A_2, A_3, A_4, and A_5, all coils are centered on the wire, with diameters 2, 3.5, 6, 9, and 12 cm, respectively. For B_1, B_2, and B_3, a 4-cm Rogovski coil with high efficiency and high calibration factor is employed. The distances between their centers and the wire are 8, 10, and 12 cm, respectively.

Typical oscillograms for these eight positions at 1μ are shown in Fig. 4.

The first picture shows the total current; the one on the right-hand side is a multiple exposure of the first. A_1' and A_5' are multiple exposures of A_1 and A_5 to show reproducibility. The total current repeats itself nicely. But A_1' and A_5' indicate the indefinite character of the current distribution after the transplosion, especially after the second half cycle.

ANALYSIS

All these I versus t curves, slightly idealized, are plotted with the same scale in Fig. 5. By drawing vertical lines at any given instant t_o, the average current distribution at this instant, can be found. Together with the knowledge of the slope of the I versus t curves in the vicinity of t_o, and an estimation of the rate of radial expansion V, the shape of the $\Delta I/\Delta r$ versus r curves in the vicinity of r_o can be determined approximately by

$$\frac{\Delta I}{\Delta r}(r_o + \delta r,\ t_o) = -\ V_o^{-1}\ \frac{dI}{dt}(r_o,\ t_o - \delta t)$$

where $\delta r = V_o \delta t$, and $\Delta I/\Delta r$ indicates the current flow in the cylindrical layer $r \to r + \Delta r$.

The estimation of the rate of radial expansion is as follows: draw a horizontal line in Fig. 5, which meets A_1, A_2, ... curves (corresponding to r_1, r_2, ...) at t_1, t_2, ... then

$$V \sim \frac{r_2 - r_1}{t_2 - t_1}$$

From Fig. $4A_1$ and $4A_1'$, a sharp drop of current at 1 to 2 μsec after transplosion is always observed. It can be explained as a sheath of current which contains 70% of the total current which

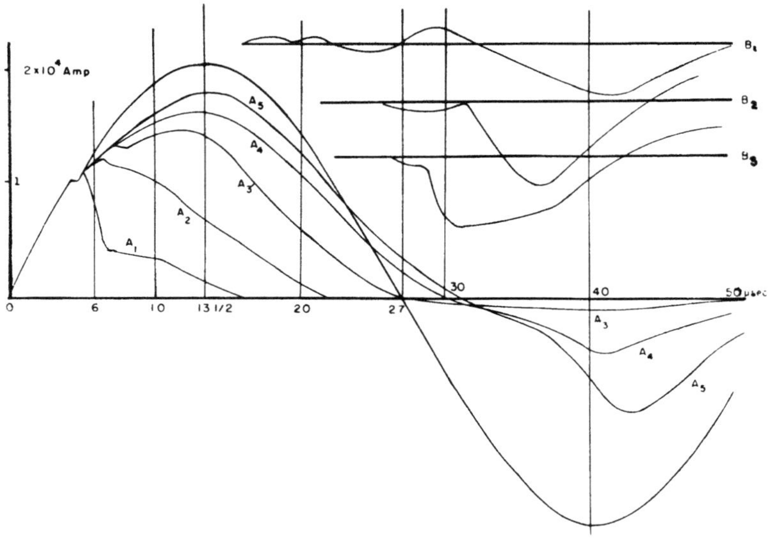

Fig. 5

passes out of the 1-cm circle at this moment with a rate of radial expansion greater than 5×10^5 cm/sec. This is shown as a sharp peak in Fig. 6a.

From Fig. 6b to Fig. 6d, the peak of the current expands with a rate of the order of 10^5 cm/sec. But in Fig. 5f to Fig. 5g, the peak of the current seems to move inward. However, due to the great uncertainty in the second half cycle, the later effect should be investigated more precisely. At present, within the probable error, the peak may actually be stationary during the second half cycle.

In Fig. $4A_3$ to Fig. $4A_5$, the partial current does not reduce to zero as the total current goes to zero. But Fig. $4B_1$ to B_3 shows that a reverse current exists in the outer region of the chamber. Fig. 6e shows the possible current distribution at this moment. It appears to be an "inverse skin effect" [7]; however, the mechanism of this phenomena is not fully understood.

COMPARISON

If the pressure is increased to around 40 μ, the same procedure characteristically yields the same results, with the exception that the rate of current expansion is considerably reduced. [See Fig. 7(1).] The peak of the current distribution seems to stop within a cylinder of radius 2 cm approximately 15 μsec after the transplosion. Further increase in pressure will reduce the expansion further. The magnitude of the initial sharp peak of the current sheath also decreases as the pressure increases. But the initial rate of radial expansion still retains the value of 5×10^5 cm/sec up to the pressure of 1 mm Hg.

As the pressure is increased above 10 mm Hg, the situations change drastically. The I versus t curves at 15 mm Hg [Fig. 7(2)] are characteristically different from the previous ones in two notable points. First, the sharp drop of current in the 2-cm Rogovski coil measurement (position A_1) occurs at 4 μsec after the transplosion instead of 1 μsec. Second, the indefinite character after transplosion disappears. Similar analysis shows that a cylindrical sheath of current moves out with a rate of radial expansion of the order $10^5 \times 10^4$ cm/sec.

The phenomena seems to be explained as follows. It has been observed by several researchers [8–10] that a large-magnitude,

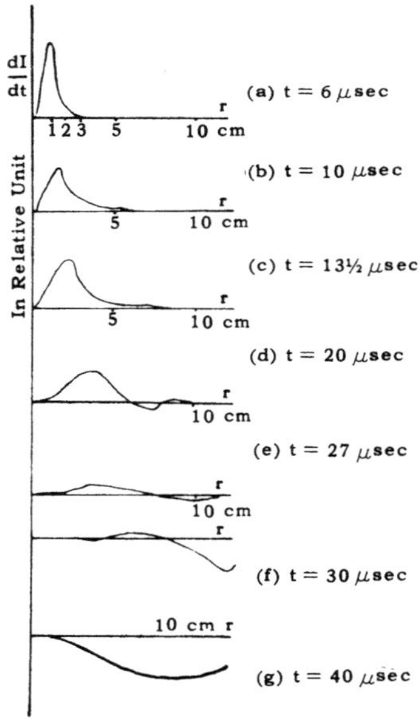

Fig. 6

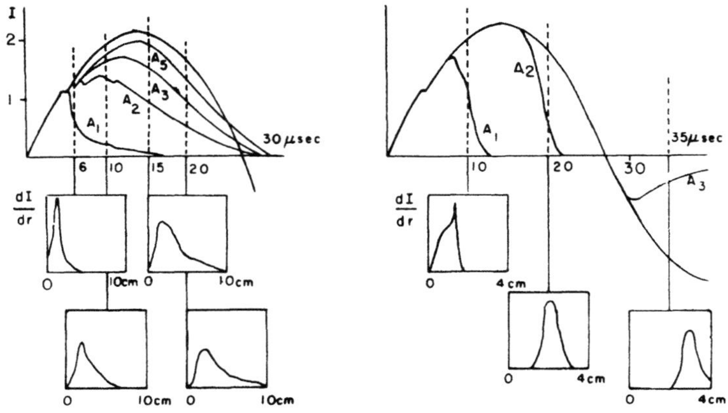

Fig. 7

short-duration voltage pulse occurs across the wire simultaneous with the transplosion. It is thought to be due to the sudden change of the inductance of the wire when exploded:

$$\text{Voltage (wire)} = iR + L\frac{di}{dt} + i\frac{dL}{dt}$$

i.e., the third term of the above equation.

With this pulse, a strong *E* field is built up in the vicinity of the wire at the moment of transplosion. For low pressure, it produces a breakdown in the ambient atmosphere. An intense sheath of current appears outside the wire immediately after the transplosion. It diffuses rapidly into a widespread cloud. The current expands much faster than the copper vapor. For higher pressures, the breakdown is not realized and the current passage is restricted near the surface of the copper vapor.

This phenomenon has been studied with a high-speed framing camera [11]. The copper vapor is seen as a well-defined image expanding with a rate of $10^5 \sim 10^4$ cm/sec. At low pressure, a long-range, irregular, dim glow is seen outside the vapor cloud. But, at pressures higher than 10 mm Hg, the glow becomes very short-range and has a clear-cut edge.

CONCLUSION

In this investigation, there are several things left to be desired. The mechanism of the phenomena is not fully understood. Although the Rogovski coils have an efficiency as high as 95%, they still need to be improved in order to achieve better quantitative descriptions, especially for the "inverse skin effect." An asymmetry of the field in the outer region of the glass chamber is also detected, which suggests that for the investigation of the second half cycle and later development, a coaxial chamber should be used.

The results, however, can be summarized as follows:

1. There is no expansion of current before the transplosion point which is indicated by the "notch" of the total current wave form.
2. The current distribution of a copper wire ($l = 4$ cm, $r = 0.02$ cm) exploded under a pressure of 1 μ expands radially with a rate of the order of $10^6 \sim 10^5$ cm/sec.
3. The rate of radial expansion reduces as the pressure increases.
4. The current distribution does not reduce to zero as the total current goes to zero. Instead, a phenomenon similar to the "inverse skin effect" is observed.
5. At low pressure (<1 mm Hg), there is an intense sheath of current, possibly an arc [1], which appears just after the transplosion. For higher pressure (>10 mm Hg), this sheath diminishes greatly and the uncertainty of later development disappears.

REFERENCES

1. I.F. Kvartskhava, V.B. Bondarenko, P.D. Meladze, and K.V. Suladze, "Electrical Explosion of Wires in Vacuum," Zhur. Eksp. i Teoret. Fiz. 31:737-744 (1956); Soviet Physics, JETP 4:637-644 (1957).
2. T. Korneff, J.L. Bohn, and F.H. Nadig, "Exploding Wire Phenomena at Reduced Pressures," Exploding Wires, Vol. 1, W.G. Chace and H.K. Moore, eds., Plenum Press, New York (1959), pp. 104-114.
3. T. Korneff, W. Chace, and H.D. Edelson, "Shunting Mechanisms in Exploding Wires at Reduced Pressures," VI Intern. Conf. Ioniz, Phenom, in Gases, Paris, Vol. II (1963), pp. 627-631.
4. H.D. Edelson and T. Korneff, "Conducting Mechanism for Exploding Wire in a Vacuum," Exploding Wires, Vol. 3, W.G. Chace and H.K. Moore, eds., Plenum Press, New York (1964), pp. 267–284.
5. F.D. Bennett and D.D. Shear, "Shockwave from Exploding Wires at Low Ambient Densities," Exploding Wires, Vol. 2, W.G. Chace and H.K. Moore, eds., Plenum Press, New York (1961), pp. 181-194.

6. O.H. Zinke, R.B. Owen, and C.K. Manka, "Translational Equilibrium of Wire Exploded in Vacuo," Exploding Wires, Vol. 3, W.G. Chace and H.K. Moore, eds., Plenum Press, New York (1964).
7. M.G. Haines, "The Inverse Skin Effect," Proc. Phys. Soc. 75(5):576-584 (1959).
8. F.H. Webb, Jr., H.H. Hilton, P.H. Levine, and A.V. Tollestrup, "The Electrical and Optical Properties of Rapidly Exploded Wires," Exploding Wires, Vol. 2, W.G. Chace and H.K. Moore, eds., Plenum Press, New York (1961), pp. 37-75.
9. I.F. Kvartskhava, A.A. Pliutto, A.A. Chernov, and V.V. Bondarenko, "Electrical Explosion of Metal Wires," Soviet Physics, JETP 3:40-51 (1956).
10. H.S. Leopold, "Effect of Bridgewire Parameters on Explosive Initiation," Exploding Wires, Vol. 3, W.G. Chace and H.K. Moore, eds., Plenum Press, New York (1964), pp. 125-152.
11. D.C. Chern, "Current Distribution for Wire Exploded in Vacuo," M.A. Thesis, Temple University (1967).

DISCUSSION

Question by H.F. Swift, University of Dayton, Dayton, Ohio.: What effect does the current sensor have on the current distribution?

Answer by Author: Framing camera studies were made of the visible interaction between plasma and Rogovski coil. There was some interaction, in the form of recombination light at the surface of the coil. We deemed it small, but not negligible.

Question by Don Warne, Sandia Corp.: As Rogovski coil is a di/dt sensor, what kind of an integration was used?

Answer by Author: An RC integrator network with a time constant of 440 μsec.

Correlated X-Ray and Optical Streak Photographs of Exploding Wires

K. S. Fansler and D. D. Shear

Ballistic Research Laboratories
Aberdeen Proving Ground, Maryland

Sequences of flash x-ray photographs of 70 nsec-duration obtained during the early expansion of copper, lead, and iron wires are correlated precisely in time with the electrical voltage pulse and optical streak photographs. Wire sizes were 10, 15, and 25 mil in diameter. The x-ray pictures show the onset and development of striations in each of these metals along the wire length during the fast resistance rise of the wire. The relatively smooth envelope of the radially expanding striations is compared with features of the optical streak pictures. Some of these optical pictures show a translucent radial cloud developing prior to the fast resistance rise, and a more opaque inner core which expands radially. Most of the striations in the x-ray pictures appear to be completely separated at the voltage maximum. After the voltage peak, the density of material appears to be greater further from the center. The observed behavior is discussed.

INTRODUCTION

The voltage-spiking mechanism for exploding wires is not fully understood. The study of this phenomenon by high-speed photographic techniques has been restricted by the brief time interval of voltage spiking and by a dense outer cloud of vaporized metal which hides the inner structures of the exploding wire. Studies [1, 2] using pulsed x-ray techniques have indicated that striations were formed during some stage of the explosion. More recently, a pulsed x-ray study [3] was conducted that correlated the x-ray pulse with the voltage and current across the exploding wire. This study indicated that transverse striations were formed during the overvoltage rise.

We felt that a study using both high-speed photographic and flash x-ray techniques might be very illuminating. Direct comparisons between the streak camera and flash x-ray photographs obtained are possible since they are correlated in time and scaled to the same size.

EXPERIMENTAL ARRANGEMENT

The 0.625-mm-diameter iron and copper wires were exploded with 63 μF of capacitance at 7.5 kV and a circuit basic frequency of 40 kHz. The 0.375-mm-diameter lead wires and the 0.25-mm copper wires were exploded at 3.5 kV and 3.0 kV, respectively, with a basic circuit frequency of 43 kHz. All wires were 2.5 cm long, with the exception of the 0.25-mm copper wires, which were 2 cm long. A Fexitron model 730-4C/235 150-kV flash x-ray system was used that emitted a 70-nsec pulse with variable delay. This pulse and the exploding wire voltage pulse were correlated on a dual-beam 555 Tektronix oscilloscope. This system is shown in Fig. 1. The x-ray tube and x-ray film were located 15 and 0.5 cm, respectively, from the wires, except for the lead wires. The lead wire to film distance was 0.25 cm. The x-ray source size was 3 mm. The x-ray film was Kodak Ultra-Speed Dental X-ray film and was used without an image intensifier. It was protected from the blast by thin

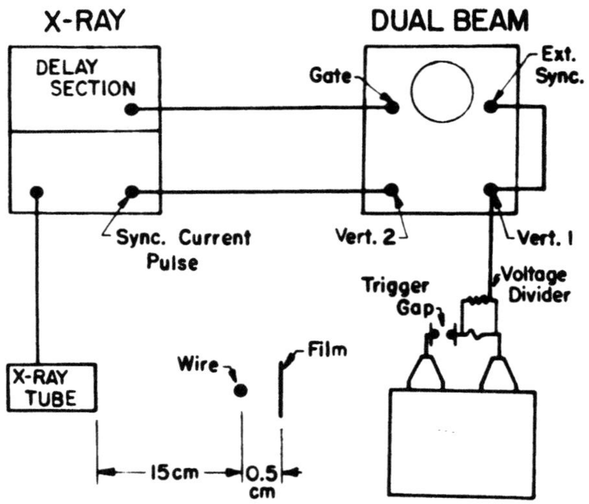

Fig. 1. Diagram of experimental setup for correlated x-ray photography.

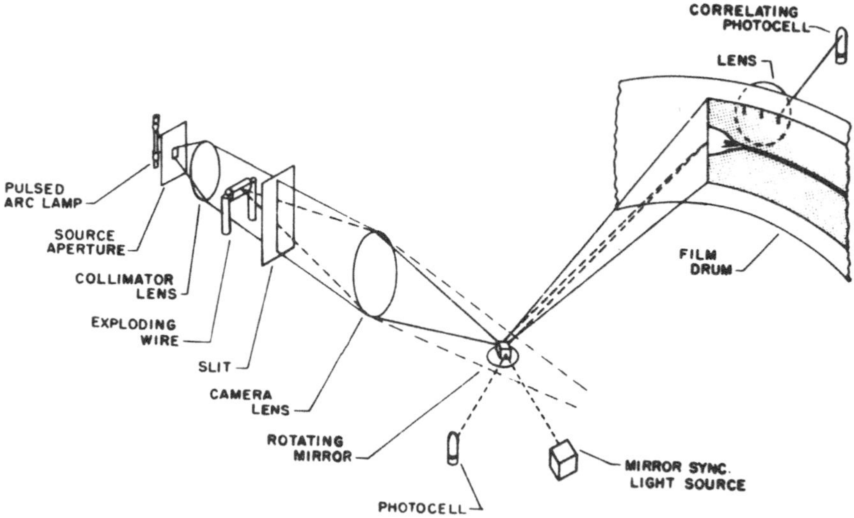

Fig. 2. Streak camera with time-correlating pulse photocell.

sheets of Mylar plastic. In an auxiliary experiment, a photomultiplier tube was used to correlate in time the x-ray pulse intensity with the current that produced it. With cable lengths being taken into consideration, ± 50 nsec was established as the total of systematic and random errors.

The basic rotating mirror camera used to obtain the streak photograph is shown in Fig. 2. The image of a short segment of the exploding wire, backlighted by a pulsed mercury vapor lamp and viewed through a perpendicular (vertical) slit is projected via the mirror onto the photographic film. Any increment of action is recorded at a position fixed by the mirror rotational velocity to its time of occurrence.

A four-sided rotating mirror, with faces 1 cm square, is located 5 in. from the lens at its focal point. With a 60-cm distance between the mirror and the film drum, a magnification of five times and a lens speed of f/64 are obtained; with the mirror rotating 1650 rps, a sweep speed of 1.25 cm/μsec is provided.

Each of three slits cut in at the edge of the photographic film transmits light from the moving slit image, through a lens, to the correlating photocell. The output of this photocell is recorded as a sequence of three spikes on one beam of a dual-trace oscilloscope. Thus, three permanent position-reference points on the film are

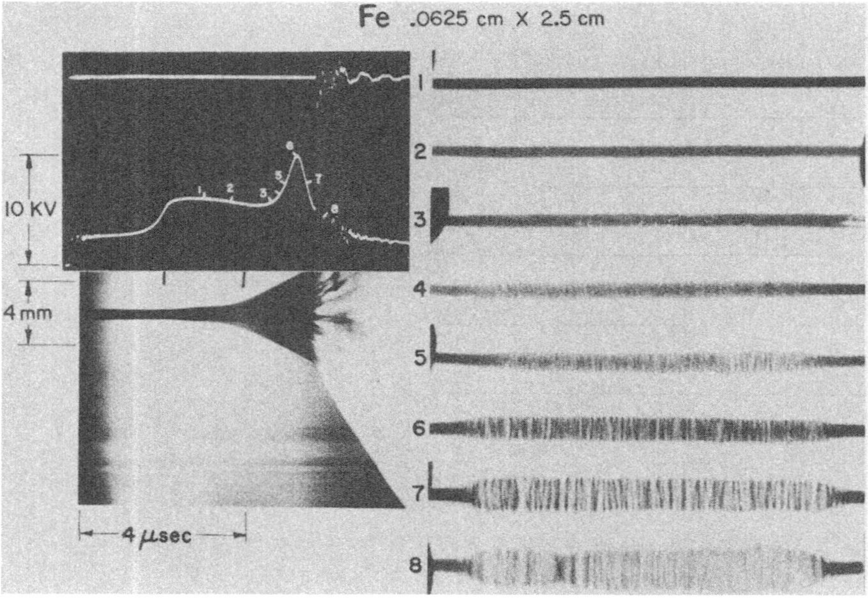

Fig. 3. Fe 0.0625 cm × 2.5 cm.

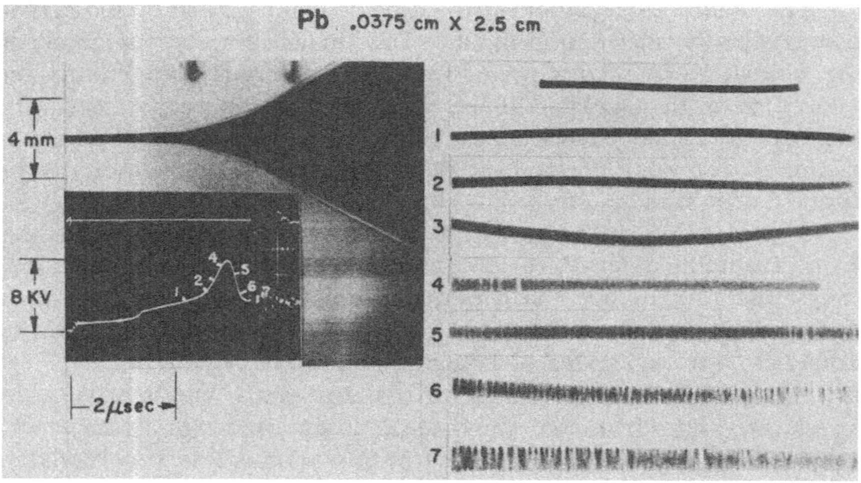

Fig. 4. Pb 0.0375 cm × 2.5 cm.

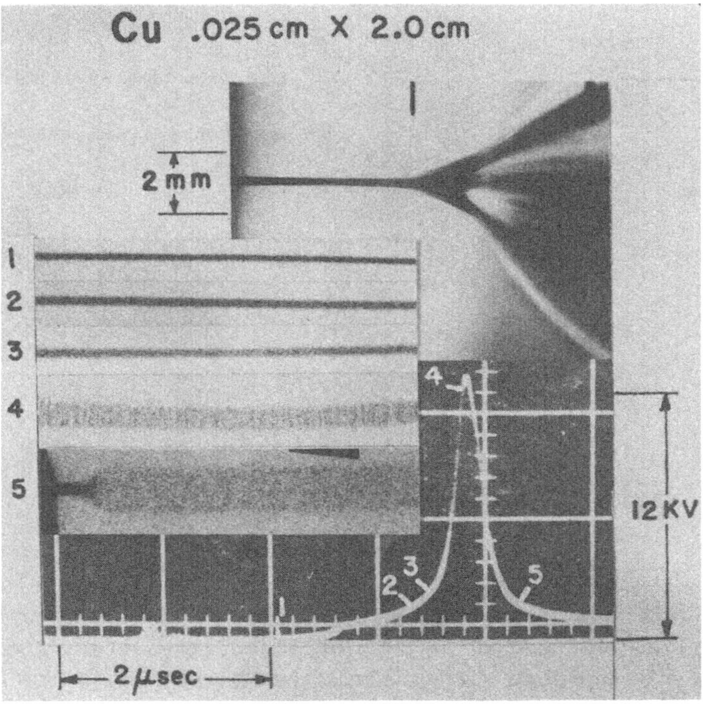

Fig. 5. Cu 0.025 cm × 2.0 cm.

correlated with time within the accuracy of the oscilloscope sweep calibration. On the second trace, an exploding wire voltage record correlates directly with the three-spike sequence. Another photocell system detects a narrow auxiliary beam of light reflected from one of the back faces of the rotating mirror and the resulting pulse triggers the time correlation scope which in turn triggers the exploding wire circuit and backlighting. Wire burst is thus preset to occur at a mirror orientation which will sweep the ensuing action across the film drum.

RESULTS AND DISCUSSION

Flash x-ray and streak photographs, together with the correlating voltage trace, are presented with the same diameter-scales for

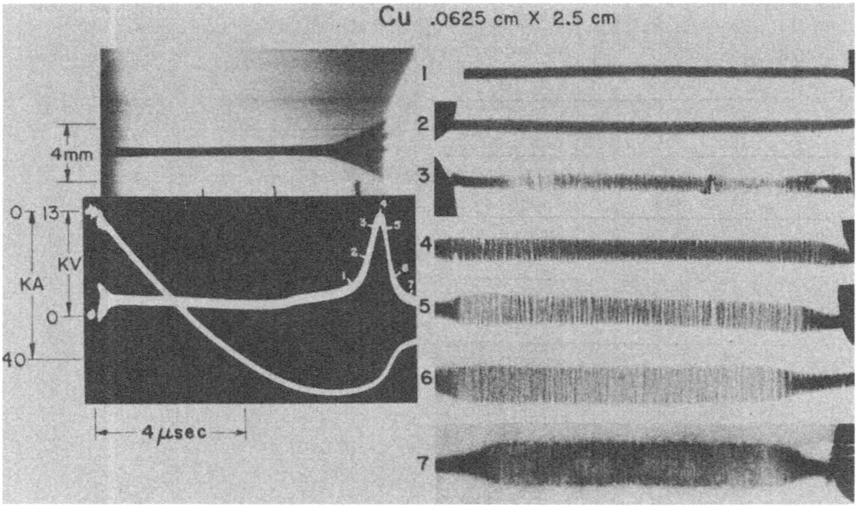

Fig. 6. Cu 0.0625 cm × 2.5 cm.

both streak and x-ray photographs. The streak photographs show an obvious increase in wire diameter before onset of vaporization for the iron and lead wires. The streak photographs of all of these wires show vaporization, starting well before the start of the overvoltage pulse, and a shock wave associated with the expansion of the vapor against the ambient atmosphere. In the streak photographs for iron and the smaller copper wires, one can observe the sharp boundary of an opaque inner core through the less dense outer vapor layer. For the smaller copper wire case, this inner core begins expanding less than 0.1 μsec before the start of the rapid upsweep of the overvoltage pulse. In the iron wire case, the rapid expansion of the denser material starts about 0.4 μsec before the beginning of the overvoltage pulse. The flash x-ray photographs reveal (small copper wire excepted) almost periodic variations in cross section along the length of the wire in the early stages of the overvoltage rise. These variations uniformly become more pronounced as the voltage peak is approached. However, the time relative to the voltage peak at which clear layers are formed through the wire varies with the wire material. The iron wires have formed into many clear striations well before their voltage peaks, while the lead and larger copper wires do not exhibit many clear portions through the metal until near the voltage peak. The number of striations at voltage peak is approximately 35/cm for the large copper wires, while it is approximately 25 for the iron wires. The number of striations diminishes with time after voltage peak. After voltage peak, both the distance between the dense disks and their diameter become larger. Throughout this expansion, the envelopes of the disks are smooth curves with slight variations in diameter. The x-ray photographs at later time for the large iron and copper wires show the center portions to be less dense [2] than the periphery. This does not seem to occur for the lead wires. It is interesting to note the smaller diameter ends of the wires. High-speed optical photography [4] has shown the ends to be quite luminous during the wire explosion. It appears that arcs forming about the ends of the wire shunt much of the current around them.

The envelopes of the x-rayed disks of the iron and small copper wires have a smaller diameter than those of the streak photographs of the inner core for the times during the overvoltage upsweep. The inner core cannot be seen in the other wires, since the outer vapor layer is too opaque.

It is felt that the vaporization wave theory [5] can offer some explanation of the phenomena observed. The expansion of the inner cores as observed in the streak photos is thought to correspond to the attainment of fast vaporization wave speeds. The vaporization wave speed as a function of energy input to the wire very slowly rises until a characteristic temperature for the metal is reached and thereafter the vaporization wave speed rises very steeply with energy input. The growth of the striations after the initiation of the striating process is caused by magnetic pinch effects and the different heating rates for the disks and the unvaporized material between the disks. Early in the striating process, the disks are assumed to be joined by bridges of liquid metal of smaller cross section. Skin effect will tend to keep the current to the outside of both—disks and bridges—but a second effect, due to the geometry, will tend to prevent the current from expanding to the outside of the disks. The effects reinforce one another to provide high current—and, thus, high energy dissipation—at the surface of the bridges, while opposing one another to produce lower surface heating in the disks. This adds further to the processes which hasten the eating away of the bridges.

It is also thought that sometime during the striation process arcs form between the transverse layers of dense metal. These too would cause localized heating in the area of the bridges. These arcs cannot be seen optically because of the dense vapor cloud surrounding the disks.

CONCLUSION

There are certain characteristics common to all of the correlated photographs. The streak pictures indicate the vaporization processes starting well before the overvoltage pulse and show the outer vapor boundary appearing as a nearly straight line with a shock wave coming off it. Inner cores appear in the iron and smaller copper streak photographs and are significantly larger than the disks appearing in the x-ray shadowgraphs.

The x-ray shadowgraphs show the approximate onset and growth of striations, the almost regular positions of the disks, and the uniform envelope of disks.

In the future, we will attempt to see through the outer vapor envelope with the streak camera and to obtain x-ray shadowgraphs with better resolution. We will also obtain these photographs correlated with current, power, and energy curves.

REFERENCES

1. W. Chace, R. Morgan, and K. Saari, in Exploding Wires, Vol. 1, W.G. Chace and H.K. Moore, eds., Plenum Press, New York (1959), p. 59.
2. W. Muller, in Exploding Wires, Vol. 1, W.G. Chace and H.K. Moore, eds., Plenum Press, New York (1959), p. 186.
3. V.M. Kul'gavchuk and G.A. Novoskol'tseva, Zh. Tekhn. Fiz. 36(3):549 (1966) [Soviet Phys. - Tech. Phys., Vol. 11, p. 406].
4. W.G. Chace, Phys. Fluids 2:230 (1959).
5. F.D. Bennett, G.D. Kahl, and E.H. Wedemeyer, in Exploding Wires, Vol. 3, W.G. Chace and H.K. Moore, eds., Plenum Press, New York (1964), p. 65.

DISCUSSION

Question by O.H. Zinke, University of Arkansas, Fayetteville, Ark.: Would you speculate on the peculiar behavior in the iron—voltage curve?

Answer by Author: Between the voltage rise that is thought to correspond to melt and the beginning of the overvoltage pulse, the voltage decreases. We have obtained current for exploding iron wires and have calculated that the resistance decreases during this voltage decrease. It is thought that the time during decreasing voltage corresponds to the liquid phase of iron. Nickel also shows this peculiar behavior. This is all we know about this at the present time.

Question by M. Levine, Air Force Cambridge Research Laboratory, Bedford, Mass.: The early development of striations and the slow current pulse seem to indicate that an early magnetic instability of the "sausage" type has occurred and this explosion can be characterized as slow in the Chace and Levine classification system. This seems to be confirmed by the fact that the instability develops later in lead and that the striations have spacings which correspond to a wire radius at the time it develops again in accordance with theory.

Answer by Author: We have not yet ascertained the basic cause of the initiation of striations but in all our studies so far the striating process seems to begin near the start of the overvoltage upsweep. Much faster explosions than ours exhibit similar voltage spiking characteristics. I am convinced that the growth of these striations causes the fast resistance rise in our experiments. I therefore feel that the growth of striations may also cause the fast resistance rise for explosions much faster than ours. We intend experimentally to find this out since no x-ray studies with faster circuit parameters have yet been made.

Indeed, according to the Chace–Levine classification, our x-ray shadowgraphs show our explosions to be classified as slow. Our basic circuit frequency is 40 kC. The length of time for most of our energy dump is approximately 4 μsec in the 0.025-in. copper and 0.025-in. iron. Since the total energy in the condenser is 1750 J and approximately 40% is deposited in the wire, i.e., 700 J in 4 μsec, I cannot consider this a slow explosion. Also note the current for the 0.025-in. copper wire has a di/dt equal to 10^{10} A/sec.

It is not at all certain where the striating processes start for the lead wire, but there seems to be evidence of the process starting very near the start of the overvoltage upsweep. Copper wires behaved similarly in this respect. Harder x-rays might have been better to use with the lead wires.

We will agree that the distance between striations for copper and iron are near the radius of these wires, but for lead these distances seem to be quite large in comparison to the lead wire radius. Kul'gavchuk, using an exploding wire circuit with an associated frequency one-half of ours, indicated that the distance between striation was about two times greater than the original radius of the wire near overvoltage peak. I feel that, although the radius may be a consideration in the striation distances, the associated circuit frequencies, the wire materials, and pressures should also merit much attention.

Time-Resolved Spectroscopy of Exploding Wires*

Roger S. Case, Jr.
Air Force Institute of Technology
Wright-Patterson Air Force Base
Dayton, Ohio

and

Arthur H. Guenther
Air Force Weapons Laboratory
Kirtland Air Force Base
Albuquerque, New Mexico

A continuous-writing, image-streaking spectrograph-camera instrument was constructed and calibrated. Time-resolved spectral emulsions were obtained, in the visible spectrum, using exploding wire plasmas of five elements (Al, Au, Ni, Cu, Ag) as a light source. Wires were exploded both in vacuum and at ambient atmospheric conditions. The time-resolved spectrograph consisted of a $f/6.3$ plane grating spectrograph coupled with a 70-mm AVCO Model 300 camera. This combination featured a time resolution of $\sim$0.26 μsec and a reciprocal dispersion of approximately 35 Å/mm, and was calibrated so that relative spectral intensities could be easily determined. The energy discharge system incorporated a 2-μF 50-kV capacitor, pressurized switch of the trigatron type, and an environmental chamber with electrical diagnostics. The ringing frequency of the system on a dead short was $\sim$330 kHz. Preliminary estimates of the temperature were calculated using measured relative spectral intensities, and a theoretical description of the expected plasma radiation. Results indicated that at late times in the discharge the plasma was not optically thin as assumed, and that meaningful temperatures could not be calculated without recourse to a detailed investigation of the plasma opacity. At early times, the assumption that the plasma radiated as a black body was applicable, and reasonable temperatures (0.1 to 1.0 eV) were calculated.

INTRODUCTION

At the present time, there is an incomplete description of the processes involved in high atomic number, high-density, transient

*This work was abstracted in part from a Master's thesis submitted by First Lieutenant Roger S. Case, Jr., to the Graduate School, Air Force Institute of Technology, June, 1966.

plasmas because of the lack of sufficient fundamental data concerning the radiative characteristics of these plasmas and the inadequacy of the untried theoretical descriptions. Exploding wire research can be used to determine physical quantities necessary for improving our ability to predict the characteristics of this class of plasmas.

Previous related work has been reported in detail [1, 4, 5]. Efforts in this area at the Air Force Weapons Laboratory (AFWL) have been confined to the exploding of wires of various materials at atmospheric pressure and have resulted in the design and installation at the AFWL of the High Energy Density Pulsed Power System for exploding wire research [2]. The research reported herein was aimed at developing optical diagnostic techniques capable of being extended to the study of much higher energy-density exploding wires using the High Energy Density System.

One of the most useful diagnostic techniques at the disposal of the plasma physicist is that of spectroscopic investigation. It is desirable that this spectral measurement be accomplished in a time-resolved manner, particularly when dealing with highly transient phenomena such as an exploding wire. When initiating a program to determine in detail the radiative characteristics such as various particle temperatures and densities, preliminary investigations always involve a survey of the spectral content in emission as well as absorption. This initial investigation will aid in the selection of the most prominent radiative features which can be fruitfully studied in detail to elucidate plasma characteristics such as temperature, density, stage of ionization, and degree of equilibrium. This paper concerns itself with a description of the construction and calibration of a high optical speed, continuous-writing, streak spectrograph-camera system. In addition, preliminary checks on the consistency of the instrument were made by applying appropriate formulas to obtain "temperatures" for the exploding wire plasmas; emphasis was placed upon development of the instrumentation and a determination of the instrumental corrections required to accurately calibrate the optical system.

OPTICAL INSTRUMENTATION AND SYSTEM CALIBRATION

The optical instrumentation consisted of a Jarrell-Ash $f/6.3$ grating spectrograph used in conjunction with an AVCO Model 300

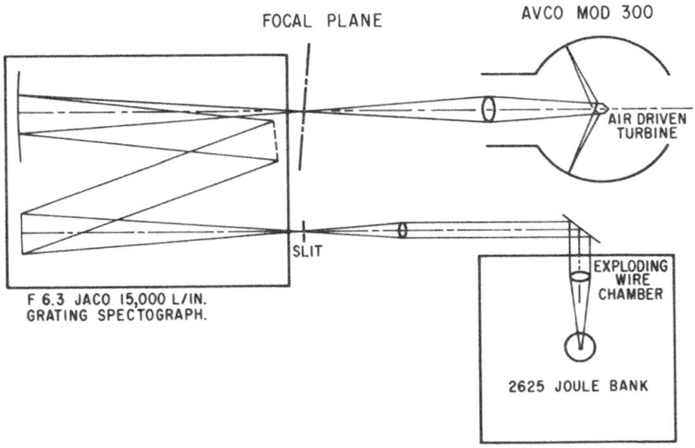

Fig. 1. Spectrograph–camera system.

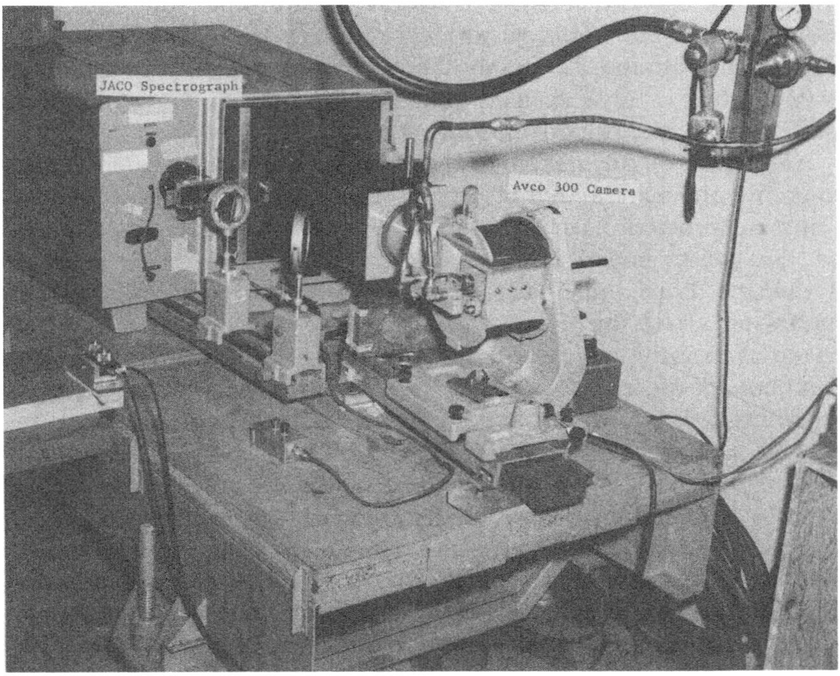

Fig. 2. Schematic of optical diagnostics.

rotating turbine camera to obtain the desired time and spectral resolution (see Figs. 1 and 2). Employing 70-mm film in the camera, the instrument is excellent for survey work over a broad range of wavelengths, as a useful wavelength coverage of > 1000 Å is obtained on one spectral emulsion. The reciprocal dispersion of the optical system at the film plane was 34.7 Å/mm, with a time resolution of 0.26 μsec at a maximum turbine speed of 3000 rps. The time resolution available with this system is much greater than that available with drum-type cameras, and matches well the 15- to 50-μsec time interval during which the plasma radiates intensely enough to be recorded by the system at high scanning rates.

The most time-consuming and critical portion of the experimental procedure was the determination of the spectral transmission characteristics of the optical recording system. It was necessary to calibrate the spectrograph, camera, and emulsion as a unit before this system could be used to ascertain relative spectral line and continuum intensities, necessary as inputs for plasma temperature evaluations. Necessary calibration curves are those relating the optical density of spectral emulsions to the relative intensities of the sources which produce these exposures. Two photographic emulsions supply the necessary data to construct these curves, while a third is used to ensure that all photographic plates were developed uniformly (equal γ).

The first plate necessary for this calibration is a streaked image of the exit aperture of the spectrograph. One coordinate of this calibration plate is the time axis, lengthwise along the plate and the other coordinate axis, across the plate, represents the wavelength dispersion. The exposure, $E(t) = I(t) \times t$, is then constant lengthwise along the emulsion of any given wavelength. The decrease in density measured along the film is caused by the changing f/number of the camera, which results from the decreasing effective writing area of the turbine face as it rotates (the turbine face acts as an aperture stop under these conditions). The variation of the f/number was assumed to be independent of wavelength. By scanning the plate lengthwise, calculating the quantity $D(x)/D(X_o)$, where x is the distance along the plate from a position X_o and D is the optical density ($D = -\log T$, where T = percent transmission), and then plotting this as a function of x, one obtains a quantitative measure of the f/number variation as a function of the position of the exposure along the film if the appropriate characteristic curve for this emulsion is known.

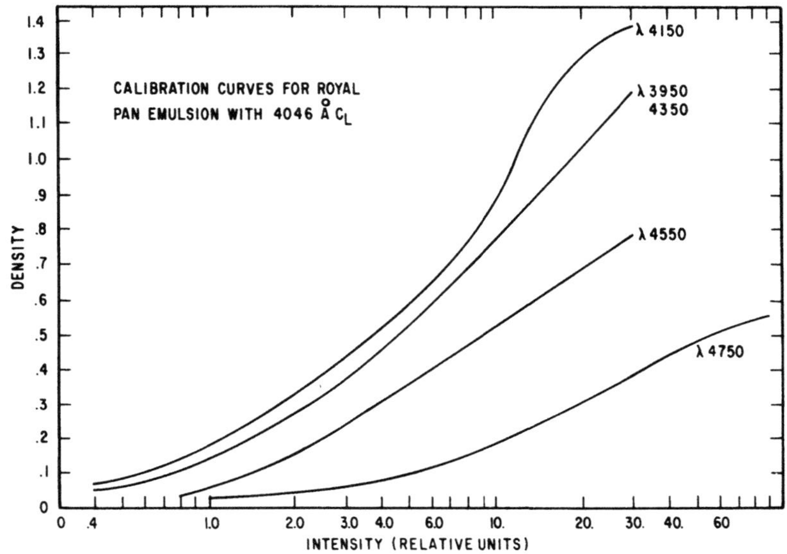

Fig. 3. Typical calibration curves for various wavelengths.

On a second emulsion, a series of static exposures are made (at different positions along the film) with the exposure time being varied from one exposure to the next. This plate is scanned lengthwise, at various wavelengths, to obtain values of the optical density of the film as functions of wavelength and exposure ($I \times t$). The f/number correction discussed above is then applied to correct the observed densities for their film position, with the wavelength a parameter. The plotting of optical density against the relative exposure yields the calibration curves, which can then be used to measure relative spectral intensities. The horizontal axis on these curves may be labeled either E (exposure) or I (intensity), since in all formulas in which these quantities appear, only the ratio of two intensities are used. In forming the ratios of two exposures E_1/E_2, the time cancels, i.e., $I_1 \times t / I_2 \times t = I_1/I_2 = E_1/E_2$. It should be noted that the curves obtained do not allow one to compare intensities at two different exposure times t_1 and t_2, but all intensities at a given time t may be compared. The calibration curves obtained for a number of wavelengths are presented in Fig. 3.

In this work, the effects of reciprocity law failure of the photographic emulsions were not taken into account [7]. It was assumed that this effect was not a wavelength-dependent function, and that any effects which did occur would cancel when intensities were compared using ratio techniques.

WIRE EXPLOSION APPARATUS

The preliminary checkout of the total optical system was accomplished using a capacitor bank with the following parameters:

charging voltage	50 kV (maximum)
ringing frequency	327 kHz
energy stored	2625 J (maximum)
peak current	6×10^4 A

The explosion chamber was designed so that wires could be exploded in either vacuum or atmospheric environments. A coaxial arrangement was employed to assure a rapid current rise time and a rapid energy delivery rate (Figs. 4 and 5).

Electrical diagnostics were those conventionally used in pulse power work: a current-viewing resistor monitored the current, and capacitive voltage divider (1015:1) was used as a voltage pickup. The spark gap was a pressurized trigatron. A schematic of the electronics system is shown in Fig. 6.

Accurate synchronization of the camera with the wire explosion was necessary to assure that the particular portion of the film

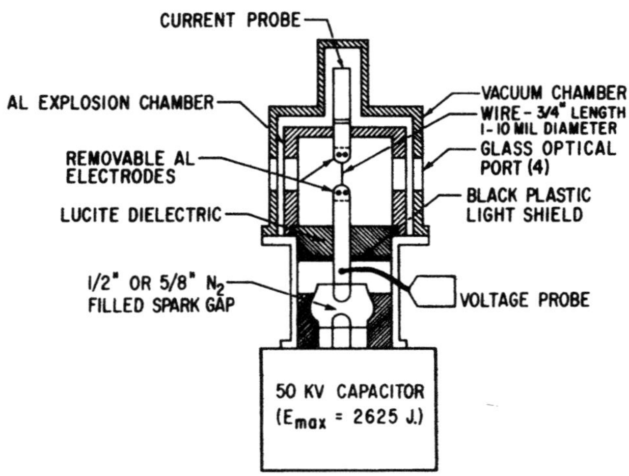

Fig. 4. Schematic of 50-kV exploding wire system (not to scale).

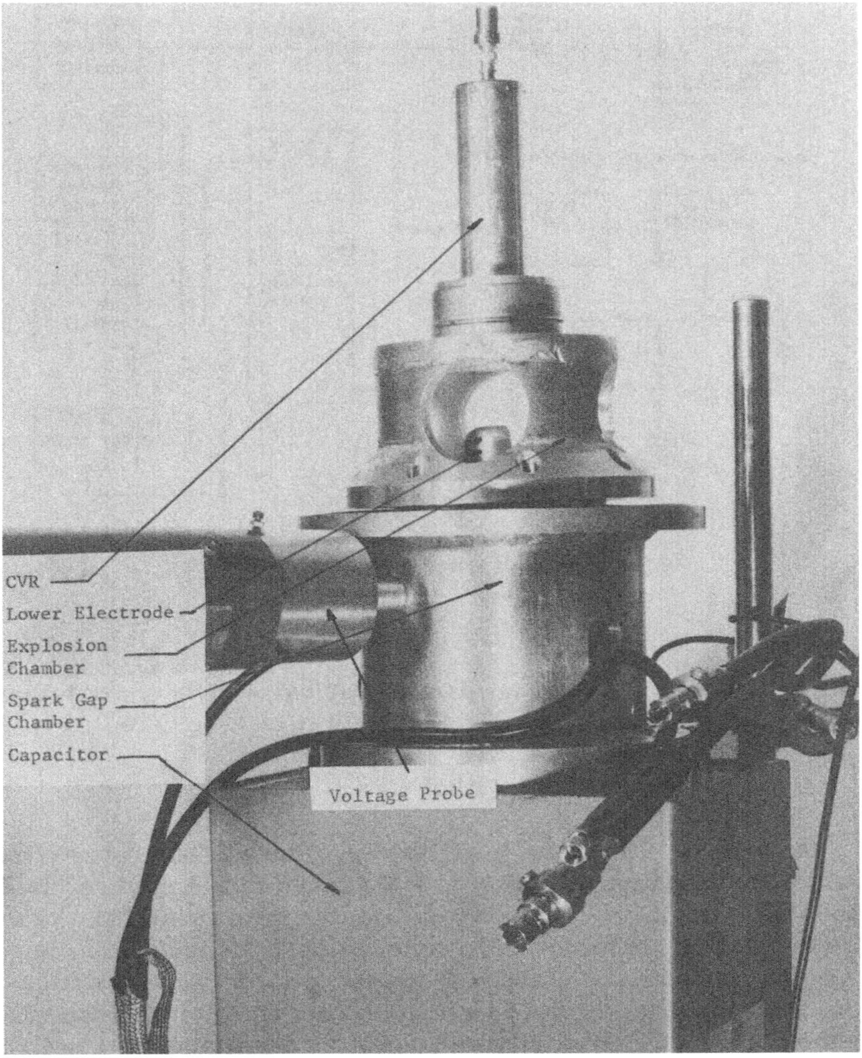

Fig. 5. Partially assembled exploding wire system.

plane on which data were recorded was that on which the non-linearity of the photographic response was minimized. To achieve this, the camera was used to initiate the discharge system at the appropriate time with respect to the turbine face of the rotating

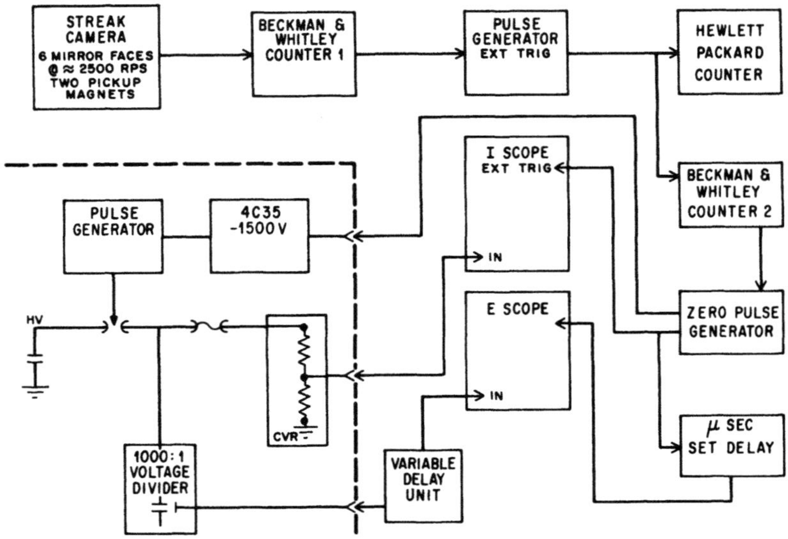

Fig. 6. Block diagram of system electronics.

mirror camera. Electrical delays (cables, switches, etc.) were taken into account.

RESULTS

A total of 49 wires were exploded during the course of the initial investigation; wires of two diameters (0.003 and 0.005 inch) and five materials (Au, Al, Ag, Cu, and Ni) were used. Most of the time-resolved records were suitable for the measurement of spectral intensities. Table I shows the general characteristics of the spectra of the five elements used. For most of these spectral plates, good electrical diagnostic results were also obtained. A good correlation existed between the time of peak current and the time of maximum light output of the plasma.

In this work, exploding the wires in a vacuum eliminated the environmental continua normally observed in atmospheric wire explosions. However, arcing was observed in the vacuum region employed (10^{-4} torr), thus decreasing the delivery of energy to the wire. Because of this, a significant portion of the energy was shunted to the chamber walls and around the dense wire plasma.

CONCLUSIONS

The instrument developed during this investigation is excellent for use as a survey instrument to provide time and wavelength resolution. By replacing the course grating employed with one of higher resolution and working in higher orders, line-broadening studies could also be accomplished. However, the major usefulness of this instrument is for the time-resolved radiative characteristic of self-luminous events. It is a reliable inexpensive instrument which produced highly reproducible results under a variety of conditions. Appendix I is a discussion of the theory and assumptions used in measuring "temperatures" with the system, and the results of the "temperature" determinations.

Table I. General Spectral Characteristics of Exploding Wire Plasmas for 5 Elements

Element	Pressure	Spectral characteristics
Al	atmosphere	Al II 3944, 3961 Å; strong and persistent. No self-absorption present. Al II lines predominant, air continua strong.
Al	vacuum	Same as at atmospheric pressure, but with environmental continua absent. 15 to 20 lines identified.
Cu	atmosphere	Electrode contamination allows Al I 3944, 3961 Å to be observed; air continua present, no reversal present. Strong lines at 4275, 4530, 3897, 4651 Å.
Cu	vacuum	Many more lines evident with environmental continua absent: 4587, 4509 Å, etc.; line structure evident at earlier times in explosion.
Au	atmosphere	Same characteristics as above. Lines include 4065, 4488, 4792, 4315 Å. As above; short-lived higher energy lines present.
Ni	atmosphere	Same as above; strong lines from neutral atom (plate badly fogged); no spectra obtained in vacuum.
Ag	atmosphere	Same as above, but no line pair available with *gf* values; no spectra obtained in vacuum.

APPENDIX I

Theory and Assumptions

In analyzing experimental measurements of exploding wire plasma parameters, certain assumptions must be made. The streak camera–spectrograph system permits one to resolve the time- and wavelength-dependent radiation characteristics of wire explosions. Applicable results of the quantum theory of radiation may then be used to deduce the temperature and opacity of the plasma from the radiation characteristics recorded on the spectral emulsions. In this research the local electron temperature was the parameter of interest, since this species dominated the energy distribution processes within the plasma. For our preliminary study of the reproducibility of the system, the plasma was assumed to be in local thermodynamic equilibrium (LTE), an assumption which permits important simplification in the complex theory involved. The physical basis for this assumption is that energy transfer within plasma occurs primarily by collisional processes. According to Griem [3]

> Whenever LTE prevails, the densities in specific quantum states are those pertaining to a system in complete thermodynamic equilibrium which has the same total mass density, temperature, and chemical composition as the actual system. The relevant temperature is that which describes the distribution function of the species dominating the reaction rates, i.e., in dense laboratory plasmas that of the electrons.

Two regions of interest are defined in the plasma produced by the exploding wire. The initial radiation ($< 2\ \mu$sec) from the plasma is predominantly continuum radiation, and

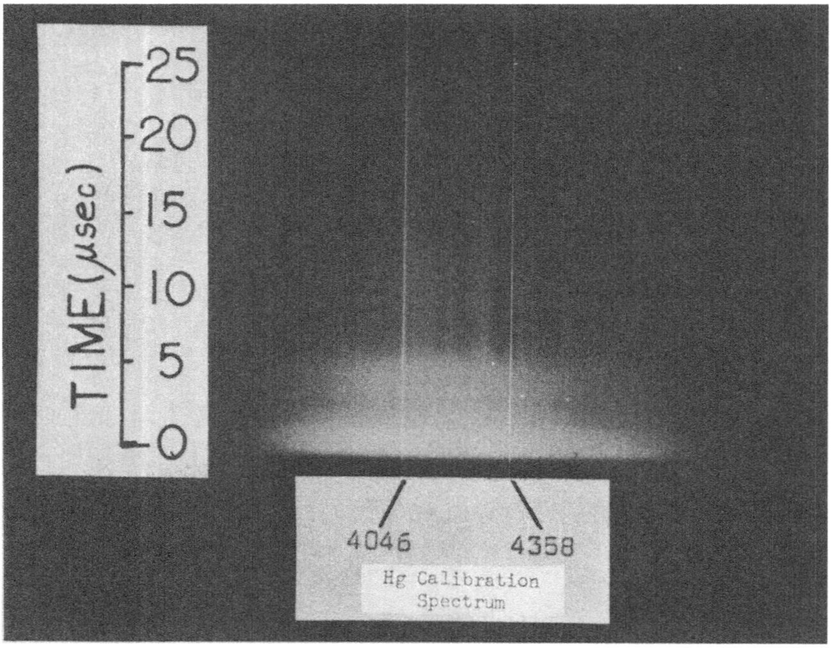

Fig. 7. Spectrum of copper plasma from wire exploded in vacuum.

the plasma at this state is assumed to be optically thick (black body radiator) [2]. As the plasma expands, there is a transition region, and at later times line radiation dominates the spectrum of the plasma. In this region, the plasma was treated as being optically thin. Two formulas derived from quantum considerations were applied [3, 4], each valid only for certain ranges of plasma opacity.

It should be emphasized here that the plasma opacity was not experimentally measured, but that values were assumed. These were compatible with results obtained by other investigators [9].

Using Wien's approximation to Planck's radiation law, the temperature of an optically thick ($\tau_\nu \sim 1$) exploding wire plasma where $\tau_\nu = \kappa \rho dl$ ($\kappa \equiv$ linear absorption coefficient, $\rho \equiv$ density, $dl \equiv$ elemental linear length) can be calculated from the measured relative intensity of the continuum radiation plasma produced by an exploding wire. The temperature is given by

$$T = \frac{hc(\lambda^{-1} - \lambda'^{-1})}{K \ln (I'\lambda'^5 / I\lambda^5)}$$

where the wavelengths referred to are the wavelengths of interest and I and I' refer to the relative spectral intensities determined at these wavelengths. The other symbols have their usual meaning. Plasmas must be optically thick (strongly absorbing) and in LTE for this formula to be applicable. As noted above, continuum radiation is predominant only at early times (< 2 μsec) in the explosion, and because of shock reheating this formula could be applied only to intensity measurements from wires exploded in a vacuum environment.

Line radiation dominates the spectrum of the plasma at times late (> 2 μsec) in the explosion. For a plasma which is in LTE and optically thin ($\tau_\nu < 1$), a second temperature-sensitive formula is applicable. Considering two spectral lines which arise from the same ground state within a given multiplet in a given stage of ionization, an applicable temperature relationship is as follows:

$$T = \frac{E' - E}{k \ln (I\lambda^3 g'f' / I'\lambda'^3 gf)}$$

where E and E' are the excitation energies of the two selected lines, I and I' the relative intensities of the lines, and the (gf) factor the product of the statistical weight and the oscillator strength of the line. One should carefully consider the choice of available lines because values of the (gf) must be available, and for nonhydrogenic atoms it is not a simple matter either to obtain or calculate reliable (gf) values. An inappropriate choice of lines may easily increase the error estimate in a calculated temperature to an unacceptable value and thus limit the results one obtains.

The measured quantities in these formulas in this work are the intensities of the continuum and line radiation. Their measurement allows calculation of plasma temperatures under the assumptions stated above.

Lines arising from the same ground state were available for temperature determinations at late times in the plasma's lifetime, but not more than one set of two (from the same multiplet) were available on the same spectral plate. This situation limited the accuracy of any temperature determination, as no cross check was available on the calculated temperature.

Figures 7 and 8 contain photographs of typical time-resolved copper spectra obtained with the system. Wavelengths present in typical gold spectra are shown in Tables II and III. Temperatures were calculated for both early and late times using the appropriate expression for temperature for the gold spectra. Using the continuum portion of the spectrum which lasted for 1 to 3 μsec, temperatures were determined using Wien's approximation. The mean temperature measured for this time was 0.456 eV. Temperature decreased with time in this region, as expected.

In attempting to apply the appropriate formula for an optically thin plasma to the spectrum at late times, "temperatures" were obtained which increased with time. This effect occurred for all pairs of lines used and was not confined to the spectrum of any one element. Previous investigations [2] showed a continual decrease from early peak tem-

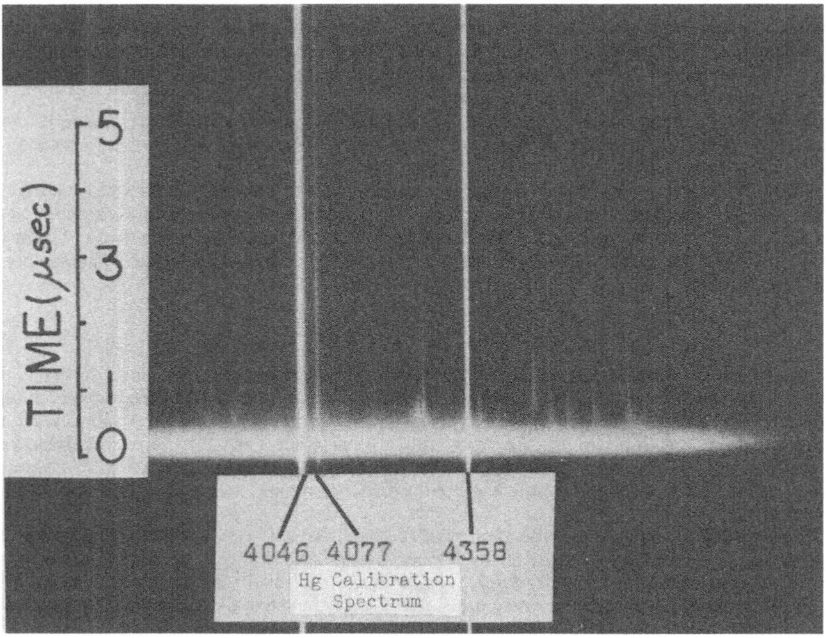

Fig. 8. Spectrum of copper plasma from wire exploded in atmosphere.

Table II. Measured Relative Line Intensities for an Au Plasma*

Time sec	I_1 4315 Å	I_2 3897 Å	I_1/I_2
1.7	0.147	0.124	1.19
3.1	0.086	0.073	1.18
5.3	0.065	00.067	0.97
9.0	0.061	0.067	0.91
12.0	0.046	0.057	0.81

*Data in Table II are for a gold plasma produced by an exploding 0.003-in. gold wire. Firing voltage: 30 kV, in a vacuum environment. Camera writing speed 1.33 mm/sec on Royal Pan emulsion. Line level values are

4315 Å—5704.9 to 32246.5 cm^{-1}
3897 Å—6598.5 to 32246.5 cm^{-1}.

Table III. Measured Relative Continuum Intensities for an Au Plasma*

Time, sec	λ_1	λ_2	I_1/I_2	T, eV
0.3	3950	4050	1.22	0.493
0.3	4050	4150	0.889	0.473
0.3	3950	4250	0.820	0.403
0.7	3950	4250	1.42	0.307
0.3			average	0.456

*Data in Table III are for a gold plasma formed by an exploding 0.003-in. gold wire. Writing speed was 3.35 mm/sec on Royal Pan emulsion in a vacuum environment.

peratures to ambient in several microseconds. Because of this anomaly, it appears quite certain that these plasmas created by explosions of wires in vacuums cannot be analyzed using the formula applicable to optically thin plasmas. The plasma opacity lies between that where black body and optically thin approximations apply, or LTE is not valid. This work was terminated before necessary time-resolved opacity measurements could be made.

In conclusion, calculated temperatures for the exploding wire plasma for early time were in the expected range of 0.1 to 1.0 eV, and lend credence to the assumptions of LTE and that the plasma was optically thick. Late time behavior throws suspicion on the assumption of an optically thin plasma and indicates that further work is needed in this area.

ACKNOWLEDGMENTS

Air Force Weapons Laboratory personnel who provided important assistance in the effort are F.O. Westfall, L.A. Johns, W.E. Kunzler, and P.W. Moore.

REFERENCES

1. W.R. Faust et al., Exploding Conductors, Final Report, June 30, 1960, to June 30, 1963, NRL Report 6009, Washington, D.C.: Naval Research Laboratory, October 31, 1963.
2. A.H. Guenther et al., "A Versatile Ultrafast Pulsed Power System," in Exploding Wires, Vol. 3, W. G. Chace and H. K. Moore, eds., Plenum Press, New York (1954), pp. 361–391.
3. H.R. Griem, Plasma Spectroscopy, McGraw-Hill Book Co., New York (1964).
4. W.G. Chace and H.K. Moore, eds., Exploding Wires, Vol. 1, Plenum Press, New York (1962).
5. W.G. Chace and H.K. Moore, eds., Exploding Wires, Vol. 2, Plenum Press, New York (1962).
6. R.S. Case, Jr., "Time Resolved Spectroscopy of Exploding Wires," Thesis, A.F.I.T., Wright-Patterson AFB, Dayton, Ohio, March 1966.
7. C.E.K. Mees, "The Reciprocity, Intermittency, and Clayden Effects," in The Theory of the Photographic Process, Macmillian Company, New York (1964), pp. 236–260.

8. F.O. Westfall, New York, private communication.
9. W.R. Faust et al., Exploding Conductors, Final Report, June 30, 1960, to June 30, 1963, NRL Report 6009, Washington, D.C.: Naval Research Laboratory, October 31, 1963.

DISCUSSIONS

Question by Jäger, University of Kiel, West Germany: Do you measure the optical thickness? Otherwise, you can't make any temperature measurements. Moreover, it must be considered the absorption of the surrounding cloud.

Answer by Author: No, the optical thickness of the exploding wire plasma was not experimentally measured, though a preliminary attempt was made at employing a re-imaging mirror behind the exploding wire in line with the optical path. Results were inconclusive. The emphasis of this paper was on developing survey instrumentation and calibration procedures rather than actual temperature calculations; the temperature calculations were an additional check on system reproducibility, and insufficient time remained to measure such parameters as the plasma opacity experimentally.

The effects of the surrounding air media were eliminated by exploding the wire in a vacuum environment, producing spectral emulsions without the undesirable air continua. Under these conditions, the lines attributable to atmospheric constituents were not recorded in strength.

Question by W. Lochte-Holtgreven, University of Kiel, West Germany: If you wait long enough, won't the plasma become (optically) thin?

Answer by Author: Yes, this was the basic assumption made in the portion of the work concerned with temperature determination of late periods in the plasma's lifetime. It appears, however, that either we did not wait long enough, as you did, or that other factors which were not considered entered the picture and lead to the unrealistic situation of an apparent increase toward infinity of the temperature with increasing time in the plasma's life history.

Question by C. C. Gallagher, Air Force Cambridge Research Laboratories, Bedford, Mass: What type of instrumental corrections did you make on your spectrograph measurement?

Answer by Author: The corrections referred to here are those associated with the calibration procedure. We had to take into account the varying f/number of the camera and the nonlinear wavelength response of the total optical system. We assumed that reciprocity law failure would be small, since we dealt with the ratios of intensities taken at similar times (exposure). We also attempted to work in the most linear portion of the calibration curves relating the spectral intensity to film density. In addition, corrections for the nonlinear wavelength and intensity response of the photographic emulsion were included.

Question by J. L. Wiza, Ameriçan Science and Engineering: Could your technique be used to establish a temperature base from which one could measure unknown oscillator strengths? What magnitudes of temperatures did you measure?

Answer by Author: Yes, this technique can be used to obtain unknown oscillator strengths, providing one knows in detail the composition of the radiating plasma and parameters such as 1) the plasma opacity as a function of time; 2) the plasma radiation characteristics have been accurately spatially resolved so that one knows the portion of the cylindrical column which is being viewed; 3) with the formula employed in this work, one of the two oscillator strengths must be known, otherwise only a ratio may be obtained. In the more general case of the comparison of the relative intensities of two spectral lines, other more complex formulas are applicable, and these in general require more information than is available using just the techniques described and used in this work.

The temperatures which were measured were in the range 0.4 to 0.75 eV at times less than 2 μsec after the plasma radiation is first recorded on the photographic emulsion.

Exploding Wire Detonators: Threshold Burst Current Dependence Upon Detonator and Environmental Parameters*

T.J. Tucker

Sandia Laboratory
Albuquerque, New Mexico

Results obtained in an experimental study of the dependence of threshold burst current I_{bth} of exploding wire (EBW) detonators upon major detonator parameters are presented. It was found that, over the range of parameter values of practical interest, I_{bth} of detonators constructed with gold or gold alloy bridgewires and loaded with granular PETN explosive, could be related to bridgewire length l and diameter d, explosive specific surface S_o^p, and density δ, and firing temperature T, by the empirical relationship,

$$I_{bth} = \frac{d}{\sqrt{l}}\left[850 + 35.5\,\frac{(ldS_o^p \times 10^{-3} - 120)^2}{(ldS_o^p \times 10^{-3})^{3/2}}\right]\left[\frac{1}{(1.88-\delta)^3}\right] \times \left[1 - \frac{2 \times 10^{-2}\,(T\text{-}24)}{\sqrt{l}}\right]$$

where I_{bth} is in amperes, d and l are in mils, S_o^p is in cm²/g, δ is in g/cm³, and T is in °C.

It is shown that within the limits of the parameter range thus far tested, the equation in general predicts I_{bth} values to within 10% of those experimentally observed. Utilization of the equation in such applied problems as EBW detonator and detonator-firing set system design is discussed, as well as the relationship of the equation to more fundamental questions concerning the mechanisms of PETN initiation by exploding wires.

INTRODUCTION

In the design and analysis of an exploding bridgewire (EBW) detonator-firing set system it is highly desirable from the stand-

*This work was supported by the U.S. Atomic Energy Commission.

point of system optimization to be able to separate the system into its major components and to specify the performance of each component in terms which are not strongly dependent on the details of a particular system. Although this is not an obvious task because of the highly nonlinear character of EBW detonator systems, the subject has been dealt with in two earlier papers—one by Tucker [1] and the other by Cnare [2].

In the paper by Tucker [1], it was demonstrated that the burst current I_b defined as the current flowing in the bridgewire at the time of peak voltage across the wire, could be used to define the initiation sensitivity of an EBW detonator. In an EBW detonator system, a continuum of burst currents can, in principle, be obtained by a suitable choice of system parameters. This continuum will, however, contain one burst current of particular importance, the threshold burst current I_{bth}, which is defined as that value of I_b which produces 50% firing of a given sample of EBW detonators. The importance of this threshold burst current is that within practical limits it is system-independent for a given EBW detonator configuration in the following sense. Independent of firing set properties and the detailed profile of the current waveform, 50% firing of a lot of detonators will result if they are subjected to a burst current equal to the threshold burst current. The paper by Cnare [2] then utilized this burst current concept in conjunction with the action criterion [3] to relate detonator initiation to firing set performance.

Although this earlier work established a technique for analyzing EBW detonator systems, it was incomplete in that the functional dependence of the threshold burst current I_{bth} had been determined for only a few of the major detonator parameters. In principle, I_{bth} values for a number of EBW detonator configurations could be measured experimentally, tabulated, and utilized as a guide for detonator selection. Considering, however, the relatively large number of parameters involved, a reasonably complete tabulation of I_{bth} values would be extremely laborious to obtain. More importantly, however, such a simple cataloging scheme does not offer an efficient method for EBW detonator optimization. A much more useful alternative would be an empirical relationship which would permit I_{bth} to be predicted from given detonator parameters.

It is the purpose of this paper to present results obtained in an experimental study of the dependence of I_{bth} on detonator parameters and to develop an empirical equation expressing the functional dependence of I_{bth} on these parameters. The particular parameters investigated are bridgewire length l and diameter d, explosive specific surface S_o^p and density δ (of PETN explosive), and finally one environmental variable of interest, the firing temperature T. These parameters and the ranges of their investigation are not exhaustive; they are, however, with the exception of the explosive type and wire materials, those most frequently encountered. With regard to these last parameters, only granular PETN explosive and bridgewires of gold, or gold alloys similar in exploding wire behavior to gold, are considered.

The paper is divided into three major sections. In the first section, the dependence of I_{bth} on bridgewire diameter and length and on the specific surface of the explosive is established. Next, the effect of explosive density is investigated, and, thirdly, the influence of firing temperature is determined. Finally, the empirical equation describing the functional dependence of I_{bth} on these parameters is presented and discussed in terms of possible models for initiation of granular PETN explosive.

THE DEPENDENCE OF I_{bth} ON d, l, AND S_o^p

Background

For detonators with gold bridgewires and PETN explosives having a density of 0.88 g/cm^3, it has been shown [1] for room temperature firing that the threshold burst current I_{bth} is dependent upon the wire length l and diameter d. Briefly, increasing the wire diameter or decreasing the wire length generally resulted in an increase in I_{bth}, but subsequent studies indicated that the magnitude of this increase was dependent upon the specific surface area S_o^p of the explosive. Further, it was observed that small-diameter bridgewires usually required explosive of high specific surface to minimize I_{bth}, while for large-diameter wires the opposite was true. These observations led to the concept of properly matching the explosive to the bridgewire; i.e., for given wire dimensions, an explosive-specific surface exists for which I_{bth} is minimized.

In this section, a function relating I_{bth} to d, l, and S_o^p is deduced from experimental results. This function not only specifies the proper specific surface explosive for matching to a given bridgewire, but, in addition, predicts effects of explosive mismatchings, which heretofore have been a source of confusion in the evaluation of experimental data.

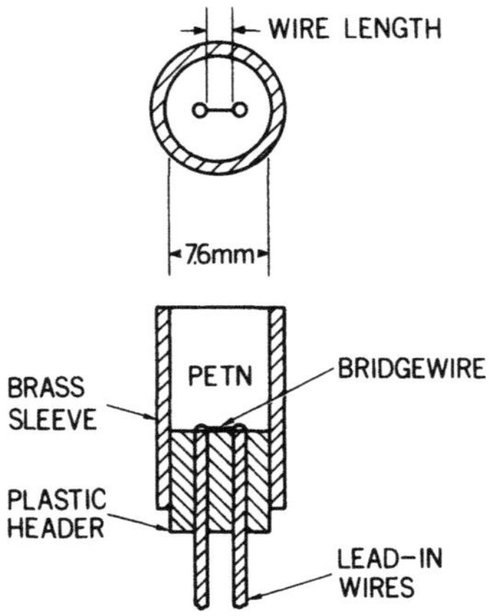

Fig. 1. Typical test detonator construction.

Experimental Procedure

All tests reported in this and subsequent sections were performed with a step-function current generator serving as a firing set [4]. It was noted in the Introduction that I_{bth} for practical firing systems is primarily dependent on the properties of the detonator and can be considered separately from the firing set. Consequently, within reasonable limits, determinations of I_{bth} as obtained with the step current generator are also applicable to arbitrary practical firing sets.

All detonators were supplied by the Mound Laboratory of Monsanto Chemical Corporation.

The physical construction of the detonators used in these studies is shown in Fig. 1. All bridgewire dimensions, explosive specific surface values, and densities are as quoted by the supplier and are assumed to be correct to within a tolerance of $\pm$ 10%. The explosive surface was determined using a permeameter, a measurement technique which will be discussed critically later.

Fifty-percent threshold firing currents (I_{bth}) were determined, using the standard Bruceton firing sequence with 10-A increments. Since, for the detonators tested, the burst-current increment producing no detonator firing to 100% detonator firing is small ($\sim$ 40 A), a sample size of 15 units was usually sufficient for each threshold determination.

Experimental Results

The results of the room-temperature detonator threshold tests with 0.88 g/cm^3 density explosive are presented in Table I. By inspection, the data of Table I suggest that I_{bth} is functionally related to d, l, and S_o^p by an equation of the form

$$I_{bth} = K_1 d/\sqrt{l} \tag{1}$$

where K_1 may be a function of S_o^p. To examine the validity of such a relationship, I_{bth} versus $d/\sqrt{l}$ is plotted in Fig. 2 with S_o^p as a parameter. For any given value of $d/\sqrt{l}$, there is a range of I_{bth} values corresponding to different S_o^p values. It is important to note, however, that a definite region exists in which no detonator thresholds were observed and that this region can be separated from that where firing occurs by a straight line of slope K_1 = 850 A · $mil^{-1/2}$ as shown in Fig. 2. It has already been noted that K_1 may be a function of S_o^p; thus the value of K_1 = 850 A · $mil^{-1/2}$ is interpreted as a minimum value of this quantity which is obtained when the specific surface of the explosive is properly matched to the bridgewire.

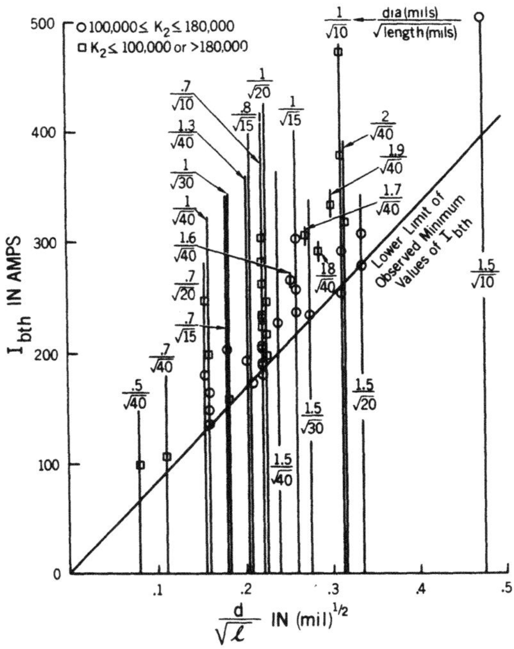

Fig. 2. Threshold burst current versus $d/\sqrt{l}$.

Table I. Survey of I_{bth} versus d, l, and S_o^p *

Item No.	Wire material	l mils	d mils	S_o^p cm^2/g	I_{bth} A	$d/\sqrt{l}$ $mil^{1/2}$	K_1 $I_{bth}\sqrt{l}/d$	K_2 ldS_o^p
1	Au	10	0.7	38,000	281	0.221	1270	266,000
2	Au	10	0.7	12,000	223	0.221	1008	84,000
3	Au	10	1.5	11,650	506	0.474	1067	174,750
4	Au	10	0.7	16,640	193	0.221	872	116,500
5	Au	10	0.7	11,870	231	0.221	1044	83,100
6	Au	10	0.7	13,390	230	0.221	1040	93,750
7	Au	10	0.7	15,550	201	0.221	908	108,850
8	Au	10	0.7	12,000	232	0.221	1049	84,000
9	Au	10	0.7	6,330	304	0.221	1374	44,300
10	Au	10	0.7	18,000	190	0.221	859	126,000
11	Au	10	0.7	10,930	259	0.221	1171	76,510
12	Au	10	0.7	15,000	189	0.221	854	105,000
13	Au	10	0.7	16,500	202	0.221	908	108,850
14	Au	10	0.7	15,000	181	0.221	818	105,000
15	Au Alloy	10	1.0	~3,800	378	0.316	1196	38,000
16	Au Alloy	10	1.0	12,000	249	0.316	788	120,000
17	Au Alloy	10	1.0	12,000	292	0.316	924	120,000
18	Au Alloy	10	1.0	~3,800	473	0.316	1496	38,000
19	Au	15	0.7	10,930	203	0.181	1123	114,750
20	Au	15	0.8	12,000	169	0.207	818	144,000
21	Au Alloy	15	1.0	13,430	236	0.258	914	201,450
22	Au Alloy	15	1.0	12,000	255	0.258	988	180,000
23	Au Alloy	15	1.0	~3,800	303	0.258	1174	57,000
24	Au	20	0.7	10,930	180	0.156	1150	153,000
25	Au	20	1.5	~3,800	307	0.335	915	114,000

26	Au	20	0.7	~3,800	247	0.156	1578	53,200
27	Au	20	1.5	4,000	279	0.335	832	120,000
28	Au Alloy	20	1.0	~3,800	218	0.224	975	76,000
29	Au Alloy	20	1.0	12,000	198	0.224	885	240,000
30	Au Alloy	20	1.0	13,430	246	0.244	1100	268,000
31	Au	30	1.5	~3,800	234	0.274	855	171,000
32	Au Alloy	30	1.0	~3,800	157	0.182	860	194,000
33	Au	40	3.0	~3,800	560	0.475	1180	456,000
34	Au	40	1.5	~3,800	227	0.237	957	228,000
35	Au	40	2.0	~3,800	318	0.316	1005	304,000
36	Au	40	0.5	~3,800	99	0.079	1252	76,000
37	Au	40	0.7	~3,800	106	0.111	957	106,000
38	Au	40	1.6	~3,800	265	0.253	1047	243,000
39	Au	40	1.7	~3,800	306	0.269	1138	258,000
40	Au	40	1.8	~3,800	292	0.285	1026	274,000
41	Au	40	1.9	~3,800	334	0.301	1112	289,000
42	Au	40	1.0	~3,800	136	0.158	860	152,000
43	Au Alloy	40	1.0	~3,800	165	0.158	1043	152,000
44	Au Alloy	40	1.0	12,000	198	0.158	1252	480,000
45	Au Alloy	40	1.3	~3,800	193	0.206	938	198,000
46	Au Alloy	40	1.0	~3,800	149	0.158	942	152,000

*All tests performed at room temperature.
Explosive density = 0.88 g/cm^3.

The general observation from Table I that for smaller values of l or d a larger value of S_o^p was required to minimize I_{bth} and, thus, K_1, suggests that a function specifying this optimum value of S_o^p might have the form,

$$ldS_o^p = K_2 \tag{2}$$

where K_2 is a constant. To test this hypothesis, values of K_1 and K_2 were computed for each test from Eqs. (1) and (2), respectively, and are tabulated in Table I. In Fig. 2, points for which $100{,}000 \le K_2 \le 180{,}000$ are designated by the symbol O, while points designated by the symbol □ were tests for which K_2 was either greater or less than the above values. Clustering of the O points at values of K_1 near that for minimum I_{bth} is apparent. To further examine this point, the values of K_2 corresponding to the lowest observed values of K_1 are listed in Table II. The data were selected from values of K_1 in Table I that were less than 900. The range of values of K_1 in Table I is from about 800 to 1600. With only two exceptions, the tabulation in Table II indicates that the lowest values of K_1 were obtained when K_2 was in the range of 105,000 to 140,000 cm^2 mil^2/g.

In Fig. 3, K_1 is plotted versus K_2 for the experimental data given in Table I. Although considerable scatter is present in the plot, an average lower limit of K_1 as a function of K_2 may be determined. The curve establishing this lower limit, in effect, establishes the minimum expected value of I_{bth} as a function of both the wire geom-

Table II. Values of K_2 Resulting in Lowest Values of K_1

Item No.	$d/\sqrt{l}$	K_1	K_2
4	$0.7/\sqrt{10}$	872	116,500
10	$0.7/\sqrt{10}$	859	126,000
12	$0.7/\sqrt{10}$	854	105,000
14	$1/\sqrt{10}$	818	105,000
16	$1/\sqrt{10}$	788	120,000
20	$0.8/\sqrt{15}$	818	144,000
27	$1.5/\sqrt{20}$	832	120,000
29	$1/\sqrt{20}$	885	240,000
31	$1.5/\sqrt{30}$	855	171,000
32	$1/\sqrt{30}$	860	194,000
42	$1/\sqrt{40}$	860	152,000

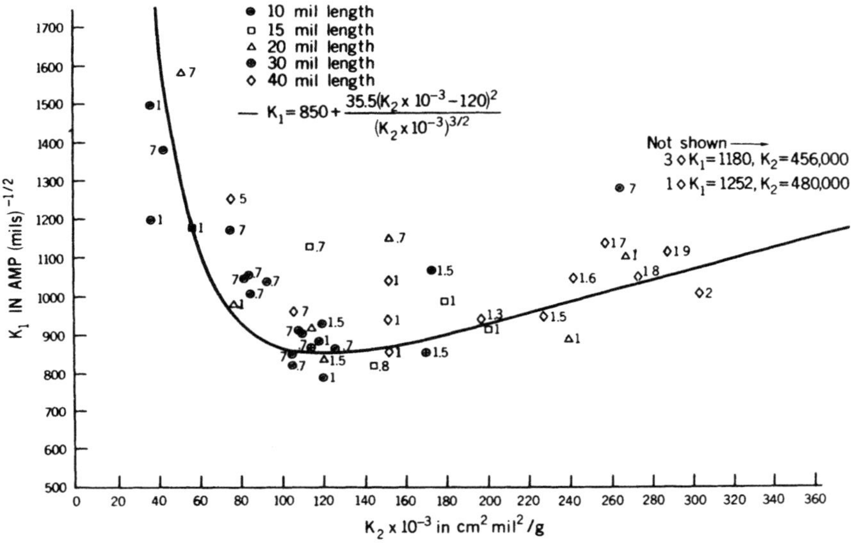

Fig. 3. Plot of $K_1 = (\sqrt{l}/d) I_{bth}$ versus $K_2 = ldS_o^p \times 10^{-3}$.

etry and the explosive specific surface. This curve can be approximately defined by

$$K_1 = C_1 + C_2 \frac{(K_2 \times 10^{-3} - C_3)^2}{(K_2 \times 10^{-3})^{3/2}} \tag{3}$$

where $C_1 = 850\ \text{A} \cdot \text{mil}^{-1/2}$ $C_2 = 35.5\ \text{A} \cdot \text{g}^{1/2}/\text{cm} \cdot \text{mil}^{3/2}$, and $C_3 = 120\ \text{cm}^2\ \text{mil}^2/\text{g}$; K_1 is in $\text{A} \cdot \text{mil}^{-1/2}$ and K_2 is in $\text{cm}^2 \cdot \text{mil}^2/\text{g}$. Combining Eqs. (1), (2), and (3), it follows that the threshold burst current I_{bth} is described as a function of wire length l, wire diameter d, and explosive specific surface S_o^p by

$$I_{bth} = \left[C_1 + C_2 \frac{(ldS_o^p \times 10^{-3} - C_3)^2}{(ldS_o^p \times 10^{-3})^{3/2}} \right] d/\sqrt{l} \tag{4}$$

Introducing numerical values of the constants into Eq. (4) gives

$$I_{bth} = \left[850 + 35.5 \frac{(ldS_o^p \times 10^{-3} - 120)^2}{(ldS_o^p \times 10^{-3})^{3/2}} \right] d/\sqrt{l} \tag{5}$$

DISCUSSION OF $I_{bth} = I_{bth}(d, l, S_o^p)$

For a given combination of d, l, and S_o^p, Eq. (5) provides, then, both an approximation of the lowest I_{bth} which can be expected as well as a method by which the selection can be made to approach this limit. From the curve of Fig. 3, however, it can be seen that some spread in K_1 values (and thus of I_{bth}) does exist for any particular value of K_2. This spread could be the result of many factors, including detonator tolerances, sampling errors, and the effects of additional uncontrolled variables (the explosive particle size distribution, for example). Clustering of points in Fig. 3 near the minimum value, however, suggests that for the studies reported here, effects which might be attributed to these uncontrolled variables are relatively small in comparison to those produced by l, d, and S_o^p, and, in general, for a given selection of the latter variables, do not exceed approximately $\pm 10\%$. Ignoring for the moment, then, these minor effects, it can be seen that the minimum value of I_{bth} is predicted to occur when the product $ldS_o^p = K_2$ is equal to 120,000 $cm^2 \cdot mil^2/g$; this value then corresponds to optimum matching between the wire dimensions and the specific surface of the PETN explosive.

The behavior of Eq. (5) at its various minimum and limiting points is of interest. Values of I_{bth} for limiting values of d, l, and S_o^p can easily be shown to be

$$\lim_{d \to 0} I_{bth} = \lim_{d \to \infty} I_{bth} = \lim_{S_o^p \to 0} I_{bth} = \lim_{S_o^p \to \infty} I_{bth} = \lim_{l \to 0} I_{bth} = \infty \tag{6}$$

and

$$\lim_{l \to \infty} I_{bth} = d^{3/2} S_o^{p\,1/2} \tag{7}$$

With somewhat greater difficulty, it can be shown that I_{bth} does not exhibit a minimum with respect to l. In contrast, I_{bth} does exhibit a minimum with respect to d given by

$$d_{\min} = \frac{53}{S_o^p\, l} \quad \text{for} \quad \frac{\partial I_{bth}}{\partial d} = 0 \tag{8}$$

where $d_{\min}$ and l are in mils, and S_o^p is in cm^2/g. Thus, for fixed values of l and S_o^p, Eq. (8) gives the wire diameter which minimizes the threshold burst current. As expected, if l and d are held fixed, then the value of S_o^p for minimum threshold burst current is given by

$$(S_o^p)_{min} = \frac{120{,}000}{ld} \quad \text{for} \quad \frac{\partial I_{bth}}{\partial S_o^p} = 0 \tag{9}$$

It is informative to consider the behavior of Eq. (5) for several different combinations of detonator parameters. For a 20-mil-long wire and various values of S_o^p, Fig. 4 illustrates the dependence of I_{bth} upon bridgewire diameter. The minimums of the curves are obvious; furthermore, these minimum values shift to lower values of both I_{bth} and d for increasing values of S_o^p. Figure 5 illustrates the effect of changes in wire length for a constant wire diameter and various S_o^p values. It can be seen that for long bridgewire lengths, e.g., l = 60 mils, I_{bth} increases with S_o^p. At short lengths, e.g., l = 5 mils, a crossover occurs and I_{bth} is predicted to decrease with increasing S_o^p.

The meaning of the constants appearing in Eq. (4) should be noted. As can be seen from Eqs. (3) and (4), the constant C_1 is the value of K_1 for the case of proper matching of S_o^p to a given d and l, i.e., the value of K_1 when the second term in the right-hand side of Eq. (3) is zero. The value of C_1 = 850, as previously indicated, represents the average minimum value of this quantity which was observed. As previously indicated, however, some spread in values of K_1, and thus of C_1, was observed ($\sim \pm 10\%$), a result which might be expected if C_1 is weakly dependent upon additional explosive

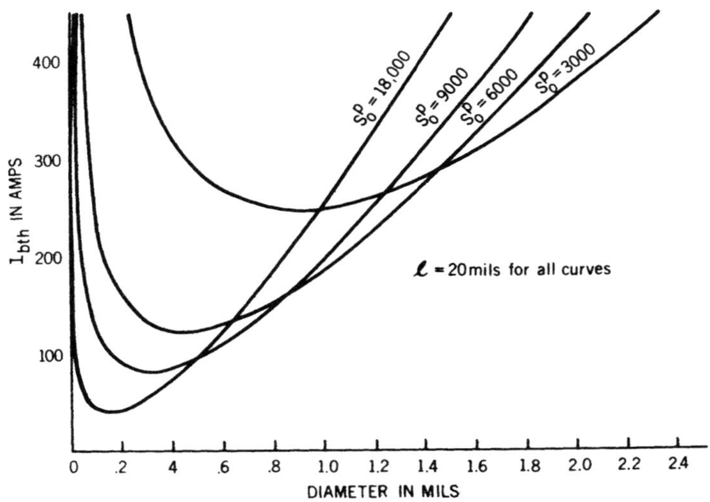

Fig. 4. Theoretical I_{bth} versus wire diameter for values of S_o^p.

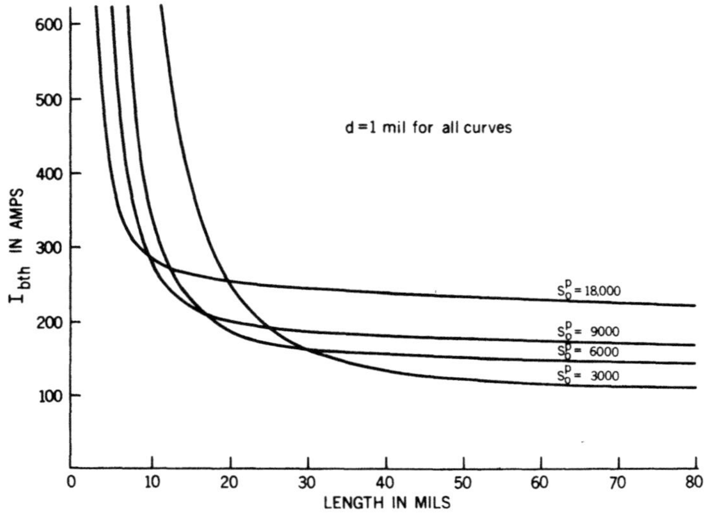

Fig. 5. Theoretical I_{bth} versus length for various values of S_o^p.

characteristics, such as purity, particle size distribution, crystal shape, and surface properties, i.e., quantities which might be expected to vary between differing explosive lots.

The constants C_2 and C_3 representing, respectively, a linear scaling factor for the explosive-to-wire matching and the optimum ldS_o^p for this matching, might also be expected to vary somewhat between explosive lots. The results reported here, however, suggest that with the PETN samples used, variations of the values of C_1, C_2, and C_3 between different explosive lots were, in general, no greater than about $\pm 10\%$.

The definition of explosive specific surface S_o^p as used in Eq. (5) and tabulated in Table I requires explanation. The values given in Table I were determined by Mound Laboratory using a permeameter. This measurement assumes that a correlation exists between the specific surface area and the permeability of a sample of particles. Such a correlation is reasonable between samples of similar particles; however, when applied to PETN crystals, several questions arise, since permeability may also be affected by crystal shape, particle size distribution, and surface and inclusion characteristics, all of which may vary slightly between differing lots of PETN. For purposes of this paper, it was assumed that, since the majority of PETN powders utilized were produced in similar fashion, they would be similar with regard to the properties indicated above and, thus, the permeameter values would yield values of S_o^p which could be correlated with detonator firing results. The existence of Eq. (5) appears to affirm this assumption for the PETN lots thus far tested. If, however, production techniques were radically changed, a direct comparison between the permeameter measurement of S_o^p of the resulting explosive to that for the PETN used in the tests reported here might be questionable.

EFFECT OF EXPLOSIVE DENSITY ON I_{bth}

Experimental Results

The effect of the PETN explosive density δ upon the threshold burst current I_{bth} has been extensively studied for only two l, d, and S_o^p combinations: a 0.7-mil-diameter by 10-mil-long gold (Au) wire embedded in PETN explosives with specific surface areas of either 16,500 or 13,400 cm^2/g. In addition, some data have been obtained using other parameter combinations, but as yet testing has covered only a small density range. The results pertaining to

Table III. Survey of I_{bth} versus Explosive Density Data

Wire length, mils	Wire diameter, mils	Wire material	S_o^p cm^2/g	PETN density, g/cm^3	I_{bth}, A
10	0.7	Au	16,500	0.7	151
10	0.7	Au	16,500	0.88	191
10	0.7	Au	16,500	0.98	268
10	0.7	Au	16,500	1.1	402
10	0.7	Au	16,500	1.2	654
10	0.7	Au	13,400	0.8	178
10	0.7	Au	13,400	0.88	230
10	0.7	Au	13,400	0.98	314
10	0.7	Au	13,400	0.98	325
10	0.7	Au	13,400	1.03	340
30	1.5	Au	~3,800	0.98	282
30	1.5	Au	~3,800	0.88	234
30	1.5	Au	~3,800	0.98	270
30	1.5	Au	4,000	1.03	272
30	1.5	Au	~3,800	0.98	281
40	1.5	Au	~3,800	0.88	221
40	1.5	Au	~3,800	0.98	331

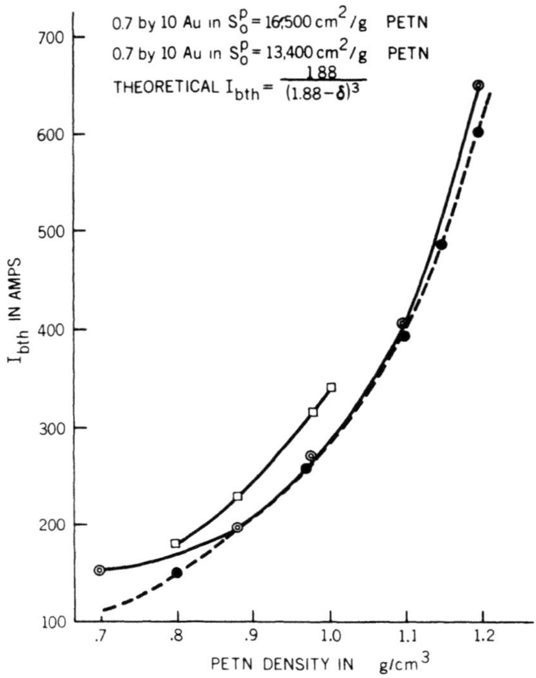

Fig. 6. I_{bth} versus explosive density.

density effects for all l, d, S_o^p, combinations thus far tested are tabulated in Table III. General characteristics of the effect of explosive density upon I_{bth} is shown graphically in Fig. 6, a plot of I_{bth} versus δ for the two 0.7-mil-diameter bridgewire studies. A relatively good approximation to the density dependence of I_{bth} for detonators loaded with 16,500 cm²/g explosive I_{bth} is given by

$$I_{bth} = \frac{C_5}{(C_4 - \delta)^3} \tag{11}$$

where $C_4 = 1.88$ and $C_5 = 188$; δ is in g/cm³ and I_{bth} is in amperes. Equation (11) is shown as a dashed line in Fig. 6 and can be seen to be in good agreement with the experimental results for densities between 0.88 and 1.1 g/cm³.

Discussion of Results

It would be desirable to incorporate the density effect expressed by Eq. (11) into Eq. (4), which characterizes the effects of d, l, and

S_o^p. Present data are insufficiently complete for the development of such a general relationship; however, all work to date suggests that, in general, all exploding wire detonator configurations exhibit decreased sensitivity with increasing explosive density. As a first approximation, then, it is reasonable to consider that the density dependence given by Eq. (11) might be introduced as a simple multiplicative term in Eq. (4). If this were true, c_5 would be equal or proportional to the right-hand side of Eq. (4). It would then follow for two detonators having the same values of l, d, and S_o^p, but loaded with PETN of 0.88 and 0.98 g/cm³ density, respectively, that

$$\frac{I_{bth}(\delta=0.98)}{I_{bth}(\delta=0.88)}=\frac{(1.88-0.88)^3}{(1.88-0.98)^3}=1.37 \qquad (12)$$

The following ratios were experimentally observed.

Detonator parameters	$\frac{I_{bth}(\delta=0.98)}{I_{bth}(\delta=0.88)}$
0.7-by-10-mil Au wire $S_o^p = 16{,}500\ cm^2/g$	1.40
0.7-by-10-mil Au wire $S_o^p = 13{,}400\ cm^2/g$	1.37
1.5-by-30-mil Au wire $S_o^p \cong 3800\ cm^2/g$	1.15
1.5-by-40-mil Au wire $S_o^p \cong 3800\ cm^2/g$	1.5

In all tests utilizing the 1.5-by-30-mil Au bridgewire, the actual explosive loading density in the bridgewire region was questionable. For this reason, the low value of $I_{bth}(\delta = 0.98)/I_{bth}(\delta = 0.88)$ for these tests is questionable and will be ignored. It can be seen that the assumption that the density function given by Eq. (11) enters as a multiplicative factor in Eq. (5) yields reasonable agreement between observed and predicted I_{bth} behavior. Thus, inclusion of the density factor in Eq. (5) yields

$$I_{bth}= d/\sqrt{T}\left[850+35.5\,\frac{(ldS_o^p\times10^{-3}-120)^2}{(ldS_o^p\times10^{-3})^{3/2}}\right]\left[\frac{1}{(1.88-\delta)^3}\right] \qquad (13)$$

EFFECT OF FIRING TEMPERATURE ON I_{bth}

Experimental Technique

Temperature control between -74°C and +84°C was obtained with a Bowser environmental test chamber in which a special detonator firing box was installed. In addition to chamber temperatures as indicated by the Bowser recorder, temperatures were determined at the firing box by a Rubicon thermocouple bridge to an accuracy of ±1°C. Before testing, all detonators were preconditioned by storage in the Bowser chamber at the desired firing temperature for a period of at least two hours.

Temperature Test Results

The results of the firing temperature tests on detonator threshold burst currents are tabulated in Table IV for detonators having 0.7-by-10-mil Au wire loaded with PETN of 16,500 cm^2/g specific surface area and 0.88 g/cm^3 density, for detonators having 1.5-by-40-mil Au wire and PETN of about 3800 cm^2/g specific surface area and 0.88 g/cm^3 density, and for detonators having 1.5-by-20-mil Au wire and PETN of 4000 cm^2/g specific surface area and 0.88 g/cm^3 density. The data are plotted as I_{bth} versus temperature T in Fig. 7.

Table IV. Survey of I_{bth} versus Firing Temperature Data

Wire length, mils	Wire diameter, mils	Wire material	S_o^p cm^2/g	Temp., °C	I_{bth}, A
10	0.7	Au	16,500	+84	151
10	0.7	Au	16,500	+24	191
10	0.7	Au	16,500	- 4	227
10	0.7	Au	16,500	-20	240
10	0.7	Au	16,500	-40	265
10	0.7	Au	16,500	-60	291
10	0.7	Au	16,500	-73.5	308
20	1.5	Au	4,000	+74	245
20	1.5	Au	4,000	+24	279
20	1.5	Au	4,000	-54	382
40	1.5	Au	~3,800	+84	203
40	1.5	Au	~3,800	+24	220
40	1.5	Au	~3,800	- 5	249
40	1.5	Au	~3,800	-40	274
40	1.5	Au	~3,800	-69	302

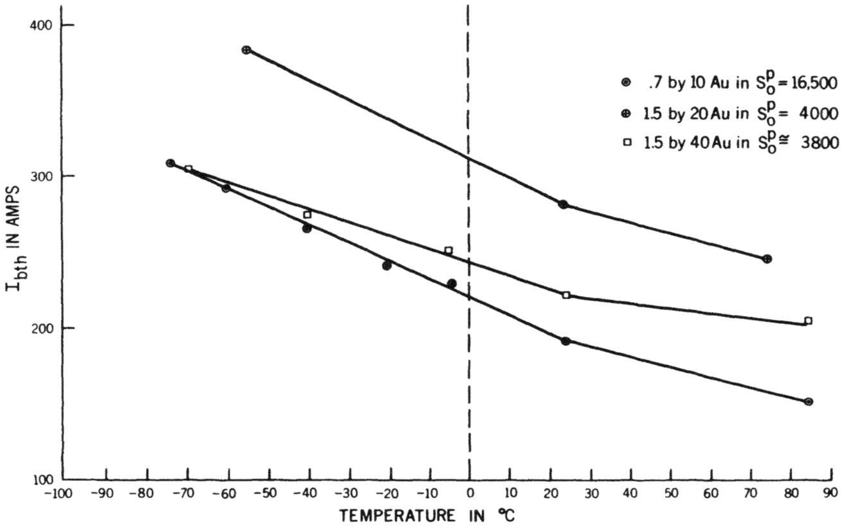

Fig. 7. I_{bth} versus firing temperature for various detonators.

Discussion of Results

It is evident from Fig. 7 that I_{bth} increases with decreasing temperature for all the detonators tested. Within experimental accuracy the functional dependence of I_{bth} for temperatures below room temperature appears to be linear. Above room temperature, a change in slope occurs; however, sufficient data are not available at present to determine whether an actual discontinuity in slope exists. In general, the behavior of detonators at temperatures below 24°C will be of greatest importance, since, in firing-set design, any increase in I_{bth} requires greater output from the firing set. Considering only this region, the test data for 0.7-by-10-mil Au wire in PETN of 16,500 cm^2/g specific surface area and 0.88 g/cm^3 density can be approximated by the function,

$$I_{bth} = 188 - 1.19(T - 24) \tag{14a}$$

where T is firing temperature in °C and I_{bth} is in amperes. Similarly, for the $d = 1.5$, $l = 40$ wire in $S_o^p \cong 3800$ PETN,

$$I_{bth} = 220 - 0.88(T - 24) \tag{14b}$$

and for the $d = 1.5$, $l = 20$ wire in $S_o^p = 4000$ PETN,

$$I_{bth} = 280 - 1.32(T - 24) \tag{14c}$$

Assuming temporarily that the temperature dependence of I_{bth} can be expressed as a simple multiplicative factor in Eq. (13), the dependence of I_{bth} on l, d, S_o^p, δ, and T would take the following form for a linear temperature dependence:

$$I_{bth} = d/\sqrt{T}\ [f_1(l,\ d,\ S_o^p)]\,[f_2(\delta)]\,[N_1 - M_1(T - 24)] \tag{15}$$

Using Eq. (13) to evaluate Eq. (15) for the 0.7-by-10-mil Au wire in PETN of $S_o^p = 16{,}500$ and $\delta = 0.88$ yields,

$$I_{bth} = 188\,[N_1 - M_1(T - 24)] \tag{16a}$$

Comparison of Eqs. (14a) and (16a) gives $N_1 = 1$ and $M_1 = 6.33 \times 10^{-3}$. If Eq. (15) is evaluated for the detonators having 1.5-by-40-mil and 1.5-by-20-mil wires, and the values for N_1 and M_1 of 1 and 6.33×10^{-3}, respectively, are utilized, the following relations are obtained:

$$I_{bth} = 230\,[N_1 - M_1(T - 24)] = 230 - 1.46(T - 24) \tag{16b}$$

for

$$d = 1.5,\ l = 40,\ \delta = 0.88,\ \text{and}\ S_o^p \cong 3800.$$

$$I_{bth} = 285\,[N_1 - M_1(T - 24)] = 285 - 1.80(T - 24) \tag{16c}$$

for

$$d = 1.5,\ l = 20,\ \delta = 0.88,\ \text{and}\ S_o^p = 4000.$$

Although the values of the temperature-independent terms appearing in Eqs. (16b) and (16c) are in reasonable agreement with the experimental results given by Eqs. (14b) and (14c), the coefficients of T are in error. In fact, in contradiction to the observed behavior expressed in Eqs. (14a) and (14b), the temperature coefficient in Eq. (16b) is greater than that in Eq. (16a). This observation suggests the following modification to the form of Eq. (15):

$$I_{bth} = d/\sqrt{T}\left[f_1(l,\ d,\ S_o^p)\right]\left[f_2(\delta)\right]\left[N_2 - \frac{M_2(T - 24)}{\sqrt{T}}\right] \tag{17}$$

Based now on Eq. (17), an evaluation similar to that used in obtaining Eq. (16a) yields the following comparison of predicted and observed temperature dependencies.

Detonator parameters	Observed	Predicted from Eq. (17)
0.7-by-10-mil Au wire, $\delta = 0.88$ g/cm^3, $S_o^p = 16{,}500$ cm^2/g	188—1.19 (T-24)	188—1.19 (T-24)
1.5-by-40-mil Au wire, $\delta = 0.88$ g/cm^3 $S_o^p \cong 3800$ cm^2/g	220—0.88 (T-24)	230—0.73 (T-24)
1.5-by-20-mil Au wire, $\delta = 0.88$ g/cm^3, $S_o^p = 4000$ cm^2/g	280—1.32 (T-24)	285—1.27 (T-24)

Although differences are evident, the agreement is remarkably good, considering the limited amount of data available. Thus, Eq. (13) may be extended to include the effect of firing temperature on I_{bth}, yielding

$$I_{bth} = d/\sqrt{l}\left[850 + 35.5\,\frac{(ldS_o^p \times 10^{-3} - 120)^2}{(ldS_o^p \times 10^{-3})^{3/2}}\right]\left[\frac{1}{(1.88 - \delta)^3}\right]\left[1 - \frac{2 \times 10^{-2}\,(T - 24)}{\sqrt{l}}\right] \tag{18}$$

SUMMARY AND CONCLUSIONS

It has been the purpose of this paper to report experimental data obtained in a study of EBW detonators and to develop an equation specifying the threshold burst current I_{bth} of the detonators as a function of major detonator and environmental parameters. This equation was found to be,

$$I_{bth} = d/\sqrt{l}\left[850 + 35.5\,\frac{(ldS_o^p \times 10^{-3} - 120)^2}{(ldS_o^p \times 10^{-3})^{3/2}}\right]\left[\frac{1}{(1.88 - \delta)^3}\right]\left[1 - \frac{2 \times 10^{-2}\,(T - 24)}{\sqrt{l}}\right]$$

where I_{bth} is in amperes, d and l are in mils, S_o^p is in cm^2/g, δ is in g/cm^3, and T is in °C. Over the parameter range thus far investigated,

$$0.5 \le d \le 3 \text{ mils},$$

$$10 \le l \le 40 \text{ mils},$$

$$3800 \le S_o^p \le 16{,}500 \text{ cm}^2/\text{g},$$

$$0.8 \le \delta \le 1.2 \text{ g/cm}^3,$$

$$-74^\circ\text{C} \le T \le +84^\circ\text{C},$$

this equation predicts experimental values of I_{bth} to an accuracy of approximately 10%. Although there are indications that this range may be extended, extrapolations outside it must as yet be approached with caution.

All experimental results were obtained with one particular type of firing set; however, the relative independence of I_{bth} from firing set details means that Eq. (18) is applicable in designing a system based on any practical firing set. Thus, the I_{bth} equation can be an extremely useful tool in the analysis and design of EBW detonator systems.

It should be emphasized that although the threshold burst current may be correlated with explosive initiation in EBW detonators, it is not proposed that this current is the direct cause of initiation. Rather, it is believed that over the range of parameters investigated, the physical mechanisms responsible for initiation are active concurrently with the attainment of threshold burst current.

Although many studies remain to be done, Eq. (18) has been valuable in providing a better understanding of the mechanisms of PETN initiation by exploding bridgewires. Considering for the moment only the situation of explosive-specific surface properly matched to wire dimensions (i.e., $ldS_o^p = 120{,}000$), then I_{bth} is directly proportional to bridgewire diameter and inversely proportional to the square root of bridgewire length. This type of proportionality would result if initiation were dependent on peak power input; i.e., if

$$I_{bth} = C_1\, d/\sqrt{l}$$

where d and l are in mils; then it can be shown that the threshold peak power P_{bth} in kilowatts is given by

$$P_{bth} = I_{bth}^2 R_{bth} = \frac{C_1^2 \rho_{bth}}{(2.54)\pi} \tag{19}$$

where R_{bth} and ρ_{bth} are bridgewire threshold burst resistance and resistivity, respectively. Although ρ_{bth} is weakly dependent on bridgewire diameter and length, most experimentally observed values of ρ_{bth} under threshold firing conditions are between 400 and 800 μohm-cm. Equation (19) indicates that these values of ρ_{bth} result in threshold power inputs between 145 and 200 kW. This suggests that for EBW initiation of PETN, a power input on the order of a few hundred kW must be generated. Additionally, the observed width of the power spike at the 145 kW power level (related to the time interval of high bridgewire resistance) is approximately 0.1 μsec; thus, the energy input during this period is approximately 7.5 mJ. These observations, then, lead to the hypothesis that exploding bridgewire initiation of PETN requires an energy deposition of 5 to 10 mJ at a very high power level of a few hundred kilowatts. The resultant high energy density in the region near the bridgewire will generate a shock wave, and it is reasonable to postulate that initiation occurs within this region of high pressure and temperature behind the shock front. Unfortunately, little is as yet known regarding exact details of the shock strength and profile required for initiation in the highly divergent EBW detonator geometry [5]. Consequently, studies of the characteristics of shock waves in granular PETN should be extremely useful in further understanding the initiation process [6].

Finally, the threshold dependence of I_{bth} (and thus the power) on explosive specific surface suggests that initiation is dependent upon interactions of microscopic disturbances which are due to the granularity of the PETN compacts. This process is far from being understood; however, the observation that for minimum thresholds, S_o^p must be increased with decreases in ld suggests that an optimum explosive environment exists; i.e., the bridgewire and explosive grain size must be scaled similarly. Although experimentally difficult, studies of shock interactions in granular pressings should be very valuable.

In summary, then, the functional dependence of I_{bth} upon the major EBW detonator parameters has been determined. The equation expressing this relationship has proved to be a valuable design tool. It is to be hoped, however, that these results will provide greater insight into the physical processes of EBW initiation of granular PETN explosive. The work reported here, although not directly defining these more fundamental processes, does provide a greater understanding of the mechanisms involved and, in addition, suggests directions for more definitive experiments.

ACKNOWLEDGMENT

The author is greatly indebted to H.E. Roberts for technical assistance and to O.E. Jones and G.E. Seay for their support and many helpful suggestions.

REFERENCES

1. T.J. Tucker, "Exploding Wire Detonators: The Burst Current Criterion of Detonator Performance" in Exploding Wires, Vol. 3, W.G. Chace and H.K. Moore, eds., Plenum Press, New York (1964) p. 175.
2. E.C. Cnare, "Exploding Wire Detonators: An Approximate Method of Predicting Exploding Wire Detonator–Capacitor Discharge System Performance" in Exploding Wires, Vol. 3, W.G. Chace and H.K. Moore, eds., Plenum Press, New York, (1964) p. 185.
3. G.W. Anderson and E.W. Neilson, "Use of the Action Integral in Exploding Wire Studies" in Exploding Wires, Vol. 1, W.G. Chace and H.K. Moore, eds., Plenum Press, New York (1959) p. 97.
4. T.J. Tucker, "Square-Wave Generator for the Study of Exploding Wires," Rev. Sci. Instr. 20:165 (1960).
5. J.H. Blackburn and R.J. Reithel, "Exploding Wire Detonators: Sweeping-Image Photographs of the Exploding Bridgewire Initiation of PETN" in Exploding Wires, Vol. 3, W.G. Chace and H.K. Moore, eds., Plenum Press, New York (1964) p. 153.
6. G.E. Seay, "Shock Initiation of Granular Explosives Pressed to Low Density" in Ninth Symposium (International) on Combustion, W.G. Berl, (ed.) Academic Press, Inc., London (1963).

DISCUSSION

Question by D.J. Jones U.S. Dept. of Commerce: Is it the temperature jump from the shock wave that initiates detonation? Have you exploded wires in PETN cavities?

Answer by Author: I simply do not know. PETN is a secondary explosive that requires both temperature and pressure for initiation. These conditions should result from the shock wave from the exploding wire; however, a good description of shock behavior in granular explosive has not as yet been obtained.

Question by R. Hadge, Edgerton, Germeshausen, & Grier, Inc., Boston, Mass.: Can your formula be extended to a foil usage?

Answer by Author: The equation was, of course, obtained empirically. I think it is possible that it might be extended to include foils; however, this would be an extrapolation outside the range of parameters tested.

Question by F. Herlach, Illinois Institute of Technology, Chicago, Ill: Could your formula be extended to the series-connection of detonators by inserting the sum of the length of the individual bridgewires?

Answer by Author: I see no reason why the equation would not apply to detonators fired in series; the equation describes I_{bth} in terms of detonator parameters and is relatively independent of the external system. I_{bth} for the detonators would, however, be that associated with the wire length of individual detonators.

The Electrical Conductivity of Metals at Very High Temperatures

H. Knoepfel and R. Luppi

Laboratori Gas Ionizzati,
Euratom – CNEN,
Frascati (Roma), Italy

The use of explosively generated megaoersted fields requires the knowledge of the electrical conductivity at the extreme physical conditions to which these conductors are subjected (at 5 MOe typically, temperature up to 20,000 °K, pressures up to 1 Mbar). The physical problems are in some respects similar to those encountered in the domain of exploding wires. Two different experiments for measuring the conductivity of metals are presented. In the first, the temperature dependence of the electrical conductivity of copper has been obtained by determining the magnetic diffusion into a hollow cylindrical conductor. The results extend up to the temperature at which the conductor explodes. In the second experiment we have measured the magnetic flux penetration into a copper bar and compared the experimental data with the results obtained from a particular theoretical diffusion model. Both experimental methods strongly suggest that, under transient heating conditions, the conductivity of liquid copper decreases much less with increasing temperature than expected.

INTRODUCTION

The knowledge of electrical conductivity and its dependence on temperature, pressure, and magnetic field is particularly important in cases where large current densities are generated, as for instance in the domain of exploding wires or for the very large magnetic fields generated by explosive-driven magnetic flux compression. The temperature range of interest in connection with megaoersted fields extends from room temperature up to 100,000 °K and more (i.e., an order of magnitude larger than the critical temperature of metals). In fact, for the interaction of megaoersted fields with normal conductors it can be shown [1] that the temperature Θ on the conductor surface is of the order

$$c_v \Theta \simeq \frac{H^2}{8\pi} \tag{1}$$

where c_v is the volume-specific heat.

The relation between the domains of "exploding wires" and megaoersted fields is not limited to the electrical conductivity but includes many fundamental physical effects typical of the "physics of high-energy density," to which both belong [1]. In a certain sense, one can say that the interaction process of megaoersted fields with metallic conductors adds a third dimension to the domain of exploding wires and foils, this being an additional argument to those of Chace [2] for calling it the domain of "exploding conductors" (E.C.).

The "third dimension" is represented by the appreciable skin depth which results from the temperature-dependent diffusion of very high magnetic fields; for a typical 10-MOe pulse, the penetration depth can, in fact, be of the order of 1 mm (see [1] and Fig. 12). The interaction of these megaoersted fields with metallic conductors results in a very complicated physical process which, in general, can be treated only by numerical methods [8–10].

In Table I, a comparison between different quantities concerning a typical exploding wire experiment and the interaction of a 6-MOe field pulse with a metal is given: it is evident that with the exception of the magnetic field value itself, all other physical quantities are of the same order. For still larger field pulses, however, the surface temperature increases roughly in accordance with Eq. (1), whereas the maximum current density (which is attained not at the surface but within the skin depth) remains of the same order.

Table I. Exploding Wire (E.W.) and Megaoersted Fields

Quantity	Dimension	E. W.*	MOe†
Surface magnetic field	MOe	0.4	6
Magnetic pressure $H^2/8\pi$	Mb	0.006	1.5
Maximum current density	MA/cm^2	500	200
Maximum surface temperature	°K	5000	20,000
Typical pulse duration	μsec	0.1	5

*From [5], copper wire 2.5×10^{-3} cm diameter, just prior to explosion.
†Typical field pulse as in Fig. 12 with H_{max} = 6 MOe and $\beta = 5.15 \times 10^{-3}$ (°C)$^{-1}$.

The Electrical Conductivity in Metals

The temperature-dependence of the electrical conductivity in the condensed phase (solid and liquid) has been measured under quasi-static experimental conditions for many metals [15, 16, 24]. Extensive literature exists on this subject, e.g., [6, 18, 21, 22, 23].

The conductivity problems of interest in the domain of E.C. and MOe fields extend, of course, to higher temperatures, where liquid–vapor mixtures or the overheated liquid state have to be considered [9, 10]. At extreme temperatures, the metallic state may reduce to that of a degenerate plasma (where the conductivity maintains a constant value [11]), while at still higher temperatures (of the order of 10^6 °K) the classical plasma conductivity law

$$\sigma \propto \Theta^{3/2} \tag{2}$$

will be approached.

Also of interest, particularly in the domain of MOe fields, is the pressure-dependence of the conductivity at high temperatures. Since high pressures reduce the amplitudes of the lattice vibration, the conductivity will increase with increasing pressure. The conductivity law (13) can be extended according to [6] to include this effect in the form

$$\sigma = \frac{\sigma_o}{1 + \beta\Theta} \cdot \left(\frac{v_o}{v}\right)^{\alpha} \tag{3}$$

where v is the specific volume and for copper $\alpha = 2.7$. Note that for copper at 1 Mbar (5 MOe) $(v_o/v)^{2.7} = 2.3$ [25].

An interesting reduction of the conductivity, but probably unimportant at high temperatures, is due to the magnetic field. According to Kohler's rule, the increase in resistivity ($\rho = 1/\sigma$) due to the field H is

$$\frac{\Delta\rho}{\rho} = f\left(\frac{H}{\rho}\right) \tag{4}$$

where ρ is the zero-field resistivity at the temperature of interest. It turns out that for normal metals the effect is small at room temperatures or above, even for fields in the MOe region [11, 19]. For copper at 20°C and for a field of 1 MOe $\Delta\rho/\rho \simeq 20\%$, but at 1000°C for the same field, one already finds $\Delta\rho/\rho \simeq 1\%$.

Furthermore, it is possible that the very high current densities obtained particularly with the exploding wires (Table I) could influence the electrical conductivity. Some authors even suggest the possibility that at current densities of the order of 10^9 A/cm^2 one might attain a sort of supraconducting state at high temperatures [20].

It should be pointed out that other physical parameters play an important role in connection with the conductivity problem. Probably the most important of these is the specific heat c_v which determines the temperature of the conductor through Joule's law [Eq. (10)] and also shows a temperature dependence [15, 16]. For simplicity we will consider c_v in this paper to be a constant.

In the following sections we will describe two new measuring methods aimed at determining the temperature dependence of the electrical conductivity of copper under transient heating conditions.

THE HOLLOW CYLINDER PROBE

Consider the conductor shown in Fig. 1. Assuming that the wall thickness is small compared to the penetration depth, i.e., the current density j_ϑ constant over d, one can write (Gaussian units, c being the velocity of light)

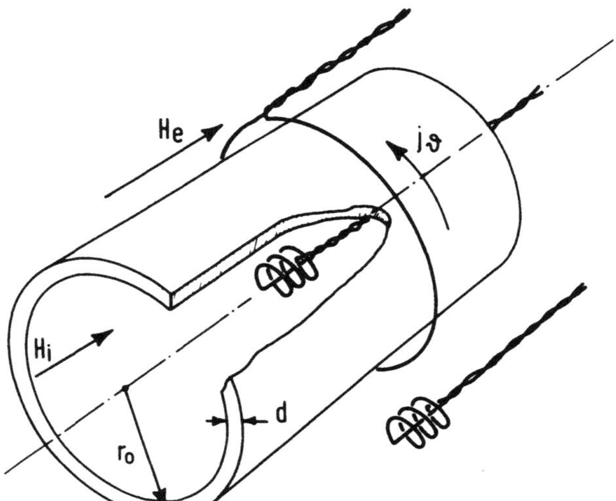

Fig. 1. Hollow cylindrical conductor. Induction probes measure outer and inner magnetic field as well as total diffused flux.

$$j_\vartheta \cdot d = \frac{c}{4\pi}(H_i - H_e) \tag{5}$$

or, from the induction law,

$$\frac{2\pi r_o}{\sigma d} j_\vartheta d = \frac{1}{c} \pi r_o^2 \frac{dH_i}{dt} \tag{6}$$

Elimination of the current results in the differential equation

$$H_i + \tau \frac{dH_i}{dt} = H_e \tag{7}$$

where we have introduced a time constant

$$\tau = 2\pi r_o d \cdot \frac{\sigma}{c^2} \tag{8}$$

From (7) we deduce

$$\frac{\sigma(t)}{c^2} = \frac{1}{2\pi r_o d} \frac{H_e - H_i}{dH_i/dt} \tag{9}$$

The temperature increase Θ is described by Joule's law

$$\frac{\partial(c_v \cdot \Theta)}{\partial t} = \frac{j_\vartheta^2}{\sigma} \tag{10}$$

which by substitution of j_ϑ and σ from Eqs. (5) and (6) and integration becomes

$$c_v \Theta = \frac{r_o}{8\pi d} \int_o^{H_i} (H_e - H_i) \cdot dH_i \tag{11}$$

When $(H_e - H_i)$ and dH_i/dt are known functions of time, it is possible to calculate from Eqs. (9) and (11) σ and Θ at any t and to deduce therefore the temperature dependence of the conductivity.

Before presenting the experimental results related to this type of probe, it may be interesting to get an insight in the system of Eqs. (7), (8), and (11) by analyzing a case often encountered in practice, namely when the outer field is sinusoidal

$$H_e = H_o \sin 2\pi \frac{t}{T} \tag{12}$$

If we assume the simple conductivity law

$$\sigma = \frac{\sigma_o}{1 + \beta\Theta} \tag{13}$$

to be valid, the temperature Eq. (10) can be integrated to give

$$1 + \beta\Theta = \exp\left[\frac{2}{\tau_o}\left(\frac{H_o}{H_c}\right)^2 \int_o^t \left(\frac{H_e - H_i}{H_o}\right)^2 dt\right] \tag{14}$$

where

$$H_c = h_c\sqrt{2\frac{d}{r_o}}, \quad h_c = \sqrt{8\pi\frac{c_v}{\beta}} \tag{15}$$

is a critical field parameter and in analogy to Eq. (8) we define

$$\tau_o = 2\pi r_o d \cdot \frac{\sigma_o}{c^2} \tag{16}$$

In Fig. 2, the diffused inner field H_i, as deduced by numerical calculation, has been plotted for the particular case $\tau_o = T/2$ and for 6 different values of the parameter H_o/H_c. The curve labeled $H_o/H_c = 0$ represents the simple diffusion problem where the conductivity remains constant ($\sigma \equiv \sigma_o$). In Fig. 3, the maximum amplitude of the diffused field $(H_i)_{max}$ is given in function of the two parameters of the problem, whereas Fig. 4 shows the temperature of the conductor at the time when the inner field is at its maximum.

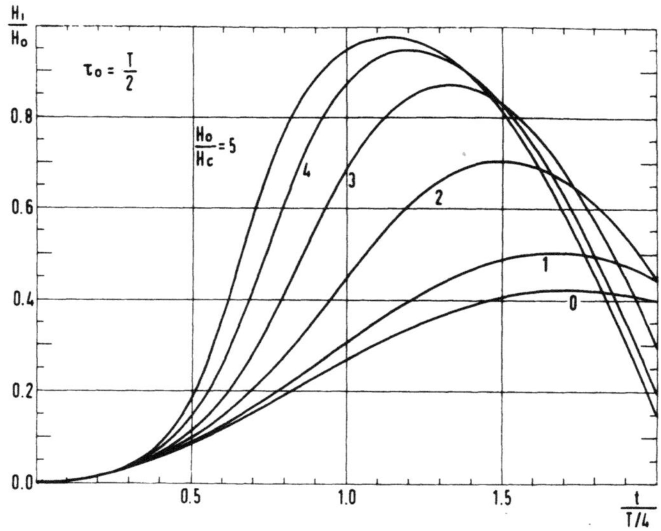

Fig. 2. Diffusion of a sinusoidal field of amplitude H_o and period T into a hollow conductor for particular case $\tau_o = T/2$; the time constant τ_o is defined in Eq. (16) and the critical field H_c in Eq. (15).

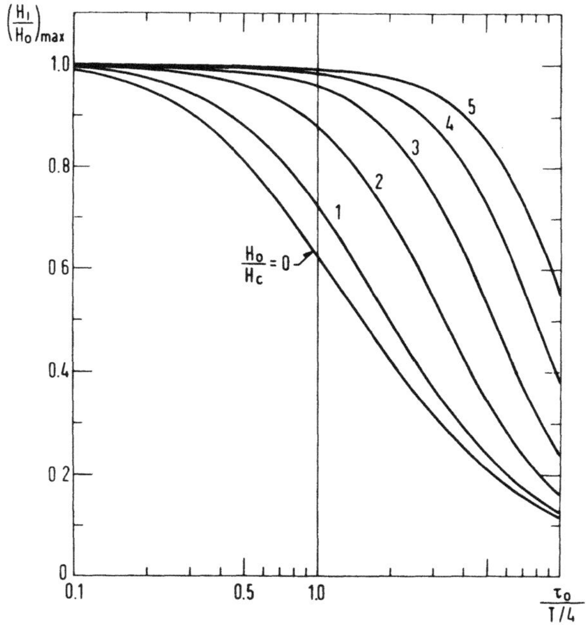

Fig. 3. Maximum of diffused inner field.

It is evident that by a suitable choice of r_0 and d, very high temperatures can be attained in this geometry. This is to be compared with the surface temperature of a thick conductor (thickness large as compared to the skin depth) which depends essentially only on the magnetic field pulse and remains limited to values as given by Eq. (1). Both Figs. 3 and 4 can be very useful in fixing the parameters of a predetermined experiment.

The experiments were performed by placing copper probe systems in a magnetic field generated by an explosive driven bellows device (Fig. 5) giving a field pulse as shown in Fig. 6, with a maximum of 1.7 MOe. The dimensions of the hollow copper cylinder, which is completely immersed in araldite, are r_0 = 2 mm and d = 0.15 mm. The magnetic field probes are built and calibrated as described in [14]. The signals of the probes (generally two pairs of probes in one conductor system) are integrated and displayed on different oscilloscopes, whose traces have blanking spots permitting to relate all signals within an error of less than 0.1 μsec (Figs. 6 and 10).

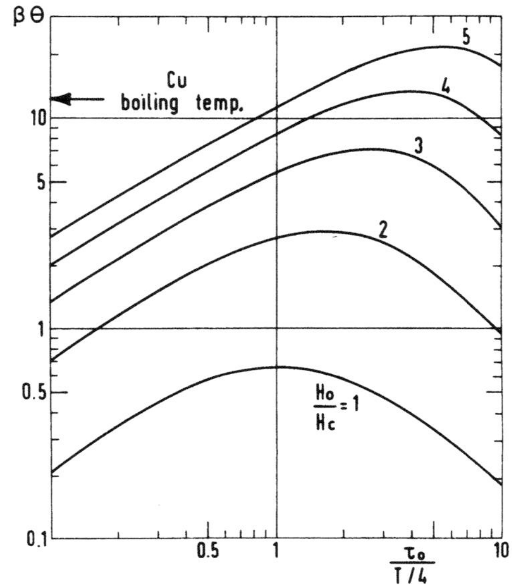

Fig. 4. Conductor temperature at maximum inner field.

The result of a typical experiment (out of a total of 4) is shown in Fig. 7, where the quantity $(2\pi r_o d\sigma)^{-1}$ [which is proportional to the specific resistivity, see Eq. (9)] is plotted in function of the inertia integral

$$I_n(t) = \int_o^t \left(\frac{j_\vartheta}{c}\right)^2 dt = \frac{1}{(4\pi d)^2} \int_o^t (H_e - H_i)^2\, dt \tag{17}$$

as is often done in exploding foil experiments [26]. Each point of the curve corresponds to a certain temperature, defined by Eq. (11); we have indicated only the melting and vaporization points (under normal conditions) which divide the curve in phase domains.

From the shape of the resistivity curve, one would expect the solid–liquid interphase to be placed at somewhat higher values. One possible explanation for this effect can be found by attributing a certain inertia to the phase transformation, considering that the experiment sweeps through the interphase within 0.3 μsec.

For the same reason, but even more because of the high pressure involved, it may well be possible that vaporization sets in at a much higher temperature than expected. For instance, the temperature point 4000°C, indicated on the curve, corresponds to the boiling point of copper at 5 kbar. The mean error related to this experimental curve can be indicated as ±15% for both coordinates.

In Fig. 8, these experimental results are formally displayed as a resistivity versus temperature curve and compared with handbook values for high purity copper [15]. Points above the melting temperature have been corrected to take the melting heat into account (equivalent to 450°C for copper). This curve

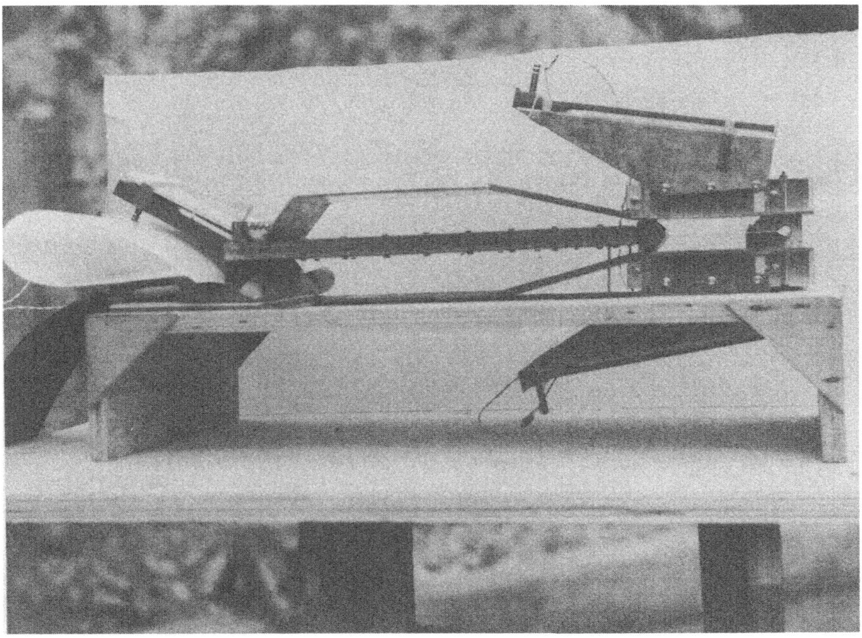

Fig. 5. Two-stage flux compression system (device N.57 [12]) on firing table. At right end of system (whose total length is about 70 cm) one looks axially into the solenoid (from which field probe has been extracted) having dimensions: diameter 1.2 cm; length, 10 cm.

has been deduced directly from Eqs. (9) and (11) with the measured $(H_e - H_i)$ and dH_i/dt values.

It is therefore necessary to discuss briefly the assumptions on which these equations rely. As can be seen from Fig. 14, the skin depth in copper for the particular magnetic field pulse used is of the order of 0.5 mm, which, compared to the conductor thickness of $d = 0.15$ mm, justifies in first-order approximation the "thin" conductor assumption. Although the specific heat c_v is not constant, its variation in the temperature range of interest, as far as known [15], lies well within 10% of the mean value given in Table II: this results in a negligible correction of the actual curve. The assumption of constant geometry is more questionable. In fact, let us suppose (as a pessimistic upper limit) that during expansion the length of the metal tube remains constant, then

$$r_o d \delta_o = r_1 d_1 \delta_1$$

and we can rewrite Eq. (9) in a form which takes the variable mass density δ_1 into account

$$\frac{\sigma}{c^2} = \frac{\delta_1}{2\pi r_o d \delta_o} \cdot \frac{H_e - H_i}{dH_i/dt} \tag{18}$$

By taking the density variation according to [15], the experimental curve is modified in the liquid phase as indicated in Fig. 8, whereas

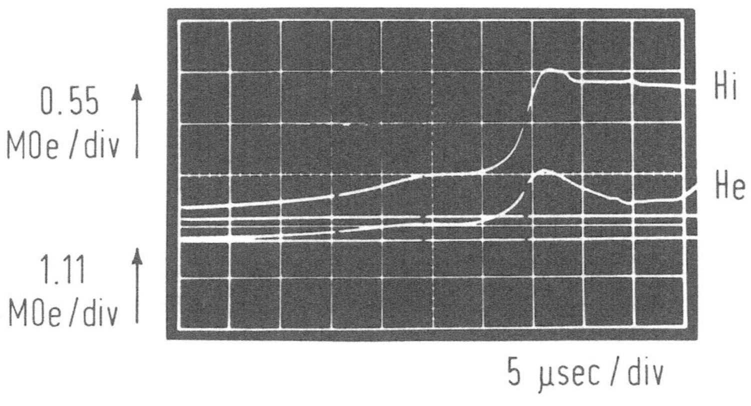

Fig. 6. Oscillogram displaying external (H_e) and inner field (H_i) related to a Cu probe system placed in a bellows generator (Fig. 5).

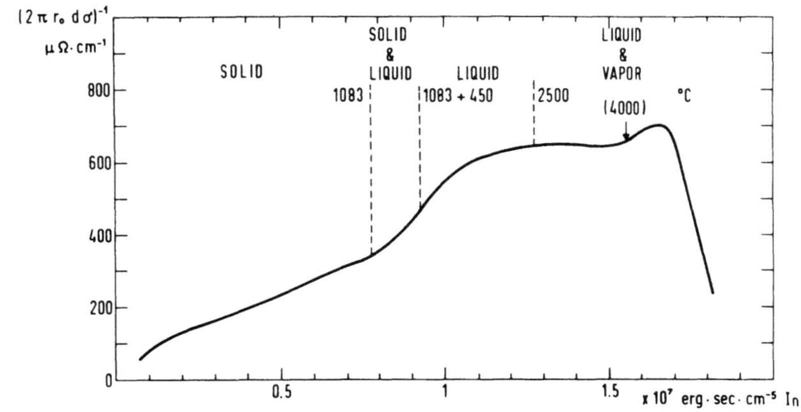

Fig. 7. Resistivity of copper plotted versus inertia integral for the hollow cylindrical conductor (shot No. 823).

in the solid phase the shift is practically negligible (and therefore not included in the figure). In any case, the deviation of the measured temperature-dependence of the conductivity from what might be expected from handbook values is evident. The slight difference between the two curves in the solid phase is most probably due to the effect of impurities on our commercial grade copper probe and has no particular meaning.

THE BAR PROBE

The probe system shown in Fig. 9 allows a direct measurement of the flux diffused into the metal, from which it may be possible, through theoretical analysis, to deduce a corresponding conductivity law. This method is therefore less precise in determining the conductivity than the previous one with the hollow conductor but can, in turn, be extended to any field level.

Since our experiments are limited to about 3.5 MOe, we assume for our theoretical calculations that the conductor is incompressible, not subjected to vaporization and obeying the simple conductivity law [13]. This problem reduces to a nonlinear diffusion problem and is described by Maxwell's cylindrical equations

Table II. Handbook Constants

	Solid					
	δ, g/cm^3	σ_0, $(\Omega\ \text{cm})^{-1} \times 10^4$	β, $(°\text{C})^{-1} \times 10^{-3}$	c_v^*, J/cm^3 °C	h_c, MOe	σ_{solid},§ $(\Omega\ \text{cm})^{-1} \times 10^4$
Cu	8.92	63.3	5.15	3.88	0.43	9.63
Al	2.7	39.2	5.85	2.79	0.30	8.10
Fe	7.87	11.28	13.6‡	19.1‡	0.60	0.76
Brass (Zn 42%)	8.44	15.7	2.1	3.81	0.67	–

*Mean value between 0°C and melting point.
†At atmospheric pressure.
‡From 0°C to the A_3 –point.
§At melting point.

$$\frac{\partial H_z}{\partial r} = -\frac{4\pi}{c} j_\vartheta \tag{19}$$

$$\frac{\partial (r E_\vartheta)}{r \partial r} = -\frac{1}{c} \frac{\partial H_z}{\partial t} \tag{20}$$

(in which we have neglected the displacement current and assumed

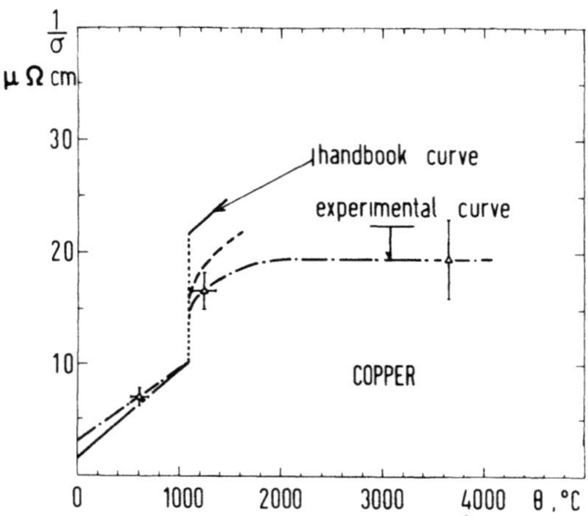

Fig. 8. Resistivity ($1/\sigma$) of copper plotted against temperature Θ. Comparison between experimental and handbook curve. The broken curve includes thermal expansion, Eq. (18).

of Metallic Conductors

Liquid			Phase parameters†			
σ_{liquid}, $(\Omega\ cm)^{-1} \times 10^4$	β_e, $(°C)^{-1} \times 10^{-3}$	c_{vl}, $J/cm^2\ °C$	Melting point, °C	Boiling point, °C	Heat fusion, J/g	Heat vapor., kJ/g
4.65	0.38	3.74	1083	2500	210	7.5
4.95	0.53	2.48	658	2057	390	6.7
0.70	–	–	1536	2450	275	–
–	–	–	900	1080	–	–

the magnetic permeability to be $\mu = 1$), by Ohm's law

$$j_{\vartheta} = \sigma E_{\vartheta} \tag{21}$$

and by the usual temperature Eq. (13).

This nonlinear diffusion problem has been solved numerically for the two particular experimental boundary fields given in Figs. 12 and 14. The results depend on the critical field parameter h_c defined in Eq. (15) and are displayed in Figs. 12 and 14 in the form of a flux skin depth s_φ which we define as

$$s_\varphi(t) = \frac{\Phi_p}{2\pi r_o \cdot H_z(r_o, t)} \tag{22}$$

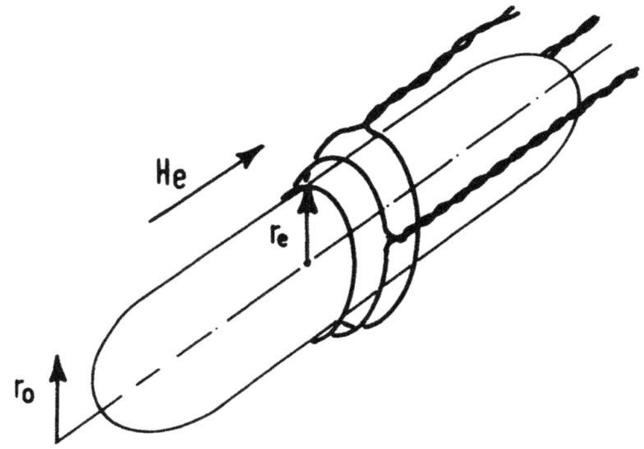

Fig. 9. Bar probe system with three concentric flux probes.

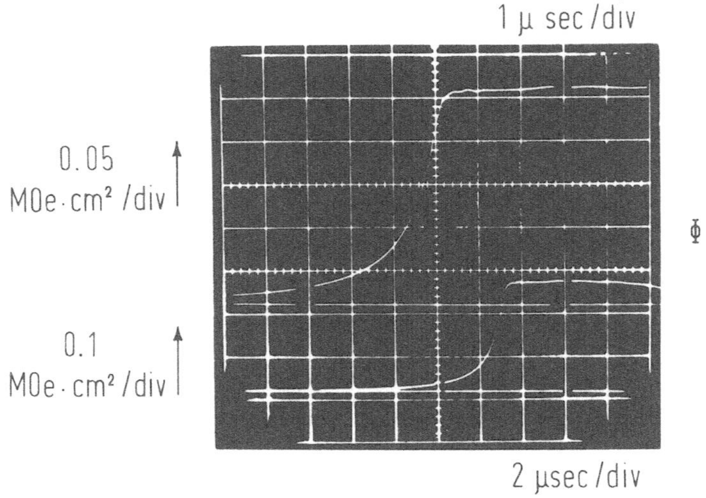

Fig. 10. Integrated signal of a flux probe.

where

$$\Phi_p = 2\pi \int_{r_o}^{o} H_z(r,\ t)\, r \cdot dr \tag{23}$$

is the total magnetic flux diffused into the cylindrical conductor and $H_z(r_o,\ t)$ is the boundary field.

Experimentally, the flux Φ_{exp} measured (through integration) by the flux probes (Fig. 10) contains two components

$$\Phi_{exp} = \Phi_p + F_e \cdot H_z(r_o,\ t) \tag{24}$$

Using the method shown in Fig. 11, we deduce the mean Φ_p as well as the surface field $H_z(r_o, t)$ from the signals of our three probes, and from Eq. (22) we find, therefore, the experimental skin depth s_φ directly. The surface between the cylindrical conductor and the concentric flux probe (Fig. 9)

$$F_e = \pi(r_e^2 - r_o^2)$$

is determined in two ways: first, by carefully measuring the radii ($r_o = 1$, $r_e = 1.15$, 1.5, 2 mm), and second, by calibrating the probes in a known magnetic field of varying frequency (from 50 to 3000 kHz) and extrapolating the results to infinite frequency, i.e., zero skin depth.

The probe system is immersed in araldite and protected by a glass tube having an over-all outer diameter of 7 mm. Due to this relatively large size, the field of the cylindrical flux compression device (type N.45, see [13]), which generally produces magnetic fields of 5 to 6 MOe [17], is limited to about 3.5 MOe.

The results of two shots are reported in Fig. 12 in the form of experimental points with an error of not less than 10%. They cross the theoretical curves and show a tendency toward high values of the temperature coefficient β of conductivity. The initial points are far from the calculated values. This is due to the fact that, in order to simplify the numerical calculations, the boundary curve has been linearly extrapolated to zero, which does not conform with the real time scale of the compressed field.

A surface temperature corresponding to the melting point is reached at nearly the same time ($t = 8$ μsec) for all curves; after this time, the liquid phase propagates in the interior with a different velocity for each curve.

The major error in the evaluation of the measurements is due to the assumption that r_o, and consequently F_e, remain constant. In fact, due to the magnetic pressure, the conductor is compressed and consequently F_e increased, so as to simulate a larger flux

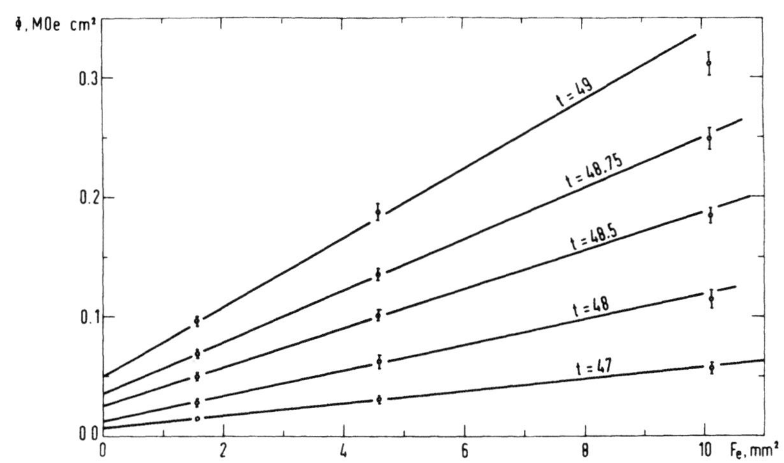

Fig. 11. Graphical method to determine diffused flux (at $F_e = 0$ ordinate) and surface field (from slope) as described by Eq. (24).

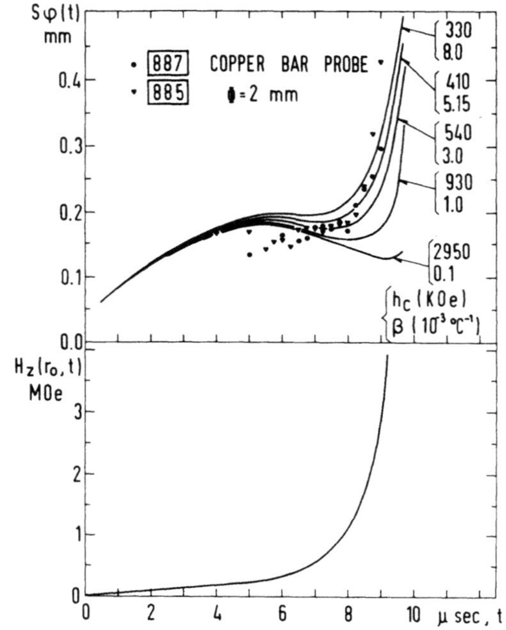

Fig. 12. Flux skin depth corresponding to a boundary field as shown at the bottom: comparison between theoretical curves and experimental points of shots No. 885 and 887.

diffusion into the conductor. We have tried to take care, in first order approximation, of this effect by calculating the compression of the core of the probe, defined by a radius $(r_o - s_\varphi)$, and subjected to the full pressure $H^2(r_o, t)/8\pi$; by using the compressibility data given in [25], a new corrected radius $r_o^*(t)$ is then determined. The experimental points of shot No. 887 thus corrected are plotted in Fig. 13. Appreciable deviations can be found only for magnetic fields higher than 1.5 MOe. This result suggests that the mean temperature coefficient is in this case $\beta = (4.2 \pm 1.2) \cdot 10^{-3}$ $(°C)^{-1}$.

The results plotted in Fig. 14 are related to a copper probe of larger diameter (5 mm) placed within a bellows generator of the type shown in Fig. 5 and capable of generating a maximum field of 1.7 MOe. Apart from the initial points, which show a similar effect as in Fig. 12, the experimental points indicate a temperature coefficient of $\beta = (5 \pm 0.5) \cdot 10^{-3}$ $(°C)^{-1}$.

DISCUSSION OF THE RESULTS

In considering the hollow conductor experiment, it should be kept in mind that it is primarily a magnetic diffusion experiment, our result being that the diffused inner field at temperatures higher than the melting point is smaller than expected.

This effect can be expressed formally by saying that the volume conductivity related to this experimental arrangement does not decrease in the liquid phase as was to be expected from handbook values. This statement is based, as we have seen, on various assumptions ("thin" conductor, constant j_ϑ, constant d, constant r_o, constant c_v) which, although they introduce not negligible errors, cannot alone be found responsible for the effect.

In the solid state, up to the melting point, no appreciable difference between the experimental results and the handbook values is found. The effects of magnetoresistance and pressure, even if not completely negligible, are hidden by the experimental error.

In analyzing results of this type of "one shot" experiment, where the interpretation is based on poor statistics, it is par-

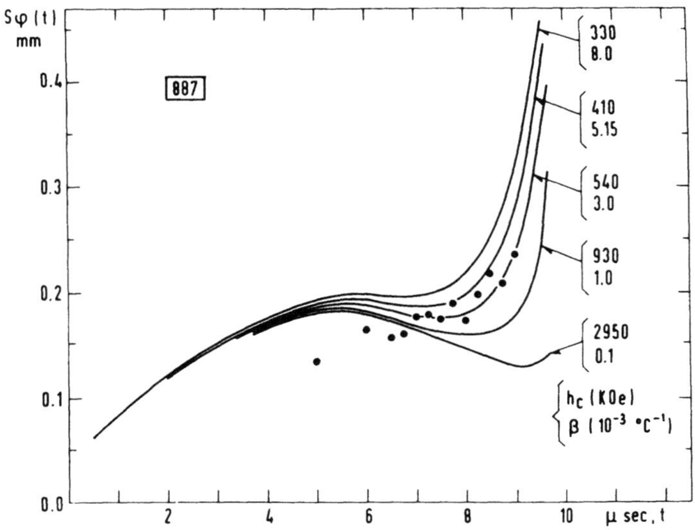

Fig. 13. Same as in Fig. 12, but corrected for compressibility effect (shot No. 887 only).

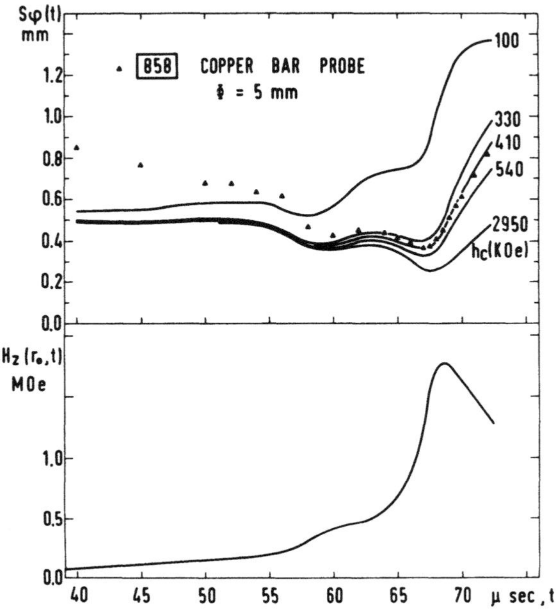

Fig. 14. Flux skin depth corresponding to a different boundary field (bellows generator; shot No. 858).

ticularly important to check the possible sources of misinterpretation. From this point of view, the only unwanted effect we can think of is represented by a mechanical distortion of the inner magnetic probe as a consequence of the pressure exerted by the conductor on its araldite case. A simple check concerning this probe distortion could be obtained by varying the geometry of the diffusion probe.

The results related to the bar probe, giving the flux diffused into the conductor, are somewhat more difficult to analyze in terms of the fundamental physical parameters involved. As we have seen, a reasonable determination of the conductivity requires, even at 3.5 MOe, a theoretical model of the diffusion process which includes the compressibility of the metal and the variation of the conductivity law at the melting point. The mean temperature coefficient determined in these experiments by assuming the simple conductivity law (3) can be given as $\beta = (4.5 \pm 1) \cdot 10^{-3}$ $(°C)^{-1}$ within the magnetic field range from 1 to 3.5 MOe.

If included in Fig. 8, this conductivity law with its uncertainty limits, represents an area which fairly well covers the experimental curve. In this sense the results of the two experiments are consistent.

In conclusion we can say that strong experimental evidence suggests that the electrical conductivity of copper in the liquid phase, obtained by a transient heating process, decreases much less with increasing temperature than expected from handbook values. On the basis of the actual experimental results, one could attribute this apparent anomaly to the fact that due to the rapid heat transfer ($\sim 1\ \mu$sec) the metal may not have attained the structure typical for its liquid phase under static conditions.

In addition, the two experimental methods used to determine the conductivity have been proved very valuable and with the improvements suggested, it will be possible to increase the precision of the measurements and to extend the range of the principal physical parameters involved.

ACKNOWLEDGMENTS

The authors are indebted to G. Di Cola, on leave from Cetis, Euratom, Ispra, who was responsible for obtaining all the numerical results reported here, and to J. G. Linhart for his constant interest and support of this work.

We would also like to thank G. B. Baratta for the evaluation of the experimental results. The explosive charges have been prepared and fired in collaboration with the Bombrini Parodi–Delfino Company, Colleferro.

REFERENCES

1. H. Knoepfel, "Megaoersted Fields and Their Relation to the Physics of High Energy Density," lecture presented at the Summer School on High Magnetic Fields, Creta 1967.
2. W.G. Chace and H.K. Moore (eds.), Exploding Wires, Vol. 2, Plenum Press, New York (1962).
3. F.H. Webb, H.H. Hilton, P.H. Levine, and A.V. Tollestrup, in Exploding Wires, Vol. 2, Plenum Press, New York (1962), p. 37.
4. W.G. Chace and H.K. Moore (eds.), Exploding Wires, Vol. 3, Plenum Press, New York (1964).
5. R.C. Maninger, in Exploding Wires, Vol. 3, Plenum Press, New York (1964), p. 47.
6. N.F. Mott and H. Jones, The Theory of the Properties of Metals and Alloys, Dover Publishing Co., New York (1958), p. 271.
7. H. Knoepfel and F. Herlach (eds.), Proceedings of the "Conference on Megagauss Magnetic Field Generation by Explosives and Related Experiments," EUR 2750.e, Euratom, Brussels (1966).

8. J.P. Somon, Proceedings of the "Conference on Megagauss Magnetic Field Generation by Explosives and Related Experiments," EUR 2750.e, Euratom, Brussels (1966), p. 235.
9. R.E. Kidder, Proceedings of the "Conference on Megagauss Magnetic Field Generation by Explosives and Related Experiments," EUR 2750.e, Euratom, Brussels (1966), p. 37.
10. A.R. Bryant, Proceedings of the "Conference on Megagauss Magnetic Field Generation by Explosives and Related Experiments," EUR 2750.e, Euratom, Brussels (1966), p. 183.
11. G. Lehner, Proceedings of the "Conference on Megagauss Magnetic Field Generation by Explosives and Related Experiments," EUR 2750.e, Euratom, Brussels (1966), p. 55.
12. F. Herlach, H. Knoepfel and R. Luppi, Proceedings of the "Conference on Megagauss Magnetic Field Generation by Explosives and Related Experiments," EUR 2750.e, Euratom, Brussels (1966), p. 287.
13. F. Herlach and H. Knoepfel, Proceedings of the "Conference on Megagauss Magnetic Field Generation by Explosives and Related Experiments," EUR 2750.e, Euratom, Brussels (1966), p. 147.
14. F. Herlach and H. Knoepfel, Rev. Sci. Instr. 36:1088 (1965).
15. $\sigma(\Theta)$ and c_v value curves for metals: Landolt and Börnstein, Zahlenwerte und Funktionen IV/2 a, b, c; IV/3, Springer Verlag, Berlin.
16. Y.S. Touloukian (ed.), Thermophysical Properties of High Temperature Solid Materials, MacMillan Co., New York (1967).
17. H. Knoepfel, Proc. of the High Dynamic Pressure Symposium, Paris 1967. ("Megagauss Fields as a Transmitting Medium for Very High Pressures").
18. G.T. Meaden, Electrical Resistance of Metals, Plenum Press, New York (1966).
19. J.L. Olsen, Electron Transport in Metals, Interscience, New York (1962).
20. R.H. Parmenter, Phys. Rev. 116:1390 (1959).
21. A.N. Gerritsen, Metallic Conductivity: Experimental Part, Encycl. of Phys., Vol. 19, S. Flügge (ed.) Springer Verlag, Berlin (1956).
22. H. Jones, Theory of Electrical and Thermal Conductivity in Metals, Encycl. of Phys., Vol. 19, S. Flügge (ed.) Springer Verlag, Berlin (1956).
23. S.F. Vonsovskii, Sov. Phys. Uspekhi (Engl. transl.) 5:219 (1962).
24. R.N. Lyon (ed.) Liquid Metals Handbook, U.S. Govt. Printing Office, Washington, D.C. (1952).
25. R.G. McQueen and S.P. Marsh, J. Appl. Phys. 31:1253 (1960).
26. G.W. Anderson and E.W. Neilson, in Exploding Wires, Vol. 1, W.G. Chace and H.K. Moore (eds.) Plenum Press, New York (1959), p. 97.

A Thermal Model of Wire Explosions in Methane

Bernard Siegel and Richard L. Johnson

Laboratories Division
Aerospace Corporation
El Segundo, California

Explosions of various metals into methane were used to develop a thermal model for the interaction of exploding wires with this hydrocarbon. This model has wider implications that are useful in predicting the behavior toward exploding wires of other gases whose endothermic pyrolyses can be initiated by this technique. Calorimetry, product analysis, and thermodynamic calculations were used to develop the model. The paper describes the role of the exploding wire in transferring energy from the capacitors to the surrounding gas, the determination of the efficiency of the energy transfer process, and the indirect measurement of the temperatures at which the principal chemical reactions occurred, as well as the temperature distributions in the overall surrounding gas. It is shown that the process is remarkably efficient in abstracting energy from the wire explosion to effect high temperature pyrolyses, and that the major reaction forming acetylene, hydrogen, and solids occurs in a narrow zone of very high temperature, while most of the surrounding gas is relatively cold.

INTRODUCTION

Most of the published experimentation with exploding wires has been concerned primarily with the mechanism of the explosion itself rather than the effect of the wire explosion upon a surrounding gas. However, the concept of instantaneously vaporizing or explosively liquefying a conductor of very high boiling point—a difficult process by steady-state methods—in the presence of various gases offers considerable promise in the field of chemical synthesis, and for this reason attention should be directed toward the study of mechanisms of interactions between exploding wires and surrounding gases. Before one is able to make reasoned deductions about the feasibility of synthesizing various classes of inorganic or organic compounds by this technique, questions such as the following must be answered. What is the mode and efficiency of energy transfer between an exploding wire and a surrounding gas, and how are these factors affected by equilibrium thermodynamics and reaction kinetics? A related question is the magnitude and distribution of temperature levels that are imparted to surrounding gases by various levels of electrical energy imparted to the exploding wires.

Cook and Siegel [1] have recently described the effect of reaction energetics upon the yields of products obtained when various metals are electrically exploded into sulfur hexafluoride. Highly exothermic reactions gave essentially quantitative yields with minimum electrical input energies, but endothermic reactions produced very low yields in relation to the imparted energies. However, in the sulfur hexafluoride explosions, the metal participated chemically in that metal-containing products were formed. It is shown in the present paper that the limitation with regard to endothermic chemical processes in wire explosions is characteristic of reactions in which the metal participates chemically, but does not necessarily apply to pyrolytic reactions of surrounding gases, in which the exploding wire is an initiator rather than a chemical participant.

The principal aims of the present study were to ascertain the temperatures and temperature distributions to which methane can be heated by wire explosions at fairly low electrical input energies where one can monitor and determine hydrocarbon pyrolysis products, and to study the efficiency of energy transfer from the exploding wire to the surrounding gas. Methane was selected because reliable thermodynamic data for its decomposition are known at temperatures up to 6000°K, as well as kinetic pyrolysis data at temperatures up to 2500°K.

EXPERIMENTAL PROCEDURES

Exploding Wire Apparatus

The wire explosions were carried out in the stainless steel bomb shown in Fig. 1. The wall thickness is $\frac{1}{4}$ in., and it was designed to consist of two sections so that the wires to be exploded can be mounted conveniently. There are four O-ring seals to keep the bomb vacuum-tight. The largest of these O-rings is readily seen on the circumference of the upper section. The other three O-rings, which are not visible, seal between the upper plate surface and the swagelock fittings. The bomb can be evacuated to a pressure of 2×10^{-5} torr, but since an initial pressure of 10^{-3} torr was found

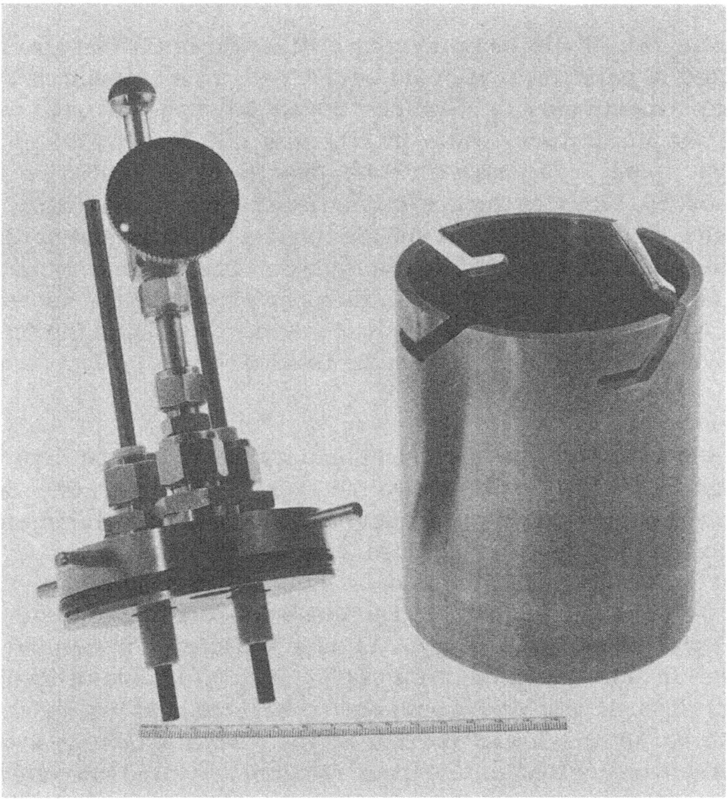

Fig. 1. Exploding wire bomb.

not to affect the results of our experiments, many of the runs were carried out after the bomb had been evacuated initially to 10^{-3} torr. Teflon insulators separate the $\frac{1}{4}$-in. copper pole electrodes from stainless steel parts. Wires to be exploded are fastened to the lower ends of the electrodes by set screws. The distance between electrodes is $1\frac{1}{2}$ in., and they are positioned to hold the wire $1\frac{3}{4}$ in. from the top of the cylinder if the wire is placed linearly between the electrodes, the only position used in our experiments. The internal diameter of the bomb is 4 in., and the internal volume is 833 cc. This volume was used in most experiments. In several runs, insertion of a Teflon liner reduced the volume to 485 cc. The entire bomb assembly was connected to a vacuum line by a steel ball joint so that the bomb could be filled with gases and so that volatile products could subsequently be transferred for analysis.

The electrical discharge system utilized three 14-μF capacitors, connected in parallel, that could be charged to a maximum of 20 kV. In many experiments, a Jennings vacuum switch was used to discharge the capacitors, while in others a GE Model 7703 ignitron tube was used. The experimental procedure consisted in filling the evacuated bomb to the desired initial pressure of methane, after which the wire was exploded and the volatile products and unreacted methane were transferred to glass bulbs for gas chromatographic analysis. Product analysis was then performed on the nonvolatile solids which could be scraped from the reactor wall, and the fraction of the wire that had been exploded was ascertained.

Calorimetry

The fraction of the electrical energy stored in the capacitors that was actually transferred to the exploding wire, rather than to the external circuitry, was measured with a water calorimeter, of which the bomb was the internal component. This is a very convenient and quite rapid method for determining the energy transferred to the exploding wire. The bomb was completely immersed in a Dewar flask containing water just sufficient for this purpose. The Dewar was enclosed at the top by several layers of insulating plastic, through which the bomb electrodes and inlet line protruded. The water in the lower portion of the Dewar was stirred with a magnetic stirrer; that in the upper portion, by a hand-operated loop stirrer. Temperature was measured on a mercury thermometer that could be read to 0.001°C.

The heat content of the calorimeter was measured by combusting benzoic acid in oxygen. The benzoic acid was contained in a small platinum boat mechanically suspended from one of the bomb electrodes. Firing a very thin platinum wire submerged in the benzoic acid produced ignition of the acid. Standard methods were used to obtain temperature–time curves. For our apparatus the calorimeter heat content was 3132 ± 65 cal/°C. We could then determine the energy transferred to exploding wires by carrying out such wire explosions in the calibrated calorimeter.

Initially, our experimentation was with a firing circuit containing the Jennings mechanical switch. The design was poor in that an excessive amount of energy was degraded in the external circuitry and only 10.6 ± 0.6% of the nominally stored energy was actually transferred to the exploding wire, when the nominal energy input was 4400 J. The firing circuit was then redesigned to reduce the inductance and resistance of connections between the capacitors, and the connections from the capacitors to the bomb. At this time, we also substituted the ignitron switch for the mechanical one. This was a very much more efficient system. At a nominal stored energy level of 2100 J, the actual delivered energy to the exploding wire was 39.2 ± 2.1% of the nominal energy, while the comparable value at a nominal energy of 4725 J was 46.3 ± 2.8%. Identical values were obtained when lead and platinum wires were exploded. The calorimetric method described here yields the desired energy transfer fraction to within 5.7%, on the average, which compares favorably with the 10% uncertainty of a vacuum aluminum calorimeter previously used for this purpose [2].

EXPERIMENTAL RESULTS

Reaction conditions and product data for the formation of volatile pyrolysis products by wire explosions into methane are shown in Table I. Despite the varying reaction parameters and the use of different metals, the reactions were all similar in that the methane decomposed predominantly according to Eq. (1).

$$2CH_4 \longrightarrow C_2H_2 + 3H_2 \tag{1}$$

In addition, minor amounts of other volatile hydrocarbon products were formed. These included propene, vinylacetylene, diacetylene, ethylene, propyne, and benzene. It can be readily seen that the magnitude of the mass of the exploding wire is a trivial parameter. Despite a 31-fold variation in mass of exploded platinum at an electrical input energy of 466 J, at identical pressures and reactor volume, we obtained identical values of 0.45 for the moles of acetylene formed per mole of methane decomposed; this corresponds to a 90% conversion to acetylene by Eq. (1). Also, the extent of pyrolysis was essentially equal in these runs, since the difference between 7.8 and 8.3% is within experimental error. The lead and aluminum explosions at 466 J produced slightly lower extents of pyrolysis and proportions of acetylene among the volatile products, but the data for these metals are essentially similar to that for the platinum explosions.

In addition to the volatile products of these pyrolyses, nonvolatile products were always formed as powdered wall deposits, admixed with unreacted metal. These nonvolatile products consisted of hydrocarbon polymers, possibly mixed with elemental carbon. In those runs for which the extent of pyrolysis to nonvolatiles was de-

Table I. Pyrolysis of Methane to Volatile Products*

Metal	Electrical energy, J†	Mass of exploded wire, mg	Reactor volume, cc	Extent of pyrolysis, %	Moles C_2H_2 formed per mole of CH_4 decomp.
Pt	466	317	485	9.6	0.45
Pt	148	341	485	3.8	0.49
Pt	466	317	833	7.8(2.5)§	0.45
Pt	148	300	833	2.4(1.5)§	0.45
Pt	466	10.2	833	8.3	0.45
Pb	466	218	833	7.0	0.40
Al	466	50.6	833	6.7	0.44
Al	2188‡	57.5	833	16.4(18.6)§	0.40
Al	823‡	55.2	833	5.5(6.4)§	0.425

*Initial methane pressures were 400 to 450 mm in each of these reactions.
†Actual energy delivered to the exploding wire, as measured calorimetrically.
‡Ignitron firing system used in these runs. In other runs the mechanical firing switch was used.
§The values in parentheses are the extents of decomposition to nonvolatile carbonaceous solids.

termined, the data are given in parentheses in Table I. It might be noted that, whereas we could determine the extent of pyrolysis to volatile products by summing all the data on accumulation of volatile species, most runs produced relatively small extents of overall conversion and the extent of conversion to nonvolatiles could not be determined accurately from the gas chromatography data. In such cases the extent of conversion to nonvolatiles was obtained by determination of the carbon contents of the scraped products. The platinum explosions for which nonvolatile data are given resulted in the formation of solids with a hydrogen-to-carbon ratio of two and presumably were $(CH_2)_x$. The aluminum explosions in which the ignitron circuit was used produced polymeric products of considerably lower hydrogen contents, with H/C ratios of 0.26 to 0.29.

A diligent effort was made to determine whether metal-containing products were formed in these reactions. However, organometallic compounds, such as alkyls of platinum, aluminum or lead, were definitely not formed. Other possible metal-containing compounds were PbH_4 or AlH_3, or the carbide Al_4C_3, but these also did not form in the explosions into methane.

DISCUSSION AND CONCLUSIONS

Role of the Wire

The inability of the metals to form chemical compounds in our experiments proves that the exploding wires were merely initiators of pyrolyses. Thus, their role is merely to transfer energy from the storage capacitors to the surrounding molecules. This accounts for the negligible effect of the mass of the exploded material and for the similar results with the three metals. The data also demonstrate that specific catalysis by an individual metal is not involved to a significant extent, and that the volatility of the metal is not a determining factor within the reaction parameters used in these experiments. Note that there is a wide range of boiling points for the metals exploded in this paper; lead, aluminum, and platinum exhibit normal boiling points of 2024, 2720, and 4100°K, respectively [3].

An important consequence of the fact that the exploding wire does not participate chemically in wire explosions into methane at these energy levels is that pyrolytic reactions can occur in principle after the original exploding wire has lost its energy. The determining factor in the reaction time is the method by which the hot methane loses temperature independently of the metal.

Conversion of Electrical to Chemical Energy

The reaction of Eq. (1) requires at 25°C the expenditure of 188,328 J of energy per mole of methane decomposed [4]. Since the proportions of volatile products found in our experiments indicate that the reaction to volatiles proceeds predominantly by this path, we can simplify the efficiency calculation by assuming that acetylene and hydrogen are the sole volatile products. Conversion of methane to $(CH_2)_x$ and hydrogen requires the expenditure of 54,250 J per mole of methane decomposed [5]. Therefore, the observed extents of reaction in the platinum explosions at 148 and 466 J have required thermodynamic input energies of 104.1 and 320.9 J. Clearly, in these experiments 70.3 and 68.9% of the imparted electrical energy had been abstracted by the system to carry out endothermic pyrolytic reactions. This is an extraordinary efficiency for a high-temperature electrical process. Chemical syntheses by electrical discharge processes typically conserve only a few percent of the imparted electrical energy and this is also true for syntheses in high-temperature plasmas; the plasma torch method for converting methane to acetylene utilizes only several percent of the imparted energy [6].

This extraordinary efficiency in utilization of electrical energy, and the almost exclusive formation of acetylene among the volatile hydrocarbon products of wire explosions in methane suggests that the exploding wire technique could have important practical applications in the synthesis of acetylene, if a way could be found to scale-up the process.

Lower efficiencies were found for those pyrolyses that were initiated by wire explosions triggered by an ignitron switch. For the experiments at an input energy level of 2188 J, we compute that 41.6% of the imparted electrical energy was abstracted by the system and used for chemical pyrolysis, while the efficiency was 36.8% at 823 J. Clearly, the more rapid explosion that occurs with the ignitron, as compared to a vacuum mechanical switch, is more wasteful of imparted electrical energy. This could result from poorer heat exchange due to the more rapid explosion, or to higher levels of radiation, since methane is transparent at wavelengths higher than the far vacuum ultraviolet.

Gas Temperatures

When one considers the physical details of a wire explosion into a surrounding gas of appreciable volume, it seems exceedingly unlikely that the gas could at any time be heated to a uniform temperature. This is substantiated by our data. If uniform heating occurred, we compute the transitory maximum temperature, for methane initially at 400 torr in an 833-cc reactor, as 503°K at an energy input level of 148 J and 823°K at 466 J. These temperatures were computed from high-temperature heat capacity data in [4], with the assumption that all the energy stored in the exploding wire had been transferred to the surrounding gas molecules. The resulting temperatures are very much below those at which we could expect appreciable rates of methane decomposition [7, 8]. The appreciable extents of methane pyrolysis found in our experiments prove that much higher temperatures were required, and that the methane must have been heated in a very nonuniform manner.

We can now turn to the question of the specific temperatures to which portions of the methane had been heated by the wire explosions. It has been reported that ethane and ethylene are major products of the methane pyrolysis in the millisecond time range at temperatures up to 2500°K [7, 8]; the reaction path to acetylene is given by Eq. (2).

$$2CH_4 \longrightarrow C_2H_6 + H_2 \longrightarrow C_2H_4 + 2H_2 \longrightarrow C_2H_2 + 3H_2 \quad (2)$$

However it can be seen from the data in Fig. 2, which were computed from thermochemical data in [4], that at very high temperatures acetylene is the predominant hydrocarbon product at equilibrium, predominating over the next important hydrocarbon, ethylene, by a factor of nearly 3000 at 2500°K. This predominance of acetylene increases further with increasing temperature. Comparable data have been computed at a pressure at 10 atm in Fig. 3. These data are given to demonstrate that higher pressures do not qualitatively change these conclusions, since pressure transients undoubtedly occurred in the methane heated by the wire explosion. Thus, the observed ethane and ethylene, in very high proportions, at temperatures below 2500°K in the previously mentioned kinetic studies, were nonequilibrium products. Since acetylene is the almost exclusive volatile hydrocarbon product in the short reaction times accompanying a wire explosion, it must be concluded that equilibrium had essentially been achieved, and that this required a temperature considerably greater than 2500°K in the volume zone where these reactions occurred.

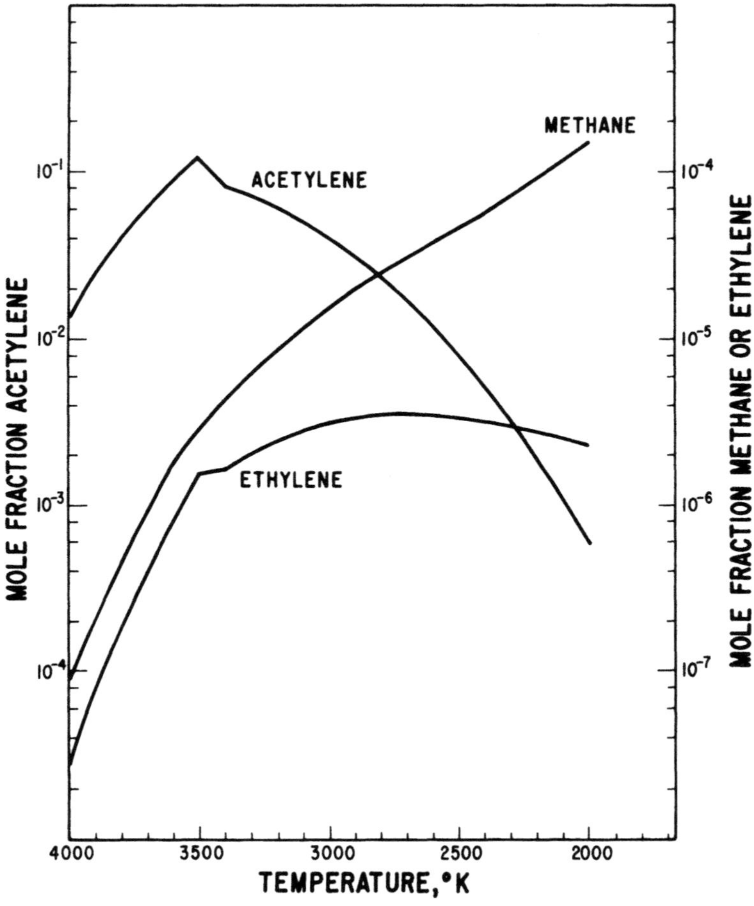

Fig. 2. Methane equilibrium at 0.565 atm.

The actual temperatures in the latter zone at the time interval when the equilibrium products must have been formed can be deduced from data in Fig. 4, which presents the temperature dependence of the equilibrium carbon to acetylene ratios among the pyrolysis products. At temperatures equal to and below 3400°K, carbon is mainly in the form of solid carbon, with minor amounts of gaseous C_n species that fall rapidly with decreasing temperature. The data in Fig. 4 have been corrected for this factor by incorporating these C_n species in the carbon/acetylene ratios, on the assumption that gaseous C_n species would mainly revert to solid carbon upon fast temperature quenching. Because solid carbon is completely volatilized at 3500°K and the pressures under consideration, the carbon at 3500°K is completely in the form of C_n molecules. For this reason, the interval between 3400 and 3500°K represents an extrapolation that we felt could not be carried to higher temperatures. The data in Fig. 4 indicate that pressure variations exert little influence over most of the temperature range under consideration.

Temperatures computed from Fig. 4 and our experimental product data are given in Table II. For the purpose of these calcula-

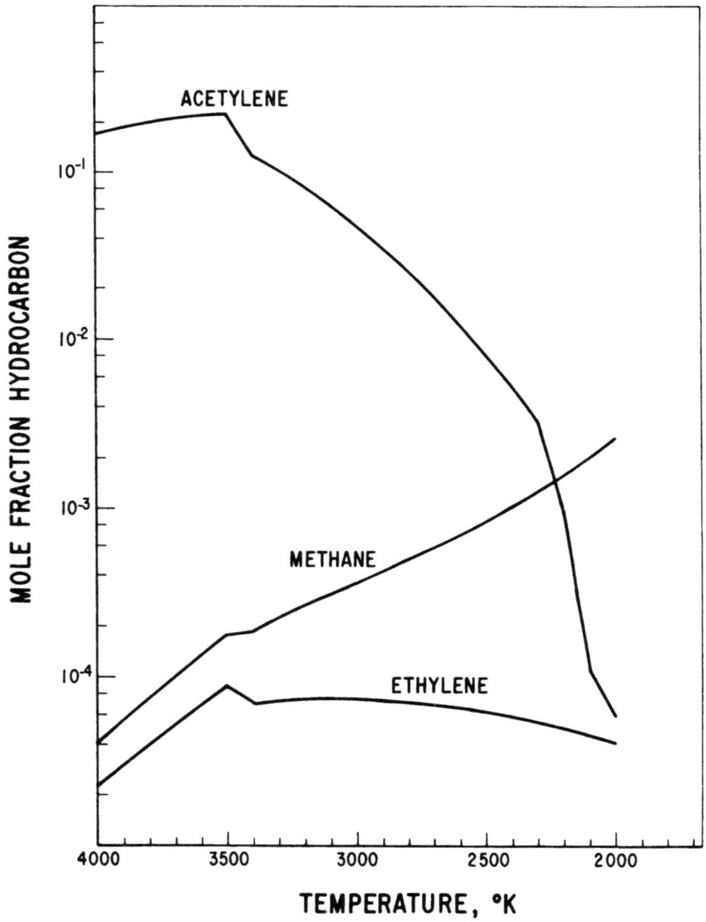

Fig. 3. Methane equilibrium at 10 atm.

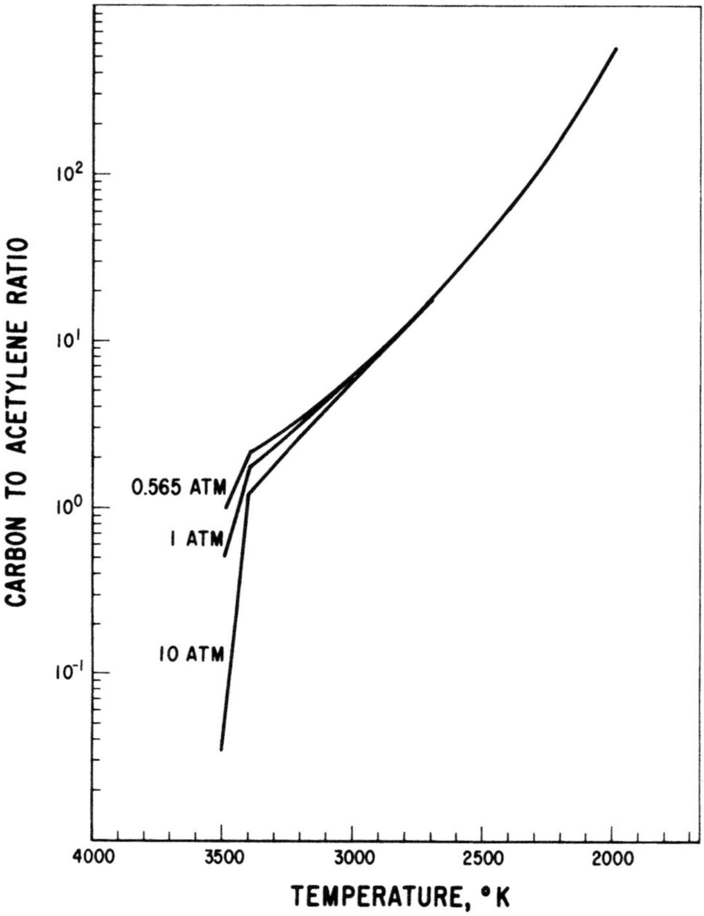

Fig. 4. Temperature dependence of carbon-to-acetylene ratio at equilibrium.

tions we have assumed that the carbon contents of the nonvolatile solids can be equated with elemental carbon. This is based on the assumption that, once solid hydrocarbon polymer particles are formed, they would continue to elemental carbon if longer time periods were provided, and the fact that, at comparable temperatures, gases pyrolyze faster than solid particles. The temperatures given in column 3 of Table II are presented as ranges rather than discrete temperatures because of the pressure effect. We have selected a maximum pressure of 10 atm for these computations.

This results in a temperature range that is probably longer than is necessary, because it is unlikely that significant portions of methane molecules are subjected to a maximum pressure as high as the selected value. From Table II it can be seen that the temperatures of the zones in which acetylene had been formed are somewhat higher when the Jennings switch was used than in the experiments using an ignitron. This is consistent with the discussion of the previous section concerning wastage of energy. In any event, the computed temperatures are near those temperatures seen in Figs. 2 and 3, at which the thermodynamic probability of acetylene formation is at a maximum.

Having determined the temperatures at which the predominant products had formed by pyrolysis of the methane molecules surrounding the exploding wire, we can ascertain the fraction of molecules heated to such temperatures. It is easily shown that this fraction is very nearly the fraction of molecules that is actually converted to acetylene and solids. Consider, for example, line 2 of Table II. At the conditions of this experiment, 9.5% of the methane molecules are converted to the predominant products, acetylene and solids. If precisely this fraction of the original methane had been heated by completely abstracting the energy contained in the exploding wire, the resulting temperature of this fraction would be 3320°K. Since 3320°K is somewhat below the temperature computed from Fig. 4, it follows that a significantly larger fraction of molecules could not have been heated by the wire explosion prior to the occurrence of the observed reactions leading to acetylene and solids. The latter reaction occurs, therefore, in a limited volume of gas approximating the fraction of methane molecules that were actually observed to form acetylene and solids. This is another manifestation of the remarkable efficiency of the methane pyrolysis during wire explosions into methane, especially when a Jennings mechanical

Table II. Temperatures from Product Data

Electrical energy, J*	Carbon-to-acetylene ratio in products	Temperature from Fig. 4, °K
148 (Jennings Switch)	1.39	3365-3450
466 (Jennings Switch)	0.71	3417-3470
823 (Ignitron)	2.74	3200-3285
2188 (Ignitron)	2.84	3185-3270

*Corresponds to runs in Table I for which product data are given.

switch is used. Virtually complete reaction is seen to occur in a narrow zone that is heated to the appropriate temperature where equilibrium is very rapidly attained, while the major portion of the methane molecules is relatively cold.

Lower Temperature Reactions

Finally, we can consider the reasons for the wasted energy, the latter representing the fraction of the imparted electrical energy that is lost to the reactor wall without effecting chemical reaction. At the high temperatures that exist in the portion of molecules that react to form acetylene and solids, equilibrium is rapidly attained, at least for the gaseous reactions. However, since Eq. (1) is highly endothermic, the progress of the reaction drastically lowers the temperature. Since methane is relatively inert at lower temperatures, little use can be made of the remaining energy at the lower temperatures and we obtained relatively small amounts of other products. This residual energy is thus mainly dissipated to the wall. If one were to substitute more reactive hydrocarbons for methane—hydrocarbons which have higher rate constants for pyrolysis and thus pyrolyze more readily at lower temperatures—one would expect to utilize a higher fraction of the imparted electrical energy. This has been substantiated by the present authors in a paper whose results will be reported elsewhere [9]. It might be noted that in the latter study virtually complete utilization of the imparted energy was made for wire explosions in ethane, using the Jennings switch and lower input energies. Under those conditions there is evidently only a small loss of energy due to radiation. If the latter is an important factor, it could only occur in the faster explosions accomplished by the use of an ignitron switch, and at higher energy input levels.

ACKNOWLEDGMENT

The authors gratefully acknowledge that Miss Cassandra Johnson obtained the gas chromatographic data, and that Mrs. Eileen Cook cooperated in obtaining the equilibrium computations of Figs. 2 and 3 by a computer program.

REFERENCES

1. E. Cook and B. Siegel, J. Inorganic Nucl. Chem., 29:2739 (1967).
2. M. Rabinowitz, Vacuum 12:217 (1962).
3. D.R. Stull and G.C. Sinke, Thermodynamic Properties of the Elements, American Chem. Soc. Advances in Chemistry Series, Washington, D.C. (1956).

4. JANAF Thermochemical Tables, Dow Chemical Co. (1964).
5. Selected Values of Properties of Hydrocarbons, National Bureau of Standards Circular 461, Nov. 1947.
6. H.W. Leutner and C.S. Stokes, Ind. Eng. Chem. 53:341 (1962).
7. G.B. Skinner and R.A. Ruehrwein, J. Phys. Chem. 63:1736 (1959).
8. H.S. Glick, Seventh Combustion Symposium, Butterworth Scientific Publications, Ltd., London (1959), pp. 98-107.
9. R.L. Johnson and B. Siegel, J. Electrochem. Soc., 115:24 (1968).

DISCUSSION

Question by A. Guenther, Air Force Weapons Laboratory, Kirtland Air Force Base, N.M.: I wonder if you have performed similar experiments on arc or spark discharges without wires since you indicated they played little part in the resultant products. If, in fact, shock hydrodynamic considerations are important, maybe a recurrent spark discharge would place this technique on a commercially sound basis.

Answer by Author: As indicated by the gentleman in the audience, the essential difference between an exploding wire technique and a continuous arc is the pulsed nature of the exploding wire process. It is the speed of the pulsed process which conserves energy and precludes significant wastage of energy conducted to the wall, with reactant gases whose reaction kinetics permit such utilization. In a continuous arc process large-scale wastage by conduction cannot be prevented, since here the process cannot be adiabatic. As we have shown, hydrocarbon pyrolysis initiated by wire explosions can be essentially adiabatic.

Comment by D.M. Wetstone, United Aircraft Research Headquarters: Arc discharges are not inefficient, as such, but only because continuous operation leads to considerable thermal losses. In contrast, exploding wire pulses are inherently adiabatic, and all free energy is used in the ambient gas.

Question by R.L. Musselman, University of California, Davis, Cal.: What happens to your hydrogen and acetylene over the platinum aerosol?

Answer by Author: There is no indication that appreciable and permanent adsorption occurs to a level that would seriously affect the product data.

Question by C.P. Nash, University of California, Davis, Cal.: The efficiencies of coupling Al and Pt to the atmosphere are very different. Do you intend to study mixtures of methane and inert carrier gases with different sound speed to insure that you have not overdriven the system in the experiments which give $\sim$100% conversion to acetylene?

Answer by Author: The role of shock waves in our experiments depends to a great extent upon the energy input. At the lowest energy input levels, it is less clear that heating is affected by a shock wave mechanism than at the higher energy levels where we have experimental proof by probe experiments not reported in the present paper that reaction follows in the wake of a reflected shockwave. For the purpose of developing a thermal model, it was deemed unnecessary to use inert gas diluents. However, we can state that inert gases can affect the mechanisms of even very high temperature regimes in our experiments. For example, we have found that titanium explosions into acetylene result in large-scale graphite formation, despite wide variations in electrical input energy, if the acetylene is undiluted. Dilution with hydrogen or argon, however, minimizes graphite formation and maximizes carbide formation.

Hydrocarbon Formation under Exploding Wire Conditions*

Charles W. Spangler, Michael J. Joncich, and Stanley K. Lott

Department of Chemistry
Michael Faraday Laboratories
Northern Illinois University
DeKalb, Illinois

Electrically exploded carbon filaments react with a surrounding hydrogen atmosphere to produce hydrocarbon mixtures in which methane, ethylene, and acetylene predominate. Explosions carried out in a mixture of hydrogen and deuterium result in essentially statistical scrambling within the residual hydrogen–deuterium mixture. All possible hydro–deutero methanes, acetylenes, and ethylenes were formed and their relative quantities determined by mass spectrometry. The data are consistent with stepwise formation of methyne, methylene, and methyl radicals. Methane formation can best be explained by four successive combinations of monatomic carbon with dissociated hydrogen. Reaction of exploded graphite filaments with a methane atmosphere is much more complicated and the product distribution must take into account the dissociation of methane in the shock or photolysis phenomena that take place during the explosion process.

INTRODUCTION

The generation and resultant chemistry of elemental carbon particles have been studied extensively during the past few years. These particles have been produced by many different techniques, ranging from evaporation from carbon electrodes in direct current arcs to the production of hot carbon atoms by nuclear bombardment. Wolfgang, McKay, and coworkers [1–3] have studied the reactions of recoil ^{11}C, produced by nuclear reaction, with various saturated and unsaturated organic target molecules. They found that monatomic carbon, assumed to be in the triplet state, reacts

*This work was supported by the Petroleum Research Fund of the American Chemical Society.

by insertion into a C—H bond, or by addition to an unsaturated linkage, followed by rearrangement to more stable species. For example, the primary products obtained from the reaction of monatomic carbon with methane are ethylene and acetylene [4].

Skell and coworkers [5, 6] have studied the reactions of both singlet and triplet carbon produced by evaporation from a carbon arc in a high-vacuum system. Their primary concern was the stereospecificity, or lack thereof, of the addition of monatomic carbon to cis- and trans-2-butenes. They concluded that the higher energy C_1 species adds stereospecifically and the lower energy species nonstereospecifically. In all probability, the temperature achieved in the generation of these carbon particles did not exceed 6000°K.

Sprung and coworkers [7] have bombarded frozen benzene with carbon vapor particles produced at temperatures ranging from 1893 to 2390°K. The vapor stream probably contained substantial quantities of C_2 and C_3 particles as well as C_1. Both toluene and cycloheptatriene were observed products, most likely arising from carbene-type insertion reactions. However, these workers were unable to provide a mechanism for the product formation. More recently this reaction has been reinvestigated by Rose et al. [8] utilizing recoil ^{11}C atoms. Vinylacetylene and diacetylene appeared as major products, as well as the products found by Sprung.

In the above reported reactions of monatomic carbon, the highest experimental temperature is 6000°K for the DC/arc. The exploding wire phenomena, previously limited to metallic elemental wires, can achieve temperatures much higher than this [9]. This paper is concerned with an attempt to explain the possible mechanisms of hydrocarbon formation under the extreme experimental conditions resulting from explosions of graphite filaments in hydrogen atmospheres.

EXPERIMENTAL

All explosions of graphite filaments (6 in. in length, 0.020 in. in diameter), purchased from Ultra Carbon Company, were performed with an exploding wire apparatus modified for the study of chemical reactions. The apparatus consists of ten 31.25 μF oil-filled capacitors connected in parallel using a copper/polyethylene/Mylar/polyethylene/copper sandwich. This bank was charged to 8 kV, and the stored energy discharged through the filament by firing a Lovotron switch with a modified Theophanis hydrogen thyratron trigger. Current–time traces were obtained with a commercially available current-viewing resistor and a Tektronix 517A oscilloscope in order to determine if an explosion had occurred. The glass reaction vessel (2 in. ID) was equipped with gas stopcocks to facilitate filling and removal of gaseous atmospheres. Initial hydrogen gas pressure was ~1.5 atm. All products were identified qualitatively and quantitatively by Gollob Analytical Service, Inc., with a CEC-21-620A mass spectrometer. Ethylene and acetylene were collected separately and analyzed individually. The isotopic methane species were calculated from the original mass spectrometer analyses.

RESULTS AND DISCUSSION

We have previously reported that reaction of electrically exploded graphite filaments with hydrogen under exploding wire conditions produces a complex mixture of hydrocarbons, with methane, ethylene, and acetylene predominating [10]. Ethane, propane, and propylene were present as minor products. Several possible reaction processes can be envisaged for exploding graphite in a hydrogen atmosphere. Our product analyses indicate that better than 95% of the hydrocarbon yield is methane and C_2 products; therefore, it is unlikely that carbon aggregates larger than C_2 survive the initial energy surge. The exploding wire phenomena, although by no means completely understood, does display certain common features in all explosions. An intense flash of light, followed by a large shock wave, usually precedes particle generation from the explosion plasma. In fact, exploding wires have been utilized as light sources for flash photolysis studies [11]. It is quite conceivable that dissociation of the gaseous atmosphere surrounding the filament by shock, photolysis, or high temperature can occur prior to chemical reaction. This paper presents evidence that this is indeed the case, even if no chemical reaction with the exploded filament occurs.

We have considered that hydrocarbons may be formed by any or all of three pathways: by direct molecular union (Scheme A), by successive hydrogen abstractions leading to free radical intermediates (Scheme B), or by direct union of carbon atoms (or polyatomic molecules) with atomic hydrogen, which also produces free radical intermediates (Scheme C).

Scheme A

(1) $C + H_2 \longrightarrow [CH_2]$

(2) $2[CH_2] \longrightarrow CH_2 = CH_2$

(3) $[CH_2] + H_2 \longrightarrow CH_4$

(4) $C + CH_2 = CH_2 \longrightarrow HC \equiv CH + [CH_2]$

Scheme B

(1) $C + H_2 \longrightarrow [CH] + H$

(2) $2[CH] \longrightarrow HC \equiv CH$

(3) $[CH] + H_2 \longrightarrow [CH_2] + H$

(4) $[CH_2] + H_2 \longrightarrow [CH_3] + H$

(5) $[CH_3] + H_2 \longrightarrow CH_4 + H$

Scheme C

(1) $C + H \longrightarrow [CH]$

(2) $2[CH] \longrightarrow HC \equiv CH$

(3) $[CH] + H \longrightarrow [CH_2]$

(4) $[CH_2] + H \longrightarrow [CH_3]$

(5) $[CH_3] + H \longrightarrow CH_4$

In Schemes B and C, one could also expect both methylene and methyl radicals to undergo coupling reactions to yield ethylene and ethane, respectively.

In an attempt to distinguish between abstraction and molecular combination mechanisms, carbon filaments were exploded in a mixture of hydrogen and deuterium, and the deuterium distribution within the products was determined by mass spectrometry. A re-

Table I

Pre-explosion composition, % of total		Theoretical scrambling	Residual composition	
			Explosion I	Explosion II
H_2	51.2	26.2	20.4	26.2
D_2	48.8	23.8	26.5	24.6
HD	trace	50.0	51.9	48.3

vealing observation was an isotope scrambling within the residual hydrogen–deuterium mixture (Table I). It might be interesting to note here that a greater yield of total hydrocarbons was obtained in Explosion I, thereby consuming more hydrogen relative to deuterium. This "isotope effect" will be discussed below. The isotope scrambling indicates that during the process initiated by the explosion, a large percentage of hydrogen and deuterium molecules must be severed. In fact, the residual compositions represent an essentially random distribution.

In order to determine if an explosion of a chemically inert wire could cause the same type of hydrogen–deuterium scrambling, a platinum filament of similar dimensions was exploded in the H_2–D_2 mixture under experimental conditions identical to those employed for carbon. These results (Table II) indicate that dissociation and scrambling of the surrounding atmosphere predominates even if no over-all chemical reaction occurs. This evidence would tend to support, at least initially, Scheme C, wherein the hydrogen atmosphere dissociates photochemically, thermally, or in the explosion shock wave prior to chemical reaction.

The isotope distribution found in acetylene led us to postulate a free radical reaction process, with acetylene arising from methyne coupling [12]. These results are shown in Table III. The predicted values are based upon initial methyne formation from monatomic carbon, followed by methyne coupling (and neglecting any possible kinetic isotope effect).

Methane formation is necessarily more complex than acetylene formation and can be explained by a variety of mechanistic path-

Table II

Pre-explosion composition, % of total		Theoretical scrambling	Residual composition
H_2	51.2	26.2	26.8
D_2	48.8	23.8	24.2
HD	trace	50.0	49.0

Table III

Product	Explosion I, % of total	Explosion II	Predicted
C_2H_2	28	29	26.2
C_2HD	47	46	50.0
C_2D_2	25	25	23.8

ways. The formation of methane from either methyl or methylene radicals is well known, and similarly, this formation can be envisaged as four successive hydrogen abstractions or combinations by monatomic carbon. We have considered several possible mechanisms for methane formation, some of the more possible of which are reproduced in Scheme D.

Scheme D

(1) $C \xrightarrow{H_2} [CH_2] \xrightarrow{H_2} CH_4$

(2) $C \xrightarrow{H_2} [CH_2] \xrightarrow{H_2 \text{ or } H} [CH_3] \xrightarrow{H_2 \text{ or } H} CH_4$

(3) $C \xrightarrow{H_2 \text{ or } H} [CH] \xrightarrow{H_2 \text{ or } H} [CH_2] \xrightarrow{H_2} CH_4$

(4) $C \xrightarrow{H_2 \text{ or } H} [CH] \xrightarrow{H_2} [CH_3] \xrightarrow{H_2 \text{ or } H} CH_4$

(5) $C \xrightarrow{H_2 \text{ or } H} [CH] \xrightarrow{H_2 \text{ or } H} [CH_2] \xrightarrow{H_2 \text{ or } H} [CH_3] \xrightarrow{H_2 \text{ or } H} CH_4$

Analysis of the methane product showed that all five possible methanes were formed. These results, as well as those predicted for each of the mechanisms of Scheme D are shown in Table IV. The predicted values were calculated neglecting any possible kinetic isotope effects.

On the basis of these results, it would seem that the only reasonable mechanistic pathway for methane formation is a four-step process resulting from four consecutive hydrogen abstractions or combinations. This mechanism has the added advantage of requiring methyne as an intermediate, which agrees with our previously postulated mechanism for acetylene formation [12].

The mode of ethylene formation is much more difficult to rationalize on the basis of a simple mechanistic model than are those of methane and acetylene. All possible hydro–deutero

Table IV

Products	Expl. I.	Expl. II.	Predicted				
			(1)	(2)	(3)	(4)	(5)
CH_4	9.3	10.1	26.2	13.4	13.4	13.4	6.9
CH_3D	24.8	25.4	0	25.6	25.6	25.6	26.2
CH_2D_2	37.8	34.7	50.0	25.0	25.0	25.0	37.5
CHD_3	22.4	23.7	0	24.4	24.4	24.4	23.8
CD_4	5.7	6.1	23.8	11.6	11.6	11.6	5.6

ethylenes are obtained. We have considered three possible mechanisms for ethylene formation consistent with our previous assignments. These are represented in Scheme E.

Scheme E

(1) $C \xrightarrow{H \text{ or } H_2} [CH] \xrightarrow{H \text{ or } H_2} [CH_2]$

$2[CH_2] \longrightarrow C_2H_4$

(2) $C_2 \xrightarrow{H \text{ or } H_2} [C_2H] \xrightarrow{H \text{ or } H_2} [C_2H_2]$

$[C_2H_2] \xrightarrow{H \text{ or } H_2} [C_2H_3] \xrightarrow{H \text{ or } H_2} C_2H_4$

(3) $C \xrightarrow{H_2} [CH_2]$

$2[CH_2] \longrightarrow C_2H_4$

Unfortunately, the H–D distributions calculated for the first two mechanisms are identical. These predicted values compared to those found are shown in Table V.

A very complex situation exists immediately after the filament explodes: chemical reaction can occur over a wide range of both

Table V

Products	Expl. I,	Expl. II,	Predicted	
	% of total		(1), (2)	(3)
C_2H_4	21	18	6.9	26.2
C_2H_3D	24	24	26.2	0
$C_2H_2D_2$	30	34	37.5	50.0
C_2HD_3	20	18	23.8	0
C_2D_4	5	6	5.6	23.8

temperature and composition. Our experiments indicate that a large portion, if not all, of the surrounding hydrogen atmosphere dissociates during the explosion process. One extreme mechanistic model would postulate that the exploded carbon filament mixes with the dissociated hydrogen atmosphere prior to any permanent bond formation. This model corresponds to the various atomic and molecular reactions displayed in Schemes A–E. An alternate over-all mechanism would postulate H-atom recombination prior to any chemical reaction with C or C_2. Both of these mechanisms would account for the isotopic scrambling found in the residual unreacted hydrogen. Although it is unlikely that either predominates, we feel the former is closer to what actually occurs experimentally. However, it is almost a certainty that reaction of carbon fragments (C and C_2) with H and D atoms is in competition with H and D recombination and carbon association.

An insight into the relative rates of hydrocarbon formation can be gained by examination of the apparent isotope effects involved in methane, acetylene, and ethylene formation. We reported a slight preference for C_2H_2 formation previously [12], and these results are briefly represented in Table III. A much more discernible isotopic preference for CH_4 formation was presented in Table IV. Ethylene formation shows an even more marked preference for C_2H_4 formation (Table V). We interpret these results to indicate the preferred reactions in the system under rapidly changing temperature conditions. Thus, in the early stages of chemical reaction, methane and acetylene are the dominant products. Ethylene formation occurs later and at a lower temperature, and is consistent with the larger observed isotope effect.

Table VI

Products	Pt filament explosion, % of total	C filament explosion, % of total
Methane	59	65
Hydrogen	35	28
Acetylene	2.5	5.3
Ethylene	2.5	0.90
Ethane	0.35	–
C_3–C_4 hydrocarbons	0.35	0.41
Benzene	0.40	0.12
Toluene	0.031	–
Xylene	0.001	–

The value of our results in investigating the mechanism of hydrocarbon formation during the explosion process hinges on the question of the relative rates of atomic hydrogen–deuterium recombination as opposed to C–H or C–D bond formation. We find it difficult to imagine that all of the dissociated hydrogen recombines prior to chemical reaction with the carbon particles generated by the explosion process. Therefore, we postulate that methane and acetylene are formed by reaction of monatomic carbon with atomic hydrogen as described above. Ethylene formation can occur either by methylene coupling or by stepwise reaction of C_2 with atomic hydrogen. However, this mechanistic scheme must be considered to be highly speculative until more information is available concerning the relative rates of hydrogen recombination and hydrocarbon formation.

The complexities of the above reactions of exploded carbon filaments with hydrogen led us to expect similar dissociation to occur in more complex atmospheres. Therefore, carbon and platinum filaments were exploded under identical conditions in a methane atmosphere. Since the chemical reactions possible in a methane vis à vis a hydrogen atmosphere are potentially more complex, we felt that a Pt–C product similarity would enhance the model of atmospheric dissociation. The results of this comparison are shown in Table VI.

The similarity between platinum and carbon filament explosions is striking, both in qualitative and quantitative make-up of the product mixture. Braun, McNesby, and coworkers [13, 14] have recently studied the vacuum ultraviolet flash photolysis of methane and have observed that ethylene is the major hydrocarbon product, although acetylene and ethane are also formed. By utilizing an isotope technique similar to ours (CH_4–CD_4 mixtures) these workers have demonstrated that CH plays a dominant role in the chemistry. It is possible that the platinum filament explosion in methane is quite similar to the flash photolysis process. The carbon filament explosion is further complicated however, due to chemical reaction of the filament with the methane atmosphere. The mechanisms leading to the observed products in both of these cases undoubtedly are related, but we cannot postulate pathways for these processes without additional information. We envisage further experiments involving CH_4–CD_4 mixtures and also carbon filaments tagged with ^{14}C to aid in the elucidation of the above mechanistic pathways.

SUMMARY

Hydrocarbon formation during filament explosion processes is extremely complex and occurs over widely varying compositions and temperatures. We interpret our results to indicate that any mechanism proposed for chemical reactions as a direct result of the explosion process must include initial atmospheric dissociation as a primary step, probably before any chemical reaction occurs. We also feel that the most probable mechanism for the formation of methane, acetylene, and ethylene involves stepwise reaction of monatomic carbon (or polyatomic molecules such as C_2) with dissociated hydrogen, followed by methyne and methylene coupling reactions. Acetylene and ethylene can also arise from the reaction of C_2 with dissociated hydrogen in a similar manner.

ACKNOWLEDGMENTS

The authors wish to express their thanks for the assistance given them by H. Giddings. We wish also to thank the donors of the Petroleum Research Fund, administered by the American Chemical Society, for support of this research (Grant No. 1814 – A1).

REFERENCES

1. C. MacKay and R. Wolfgang, J. Am. Chem. Soc. 83:2399 (1961).
2. C. MacKay, P. Polak, H. Rosenberg, and R. Wolfgang, J. Am. Chem. Soc. 84:3081 (1962).
3. M. Marshall, C. MacKay, and R. Wolfgang, J. Am. Chem. Soc. 86:4741 (1964).
4. For an excellent review see C. MacKay and R. Wolfgang, Science, 148:899 (1965).
5. P. Skell and R. Engel, J. Am. Chem. Soc. 87:1135, 1136, 2493 (1965).
6. P. Skell, L. Wescott, J. Golstein, and W. Libby, J. Am. Chem. Soc. 87:1812 (1965).
7. J. Sprung, S. Winstein, and W. Libby, J. Am. Chem. Soc. 87:1812 (1965).
8. T. Rose, C. MacKay, and R. Wolfgang, J. Am. Chem. Soc. 89:1529 (1967).
9. W.G. Chace and H.K. Moore (eds.), Exploding Wires, Vols. 1–3, Plenum Press, New York (1959, 1962, 1964).
10. C. Spangler, S. Lott, and M. Joncich, Electrochem. Tech. 5:214 (1967).
11. A. Shilman and R. Marcus, J. Chem. Phys. 39:996 (1963).
12. C. Spangler, S. Lott, and M. Joncich, Chem. Comms. 842 (1966).
13. W. Braun, K. Welge, and J. McNesby, J. Chem. Phys. 45:2650 (1966).
14. W. Braun, J. McNesby, and A. Bass, J. Chem. Phys. 46:2071 (1967).

DISCUSSION

Question by D. M. Wetstone, United Aircraft Research Laboratories: The amount of C_2H_4 formed appeared to be substantially greater than the amount of C_2D_4.

Answer by Author: We interpret this as an indication that ethylene formation occurs at a lower temperature than does either methane or acetylene formation. This is probably due to the competition with hydrogen and deuterium recombination.

Question by B. Siegel, Aerospace Corp.: I would like to comment on another question to Dr. Spangler. The discrepancy alluded to between the present data and those of our experiments is easily explainable on the basis suggested by Dr. Spangler. The pyrolysis of methane to acetylene is known to proceed via the stepwise process, $2CH_4 \rightarrow C_2H_6 \rightarrow C_2H_4 \rightarrow C_2H_2$. Shock tube studies at temperatures up to 2500°K have demonstrated that in the millisecond time range both ethane and ethylene can be more important than acetylene at relevant temperatures. Since Spangler et al. used very much higher input energies than we did, they are evidently observing nonequilibrium behavior occurring at lower temperatures. Compared to our experiments they literally swamped the system with energy. This permits an attempt at mechanism study.

Exploding Wire Detonators for Use in Experimental Physics

Fritz Herlach*

Illinois Institute of Technology
Department of Physics
Chicago, Illinois

Exploding bridgewire detonators without primary explosive are considered for application in experimental physics because of their absolute safety against stray currents. The performance of two different types of such detonators has been measured. One type is manufactured commercially on basis of standard techniques; the time jitter of this detonator can be kept to ±0.1 μsec. The other type is precision-made in the laboratory, with a resulting time jitter of ±20 nsec. The transmission time depends strongly on the crystal size of the PETN; this effect is demonstrated by a few examples. The influence of the PETN loading density on the transmission time has been measured. The over-all performance of the secondary explosive detonators depends critically on the firing circuit. Experiments have been carried out over a wide range of firing circuit parameters and are evaluated in terms of Tucker's burst current criterion. A discussion of the firing circuit is included.

INTRODUCTION

In recent years, high explosives have found several applications in experimental physics. By their use, new possibilities have been opened for performing experiments at extreme values of various experimental parameters, such as pressure and density [1, 2], MG fields, and large pulses of electromagnetic energy [3–5]. Many of these experiments require advanced techniques for the fabrication of precise explosive devices [6] and their initiation by detonators with a small time jitter. The detonation velocity of the explosives used in most of these experiments is of the order of 8 mm/μsec; thus, a time jitter of 0.1 μsec in the initiation corresponds to a relative displacement of the detonation wave of 0.8 mm, which is

*Permanent address: Euratom–CNEN, C.P.N. 65, Frascati (Roma) Italy.

still a large value in comparison to the ±0.1 mm precision that can be easily achieved in machining the explosive charges, and is often required in experiments. For example, the success of a magnetic flux compression experiment, in which magnetic flux is compressed by imploding hollow metallic cylinders from initial radii of the order of 50 mm to final ones of the order of 2 mm, depends largely on the regularity of the implosion. The most straightforward way to obtain such an implosion is by the initiation of a precise annular explosive charge with a ring of detonators.

In experiments such as flux compression, with many electrical connections to the explosive device, ordinary electrical detonators could be very hazardous. In these ordinary detonators, a fine wire (the "bridgewire") is immersed in a so-called primary explosive (e.g., lead azide, mercury fulminate, etc.) which detonates on heating of the wire. The detonation of the primary explosive, of which only the necessary minimum quantity is used, initiates in turn a

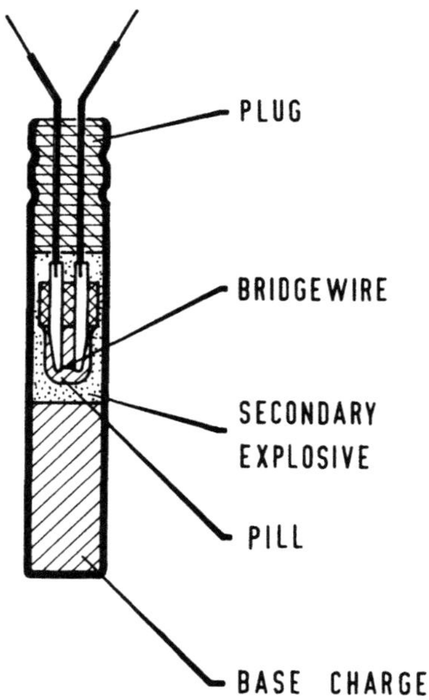

Fig. 1. Schematic drawing of ordinary electric bridgewire detonator. "Pill" contains a small quantity of primary explosive. In modified version of a safety detonator, pill is made with secondary explosive.

small charge of secondary explosive (Fig. 1). The secondary explosives are characterized by the fact that they detonate only on the impact of a sufficiently strong shock; when heated gradually they may only burn. (In exceptional circumstances large quantities, when burning, may convert to detonation.) Enough heat for the initiation of the primary explosive is developed on application of a relatively small voltage: for example 1.5 V across the 10-Ω bridgewire of a KX-1 precision detonator (manufactured by Dynamit Nobel AG, Troisdorf, Germany). The time interval from the application of the voltage to detonation is then mainly determined by the time required to heat the wire and may be in the order of milliseconds, with a correspondingly large time jitter. To obtain a small time jitter, the wire must be electrically exploded instead of being heated. Therefore, a low-inductance, high-voltage firing circuit is needed, and the performance of the detonators will depend largely on the quality of the firing circuit. As an exploding wire is necessary anyway to obtain the required precision, it would seem a natural development to replace the dangerous primary explosive by a sufficiently strong wire explosion. Such a detonator would possess a very high grade of safety, as it would never fire even on strong stray currents, but only on a very fast-rising pulse from the firing unit. It turns out, however, that secondary explosives are generally extremely hard to initiate. Only very particular arrangements of bridgewires and secondary explosives will result in proper initiation, and severe problems are encountered if we want a small time jitter.

In the United States, precise safety detonators were developed several years ago, and recently a good deal of information about their working principles and performance has been made available [7–10].

Unfortunately, in Europe these detonators are not available on the market. At our request, Dynamit Nobel AG, Troisdorf, has kindly developed a safety detonator, model PL 438 (Laboratory number). These detonators are manufactured with standard techniques as used for ordinary mining detonators. Therefore, they are relatively inexpensive, but their precision is limited. To obtain the higher precision needed for some experiments, we have manufactured detonators in our laboratory to close mechanical tolerances. In this paper, we will report on this development and also on the performance of the PL 438 detonators with regard to the particular requirements of flux compression experiments.

FIRING CIRCUIT AND BRIDGEWIRE EXPLOSION

The firing circuit used in our experiments is a RLC-combination where the energy is originally stored in the capacitor, the inductance arises from the cable connections from the capacitor to the detonator, and the resistance is given by the bridgewire. As long as the resistance of the bridgewire is approximately constant, the discharge current waveform will be a damped sinusoid,

$$I(t) = U_o \sqrt{\frac{C}{L}} \frac{e^{-(R/2L)t}}{\sqrt{1 - CR^2/4L}} \sin \omega t \tag{1}$$

$$\omega = \sqrt{\frac{1}{LC}} \sqrt{1 - \frac{CR^2}{4L}} \tag{2}$$

The initial part of this waveform is a linear current rise

$$I(t) = \frac{U_o}{L} t\, e^{-(R/2L)t} \cong \frac{U_0}{L} t \left(1 - \frac{R}{2L} t\right) \tag{3}$$

(After one microsecond, the term $t \cdot R/2L$ is of the order of 10^{-1} R, R in ohms.) It has already been pointed out that the instant of the wire explosion should fall into this linear part of the waveform to minimize the jitter of the burst time [10]. In this region, the burst time for a given wire will depend only on the slope of the current rise, $(dI/dt)_{t=0} = U_0/L$ (Figs. 7 and 14). It does not depend on the capacity, provided the capacity is large enough to allow for a sufficiently long linear current rise. Consequently, only a small part of the energy originally stored in the capacitor is used for the wire explosion. A better efficiency in this regard would be obtained with a coaxial line pulse generator, which produces a pulse of nearly constant current [11] only for the required time interval. Tucker [8] has shown that the performance of any particular lot of detonators can be fully described in terms of the burst current I_b; i.e., the current passing through the wire at the instant of the wire explosion. The detonators will function above a certain threshold burst current, and the transmission time t_x, i.e., the time interval from the wire explosion to the arrival of the detonation wave at the end of the detonator, will decrease with increasing burst current [8]. Near the threshold burst current, t_x will vary strongly with I_b, whereas at currents well above the threshold t_x tends asymptotically to a limiting minimum value. In this region, the performance of the detonators will depend less on small changes in the firing circuit and also the jitter of the transmission time will be smaller.

The design of the firing unit used in these experiments is described in [12]. A 4.5-μF capacitor is charged to 5 kV and discharged by means of a triggered spark gap. The internal inductance of this firing unit is 0.5 μH. The detonators are connected to the firing unit by cables RGU/58 A (250 nH/m), BRAND-REX V-1431 (43 nH/m) or twisted wires ($\sim$1 μH/m). For experiments with many detonators, it will be convenient to use less than one cable per detonator; i.e., to minimize the number of cables for a given detonator arrangement. Unfortunately, the possibilities for doing so are rather limited, as it turns out that a very fast current rise is required to minimize the time jitter. The rate of current rise is determined by the inductance of the firing circuit rather than by the resistance of the detonator. It is therefore unfavorable to connect detonators in parallel, as this will divide the current available per detonator by the number of detonators. With a high-voltage firing unit, a series connection of detonators may be feasible.

METHOD OF MEASUREMENT

The over-all performance of the detonator is given by the time interval t_d from the application of voltage to the bridgewire until

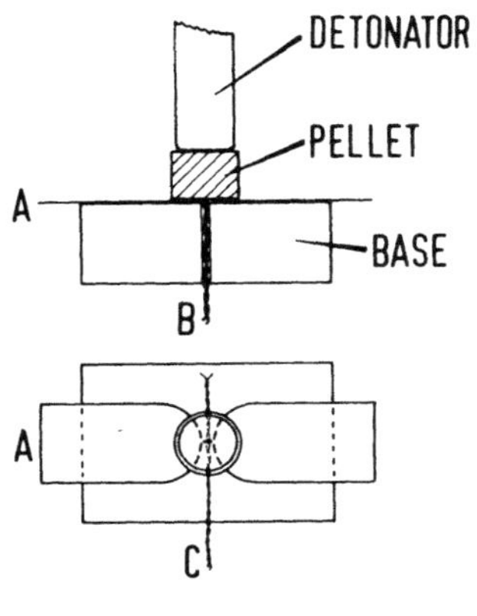

Fig. 2. Experimental setup for comparison of different ionization indicators. A tetryl pellet is placed between detonator and contacts, arranged as follows: (A) Self-adhesive copper strips, 10 mm large and 0.1 mm thick; gap distance has been varied from 0.5 to 2 mm, (B) twisted formvar-coated wires, 0.4 mm in diameter, precisely clipped at end which is in contact with explosive, (C) twisted formvar-coated wires, 0.05 mm thick, detonation wave makes contact by breaking insulation.

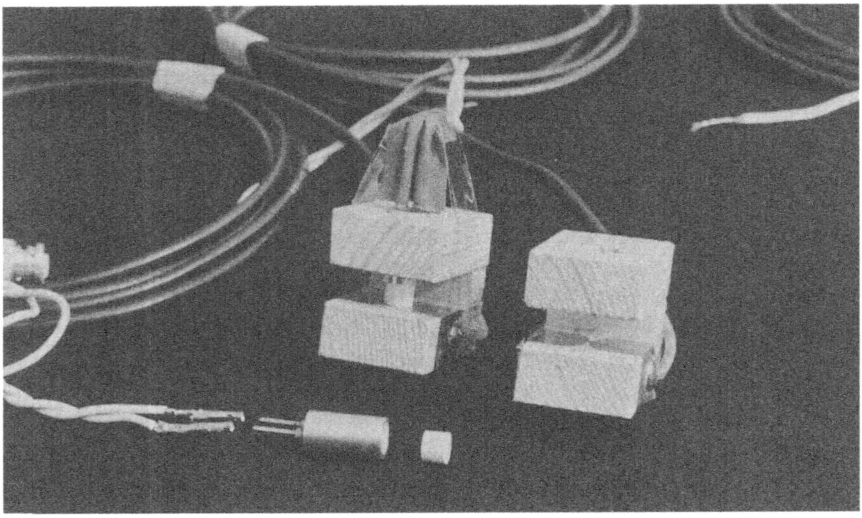

Fig. 3. Experimental arrangement for measuring detonation time of DY-1 detonators. At right, prepared contact assembly; at left, with a detonator mounted ready to fire. In foreground, a detonator DY-1 with contact leads and tetryl pellet.

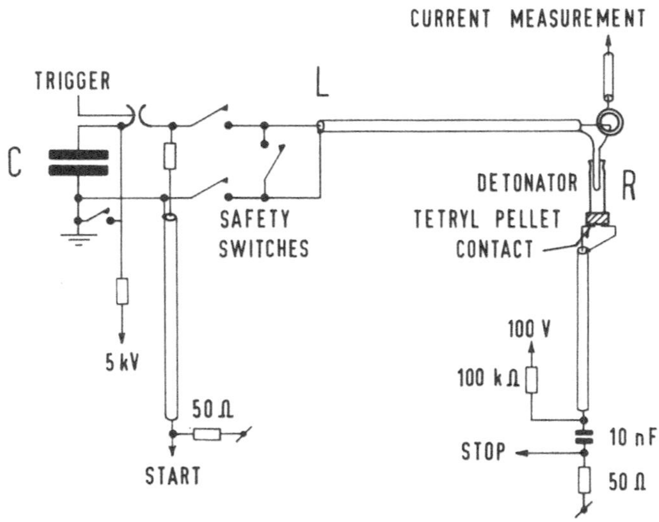

Fig. 4. Circuit diagram for measurement of detonator performance. Voltage dividing resistor in the "start" circuit is 2.5 kΩ; coaxial signal cables are RGU 58-A.

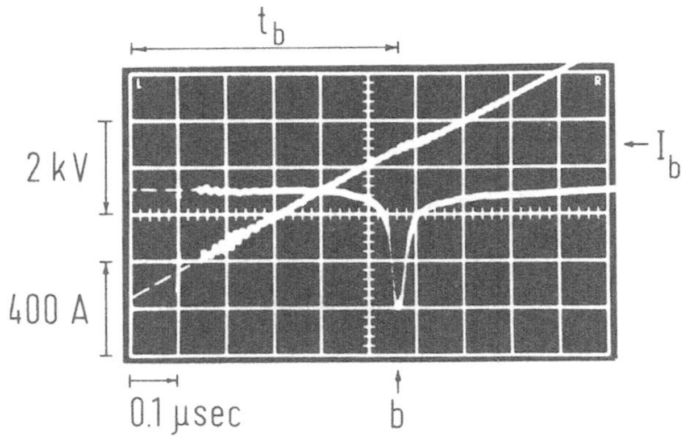

Fig. 5. Oscillogram of voltage and current across an exploding bridgewire of DY-1 detonator.

the detonation wave emerges from the end of the detonators. t_d is the sum of the burst time t_b and the transmission time t_x; it can be measured directly by starting a time interval meter with the voltage pulse of the firing unit and stopping with an ionization probe at the end of the detonator. To avoid interferences of the firing circuit voltage with the ionization probe, and to include in the measurements the initiation of an external charge by the detonator, in all measurements a precise tetryl pellet (diameter 8 mm, thickness of 6 mm, density 1.57 g/cm^3) is placed between the detonator and the ionization probe. The transmission time of these

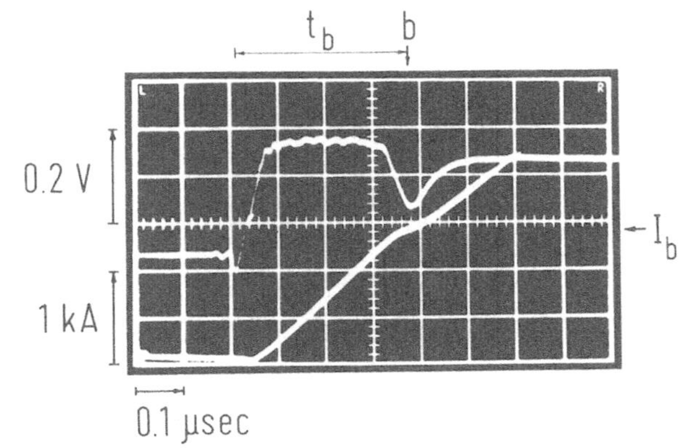

Fig. 6. Current I and its derivative dI/dt of a detonating DY-1 detonator.

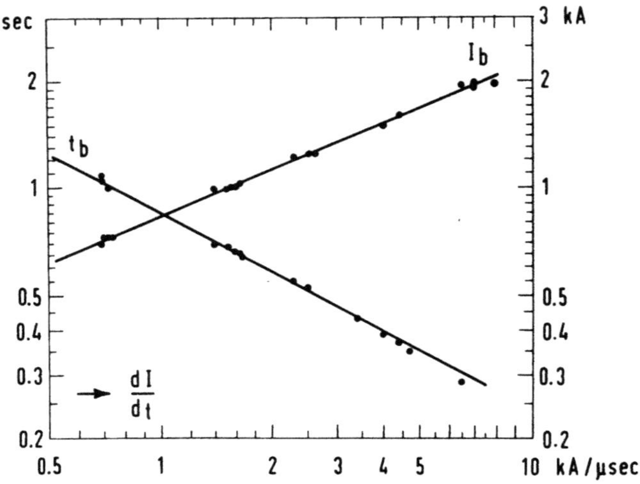

Fig. 7. Bridgewire performance of PL 438 detonators: burst current and burst time in function of initial current rise.

pellets was determined to 0.83 μsec in a separate experiment, and subtracted from the measured time interval. In preliminary experiments, several types of ionization probes have been compared (Fig. 2), and no difference was found within the limit of error of ±10 nsec. The arrangement that has been used in all experiments is shown in Fig. 3, and the circuit diagram in Fig. 4.

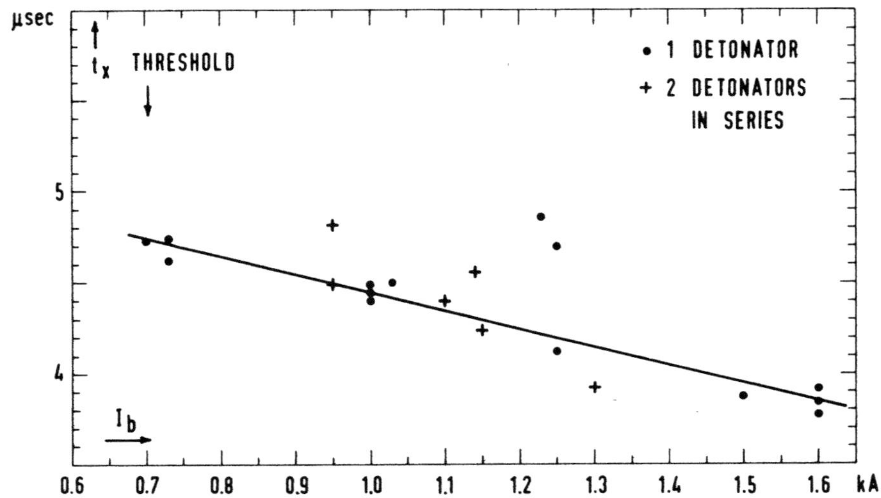

Fig. 8. Transmission time of PL 438 detonators as function of burst current.

The time interval was measured with a Hewlett Packard 5275 A time interval meter, and the current through the bridgewire with a Pearson Model 110 A current transformer. To determine the burst time, which is given by the peak of voltage across the bridgewire (Fig. 5), a measurement of this voltage would be required. Even by differential measuring techniques this measurement has turned out to be quite difficult, particularly on "live" detonators. Fortunately, the voltage peak is sufficiently well displayed in a dI/dt signal derived from the current transformer by electronic differentiation with 10 pF, 50 Ω (Fig. 6). On a 0.1-μsec/cm trace, the burst time can be determined with a precision of ±10 nsec, (±20 nsec in some unfavorable cases). t_x is then determined from $t_d - t_b$. The voltage was kept constant in all shots to 5 kV, the current rise $(dI/dt)_{t=0}$ was varied by changing the inductance of the firing circuit.

TEST RESULTS OF PL 438 DETONATORS

These detonators have been developed and manufactured by Dynamit Nobel AG, Troisdorf (Germany) to replace the primary explosive KX 1 detonators previously used in our experiments. The design is similar (Fig. 1), except that in the pill, which is normally obtained by dipping the bridgwire assembly into a mixture of glue and primary explosive, a secondary explosive is substituted for the primary. As most of the parts and manufacturing techniques can be taken from the mass production line, these detonators are rather inexpensive (less than $1 apiece). However, for the same reason, the mechanical precision is limited, and, therefore, it does not appear feasible to achieve a time jitter smaller than ±0.1 μsec. This is sufficient for many purposes, and even critical experiments can be designed in such a way that a jitter of this order of magnitude can be tolerated [6, 13].

The performance of the bridgewires can be fully described by a diagram of t_b and I_b versus $(dI/dt)_{t=0}$. Both curves are straight lines on a double logarithmic plot (Fig. 7).

In accordance with the burst current criterion, the detonator performance is given by the transmission time t_x as a function of the burst current (Fig. 8). It must be emphasized that the results given in Fig. 8 are obtained with a particular lot of detonators which had a time jitter of only ±0.1 μsec. Time jitters as large as ±0.5 μsec have occurred occasionally with other lots of these detonators, but it is felt that the appearance of such large time jitters can be eliminated in the future by more precise control of the loading process.

LABORATORY-BUILT DETONATORS, MODEL DY-1

To achieve better precision and to be able to investigate the influence of various design parameters on the performance of the detonators, we undertook to manufacture precise detonators in the laboratory on a trial basis. The design (Fig. 9) is based on the SE-1 detonator, the drawings of which were kindly made available to us by the Los Alamos Scientific Laboratory. The design of the casing was slightly modified to adapt the construction to the means of the laboratory, and instead of wire connections, contact pins are provided which fit into AMP 42993-1 connectors. The use of connectors makes it possible to prepare the complete detonator wiring

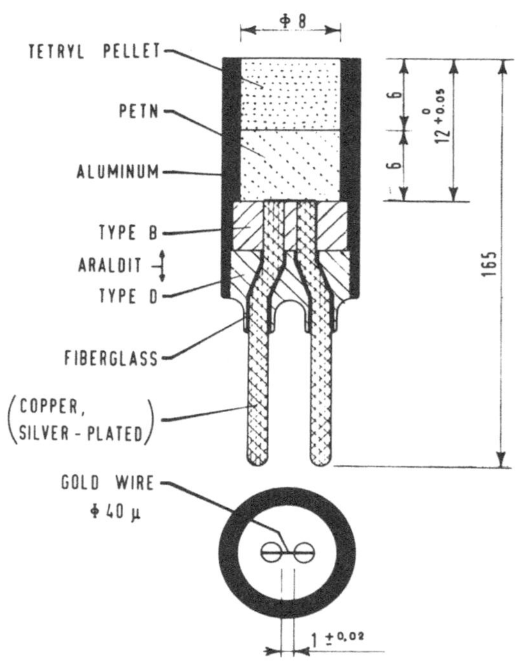

Fig. 9. Drawing to scale of DY-1 detonator. Dimensions in millimeters.

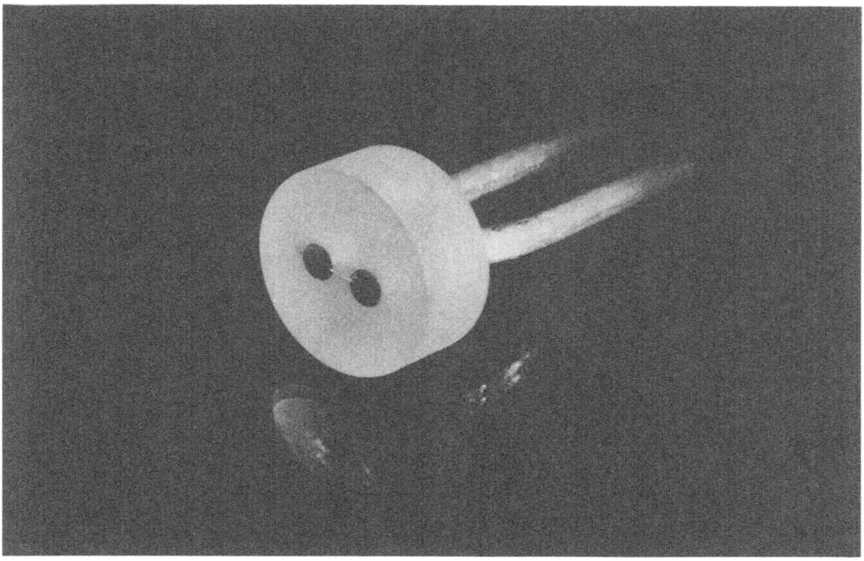

Fig. 10. Bridgewire assembly of DY-1 detonator.

in the laboratory, with consequent saving of assembly time on the firing table and the possibility of controlling more precisely the inductance of the firing circuit.

The detonators are assembled as follows: First, the contact pins are pressure fitted into the Araldit B* pellet and secured with a drop of Araldit D. The gold wires are then soldered under the microscope. For each contact, 0.8 mg of the alloy Pb 37.5%, Sn 37.5%, In 25% are used, and G. C. Electronics soldering paste No. 1207 as solder flux. The gold wire, of 99.999% purity, was obtained from Degussa, Hanau, Germany. The complete bridgewire assembly (Fig. 10) is fitted under light pressure into the outer aluminum casing. It was found possible to keep the critical distance of 12 mm between the bridgewire and the upper border of the casing within the described precision of $\pm 25\ \mu$ without further machining. (Previously, it had been envisaged that it would be necessary to retouch the casing individually after fitting the bridgewire assembly.) Two fiberglass sleeves are put around the contact pins and Araldit D is moulded into the rear end of the casing. The wall of this part of the casing is made deliberately

*Araldit is the trade name for epoxy resins made by CIBA AG, Basel, Switzerland.

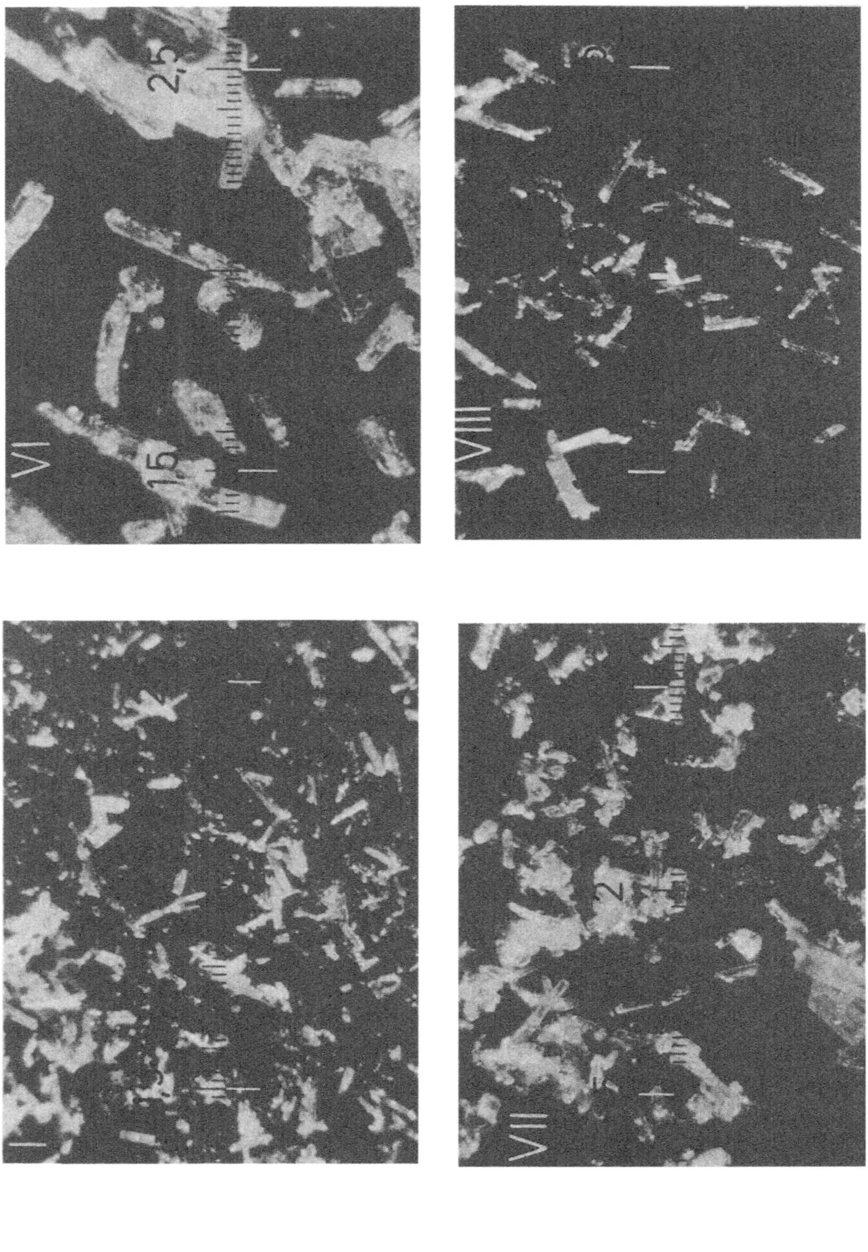

Fig. 11 Photomicrographs of recrystallized PETN used for charging DY-1 detonators. Distance between white markers is 1 mm. Number of preparation is written in left upper corner of each photograph.

Table I. Transmission Time of DY-1 Detonators, Loaded with 265 ± 2 mg of PETN. Bridgewire Diameter 40 μ, Burst Current 1.3 ± 0.05 kA, Burst Time 0.36 ± 0.01 μsec

PETN Lot (Fig. 11)	I	VIII	VI	VII
Transmission time, μsec	2.691 ± 0.006	2.692 ± 0.008	3.81 3.75	3.19 3.19
Standard deviation, μsec	± 0.020	± 0.027		
Maximum (absolute) deviation, μsec	± 0.03	± 0.04		

rough on the lathe to improve adhesion of the moulded assembly to the casing. Before loading, the resistance of each detonator is measured. With very few exceptions, the resistance was found to be in the limits 21 ± 1 mΩ. Recrystallized PETN is weighed on a precise balance and loaded by pressing it gently with a special Nylon plug to an exact thickness of 6 mm. After this, the tetryl pellet of precise dimensions and density is fitted and fastened by a small quantity of glue, making sure that the surface of the pellet coincides precisely with the upper border of the casing. The loading is checked by determining on the analytic balance the weight of the casing before loading, the weight of the tetryl pellet and the weight of the complete detonator before and after applying the glue.

The performance of the single detonators was determined in the same way as described before. As a first result, a large influence of the crystal size on the transmission time t_x was found. Figure 11 shows photomicrographs of four of the various PETN preparations. Lot VIII was prepared in the same way as lot I, to study the possibility of reproducing a lot with the same properties. Detonators were loaded with 265 ± 2 mg of these preparations and fired under identical conditions to obtain a burst current of 1.3 ± 0.05 kA and a burst time t_b = 0.36 ± 0.01 μsec. Ten detonators each were fired with the results listed in Table I. Two detonators each were fired with preparation VI and VII; and a substantial increase of the transmission time was found with increasing crystal size (Table I).

Five detonators were loaded with 280 ± 2 mg of preparation VIII (Table II).

Table II. Performance of Five DY-1 Detonators, Loaded with 280 ±2 mg of PETN Lot VIII and Fired under Identical Conditions

I_b, kA	t_b, μsec	t_x, μsec
1.4	0.35	2.58
1.3	0.36	2.58
1.3	0.37	2.59
1.3	0.37	2.59
1.3	0.36	2.58

More results regarding the dependence of the transmission time t_x on the loading density are reassumed in Fig. 12. It follows that a time deviation of ±20 nsec corresponds roughly to an error in the loading of ±4 mg.

The influence of the firing circuit parameters is displayed in the diagrams of t_x versus I_b (Fig. 13) and of t_b and I_b versus $(dI/dt)_{t=0}$ (Fig. 14). For comparison, the curve derived by Tucker [8] for similar detonators is included in Fig. 13; it has been shifted by ± 0.97 μsec to compensate for the different length of the explosive column. The curves are in fair agreement for burst currents above 800 A. Below this value, relatively large time jitters are found in our experiment. Most probably this is not the

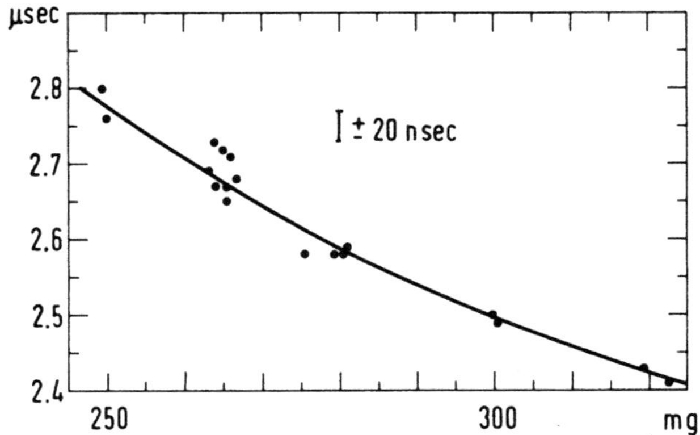

Fig. 12. Transmission time of DY-1 detonators in function of loading density. Volume occupied by the explosive is 0.3016 cm^3, thus a charge of 280 mg corresponds to a loading density of 0.927 g/cm^3.

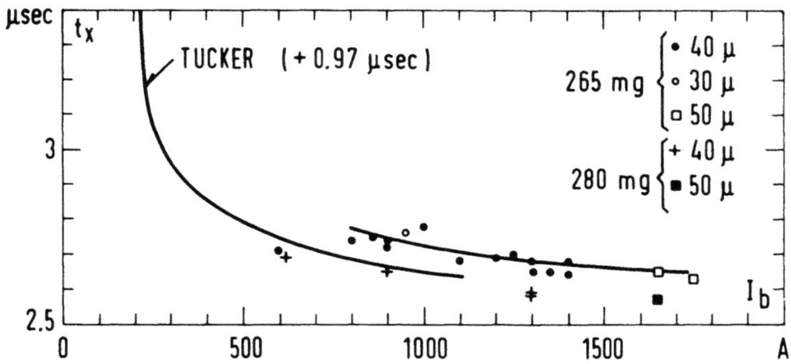

Fig. 13. Transmission time in function of burst current for DY-1 detonators with different loadings and bridgewire diameters. For comparison, curve given by Tucker [8] is included in diagram.

fault of the detonators but of our experimental setup which becomes inadequate, for reasons not completely determined, at lower values of the initial current rise. This inadequacy manifested itself also by occasional trigger failure of the time interval measuring set-up under these conditions. A few tests have also been included with 30-μ- and 50-μ-diameter bridgewires. Within the very limited statistics, these do not show any particular behavior.

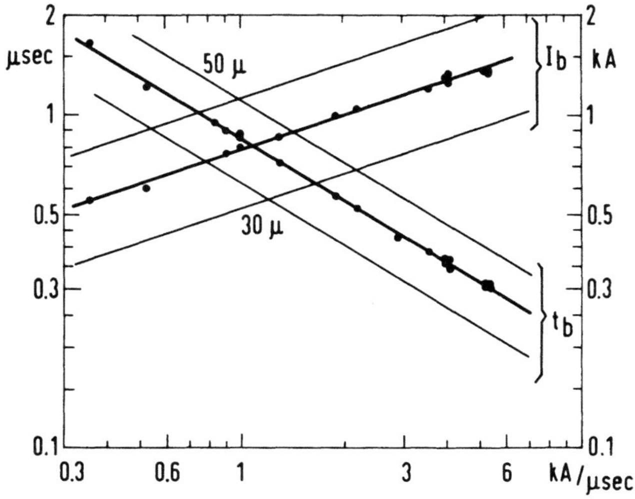

Fig. 14. Bridgewire performance of DY-1 detonators. Thick lines with indicated experimental points are for 40-μ-diameter gold wires; the upper line pair is for 50-μ wires and lower line pair for 30-μ wires.

Table III. Burst Time t_b and Transmission Time t_x of DY-1 Detonators Connected in Series. Bridgewire Diameter 40 μ, Loading 265±2 mg of PETN Lot VIII

	I_b, kA	t_b, μsec	t_x, μsec
1 detonator	1.3	0.36*	2.69*
2 in series	1.2	0.36	2.77
3 in series	1.1	0.38	2.72
4 in series	1.0	0.37	2.80

*Mean value.

To reduce the number of cables leading to an explosive device with many detonators to be fired simultaneously, it is of interest to study the behavior of the detonators when they are connected in series. In principle, this is equivalent to increasing the length of the bridgewire. In the series of experiments described in this paper, only a few detonators were left over to investigate this problem. Therefore, in every series connection, there was only one "live" detonator whose performance was measured, the others were substituted by identical bridgewires. The results are given in Table III. There is some inconsistency in these data, which calls for caution in the interpretation of the results. It follows clearly that the burst current is decreased by putting more detonators in series, without a corresponding increase in the burst time. With respect to the transmission time, something may have gone wrong with the shot where two detonators were connected in series. In the whole, one might, however, conclude that up to four detonators may be connected in series with our 5-kV firing circuit without seriously disturbing the detonator performance. As the voltage that builds up across the bridgewire is of the order of 1 kV at burst time, it is anticipated that a higher voltage firing circuit will be needed for series connections of more detonators.

GENERAL CONCLUSIONS

It has been demonstrated that safe, reliable, and precise exploding bridgewire detonators can be manufactured in a normally equipped laboratory at a reasonable price. (It was estimated that the total effort of manufacturing one complete detonator could be kept within one man-hour, when the mechanical parts are made in a series of 5000 pieces.) Besides maintaining close mechanical tolerances for the dimensions of the volume containing the explosive, it is essential to load the detonators with PETN of a homogeneous and well-defined crystal size. It seems possible to repeat the process of recrystallization sufficiently well to get reproducible results within ±10 nsec.

Once sufficiently precise detonators are available, there still remains the considerable problem of having them initiate simultaneously, as the detonator performance depends strongly on the inductance of the firing circuit, mainly through the large influence on the burst time of the wire (Fig. 15). Separate firing circuits must therefore be carefully adjusted to have the same inductance. A series connection of the detonators would have the advantage that the same current passes through all the bridgewires, apart from the economy in the connecting cables. This problem would merit a further extended study; it may be anticipated that very high voltage (10 to 20 kV) firing units will be needed for the purpose.

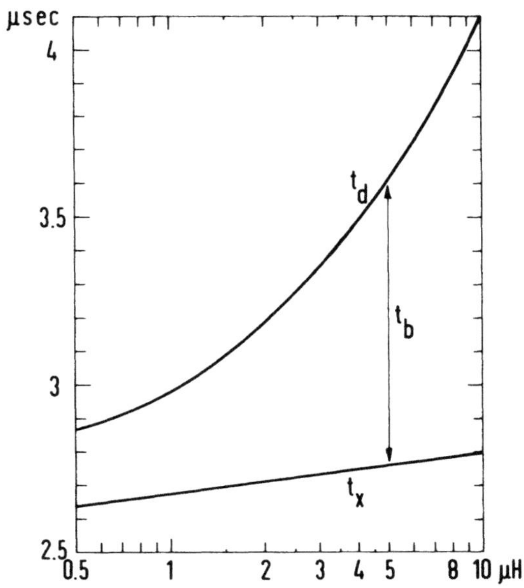

Fig. 15. Transmission time t_x and total detonating time $t_d = t_x + t_b$ for DY-1 detonators with 40-μ bridgewires and 265 mg PETN loading, fired with 5-kV capacitor discharge. Abscissa of this diagram is total inductance of firing circuit.

ACKNOWLEDGMENTS

The experimental work reported in this paper has been carried out at the Laboratori Gas Ionizzati, Frascati, Italy, under the auspices of a contract of association between the European Atomic Energy Commission and the Comitato Nazionale per l'Energia Nucleare, Rome. The DY-1 detonators were loaded in collaboration with the Bombrini Parodi-Delfino Company at Colleferro, which has also assisted the experiments on the shooting range, with particular contributions from A. Puzzuoli and O. Giovanetti. Soldering of the bridgewires and development of methods for the precise fabrication of the mechanical parts have been expertly handled by L. Bettinali at L.G.I. and the problems of mass production have been studied by G. Celentano. The author also wishes to express his gratitude to J. van Montfoort for helpful discussions and practical advice, and to A. Santececca and G. Turella for their efficient help with the experiments. The PL 438 detonators have been developed by P. Röh at Dynamit Nobel AG, Troisdorf. For a discussion of the manuscript, the author is indebted to J. E. Kennedy. The preparation of the manuscript has been partially supported by the Army Research Office, Durham, N.C.

REFERENCES

1. R.G. McQueen and S.P. Marsh, J. Appl. Phys. 31:1253 (1960).
2. L.V. Al'tshuler, Uspekhi Fiz. 85:197 (1965); English translation: Soviet Physics — Uspekhi 8:52 (1965).
3. C.M. Fowler, W.B. Garn, and R.S. Caird, J. Appl. Phys. 31:588 (1960).
4. Proceedings of the Conference on Megagauss Magnetic Field Generation by Explosives and Related Experiments, Frascati, Sept. 21–23, 1965. EURATOM 2750e, Brussels (1966).
5. F. Herlach, H. Knoepfel, R. Luppi, and J. van Montfoort, Compt. rend. du Colloque "Les champs magnetiques intenses, leur production, leurs applications," Grenoble, Sept. 12–14, 1966.
6. F. Herlach and H. Knoepfel, Laboratori Gas Ionizzati Report 65/17, Frascati (1965).
7. J. H. Blackburn and R. J. Reithel, in Exploding Wires, Vol. 3, W. G. Chace and H.K. Moore, (eds.) Plenum Press, New York (1964), p. 153.
8. T. J. Tucker, in Exploding Wires, Vol. 3, W.G. Chace and H.K. Moore, (eds.) Plenum Press, New York (1964), p. 175.
9. E.C. Cnare, in Exploding Wires, Vol. 3, W.G. Chace and H.K. Moore, (eds.) Plenum Press, New York (1964), p. 185.
10. H.S. Leopold, in Exploding Wires, Vol. 3, W.G. Chace and H.K. Moore, (eds.) Plenum Press, New York (1964), p. 125.
11. T.J. Tucker, J. Appl. Phys. 32:1894 (1961).
12. F. Herlach, H. Knoepfel, D. Nencini, and J. van Montfoort, Laboratori Gas Ionizzati Report 65/16, Frascati (1965).
13. F. Herlach and H. Knoepfel, Rev. Sci. Instr. 36:1088 (1965).

DISCUSSION

Question by H. Leopold, Naval Ordnance Laboratory, White Oak, Md.: In the first detonator shown, is there an air gap between the buttered PETN on the wire and the base charge of PETN?

Answer by Author: No. The "pill" is surrounded by loose PETN.

An Exploding Wire Self-Healing Fuse

C. A. Renton and R. J. Manco
N.A.S.A. Electronics Research Center
Cambridge, Massachusetts

INTRODUCTION

The self-healing fuse employs a liquid metal as the contacting material. It is held in a closed system consisting of two reservoirs to which leads are attached, connected via a heavy-walled insulating capillary. Liquid metal completely fills the system, the metal within the capillary tube serving as the fuse element. A mercury-in-glass fuse used in the first experiments is shown in Fig. 1.

Contact is broken when excessive current flows due to the explosive evaporation of the liquid metal within the constricted section. Forces generated by the explosion cause expulsion of the liquid from the capillary tube into the reservoirs at either end. Contact is subsequently reestablished as the liquid flows back into the capillary section.

After the initial vaporization of the liquid occurs, current still flows through the fuse in the form of a plasma discharge. This plasma column elongates as the liquid leaves the capillary section until finally a critical length is reached when the power source voltage is inadequate to maintain the discharge. As the discharge ceases, the fuse action is completed.

If the circuit fault is still present when the fuse reseals, an oscillating condition will result. A protective function is still fulfilled, since the average current flow is reduced. However, many applications require that the contact remain open until some later time that can be independently controlled. This function can be provided by a series relay which would not be subject to the usual stress of opening under load and, therefore, can be made smaller than normally used. Two materials decisions are required for the liquid metal and for the container. The choice of mercury

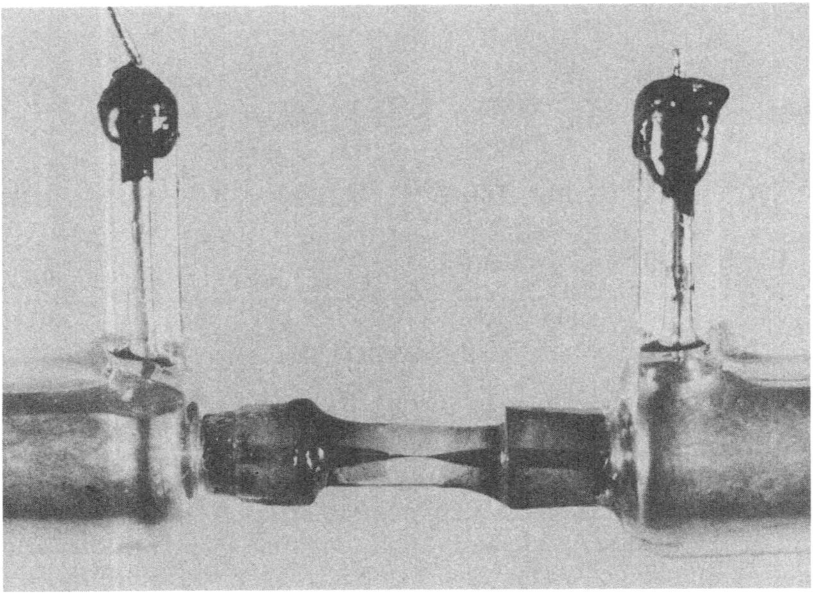

Fig. 1. Mercury in glass fuse. Constricted section of mercury forming fuse is seen in center of heavy-walled pyrex tube.

for the liquid metal is almost self-evident, combining as it does the two desirable properties of being liquid at room temperature and below (its freezing point is −39°C) and boiling at the relatively low temperature of 357°C. Another candidate for operation at a slightly elevated temperature is gallium, but this has a much higher boiling point and would create great problems in the selection of a container material.

The choice of container material introduces stringent requirements and an optimum material is by no means obvious. It should, of course, be an insulator with a melting point well above the boiling point of mercury; it must be strong enough to withstand the force of repeated occurrences of the explosion; and it must be made into a capillary tube. It is here that the ultimate limit to current and voltage capability will lie. No definitive answer to this problem exists, and current work on composites [1] promises extensions over any materials existing now.

Several factors enter into the choice of geometry, not the least of which is its realizability. In many applications it will be desirable to minimize the recovery time of the device. This will in part be dictated by the channel length of the current limiting column, and the reseal time will decrease for shorter channels. However, a minimum channel length must be maintained to insure effective current turn-off. These factors will both be affected if a multichannel configuration is employed in place of a single channel. Then, for a given cross-sectional area, the reseal time will be lengthened, since viscosity will partly determine this time, but it may be possible to compensate for this by shortening the channel.

EXPERIMENTS AND RESULTS

The first series of experiments was designed with the primary objective of demonstrating feasibility of the device. Heavy-walled pyrex tubing was drawn down to form a fine hole in the central section and then attached between two reservoirs with vacuum wax as shown in Fig. 1. Care was exercised while filling the system with mercury to completely remove all air so that the reseal was not impeded by formation of an air bubble in the capillary section. After filling, the system was pressurized with nitrogen gas to provide the driving force during the reseal part of the cycle. It was found that the fuse would sustain a 5-A direct current, while raising the level to 6.5 A caused a break after 1 msec. The time delay is reduced as the current level increases, being 300 μsec at 14 A. The reseal time in both cases was 1 msec.

In the course of these experiments, it was found that pyrex is not a suitable capillary tube material as the explosions cause internal crazing that results in changing fuse parameters and failure after about one hundred shots. Furthermore, the explosive force depends markedly on the source voltage as evidence by the more rapid degradation of the fuse at higher voltage operation. Since the delay time to explosion only depends on the current and not the source voltage, this suggests that superheating may not be significant in these explosions and that most of the energy is fed to the container walls during the plasma discharge phase. This has only been examined qualitatively so far and experiments are planned to study this in more detail.

The remaining experiments to date have concentrated on evaluating suitable capillary tube materials. The only one found to be readily available has been alumina, which can be obtained in the form of tubes of various sizes. Unfortunately, the tubes most readily available do not have a thick enough wall for the present purpose, and one of desired geometry is still under construction. Meanwhile, some insight into fuse operation has been gained with a $\frac{1}{4}$-in. O.D., $\frac{1}{16}$-in. I.D. alumina tube that has been equipped with a constricted section provided by inserting a length of pyrex rod into it. A schematic illustration of this is shown in Fig. 2. This fuse initially withstood a direct current of 55 A, but in the course of operation it changed its characteristic, becoming a 20-A fuse. The data of Fig. 3 show the delay time before fuse action of the 20-A fuse, both when the initial current is zero and when it is 20 A.

During the course of the experiments, no degradation of the alumina tube was observed, but it was found that the change of fusing current had been caused by breakage of the glass insert. The absence of damage to the alumina tube suggests that an alumina–mercury fuse in the 10-A range should work without difficulty.

CONCLUSIONS

Feasibility of the self-heating fuse using mercury as the working material was demonstrated. Alumina was found to be a suitable container for a fuse up to at least 10 A. Life, total range, reseal time, optimum geometry, and container materials must still be established. The dynamics of the conversion of electrical to mechanical energy is not completely understood. Since breakage of the capillary material is likely to prove the major limitation of the device, understanding this conversion will be crucial as the fuse is extended to higher power and inductive circuits. There is

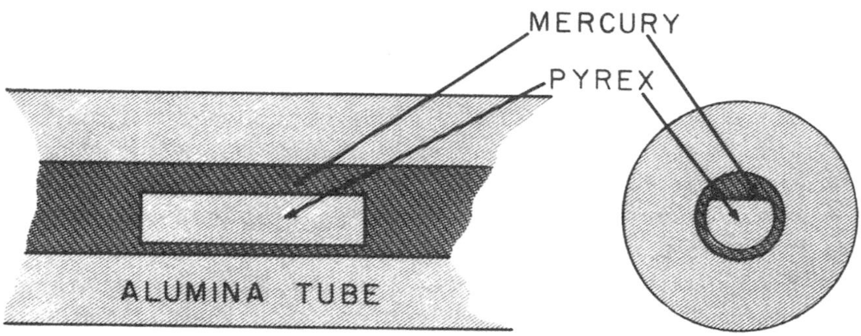

Fig. 2. Cross section of alumina tube with pyrex rod constriction inserted. Rod is milled flat on one side.

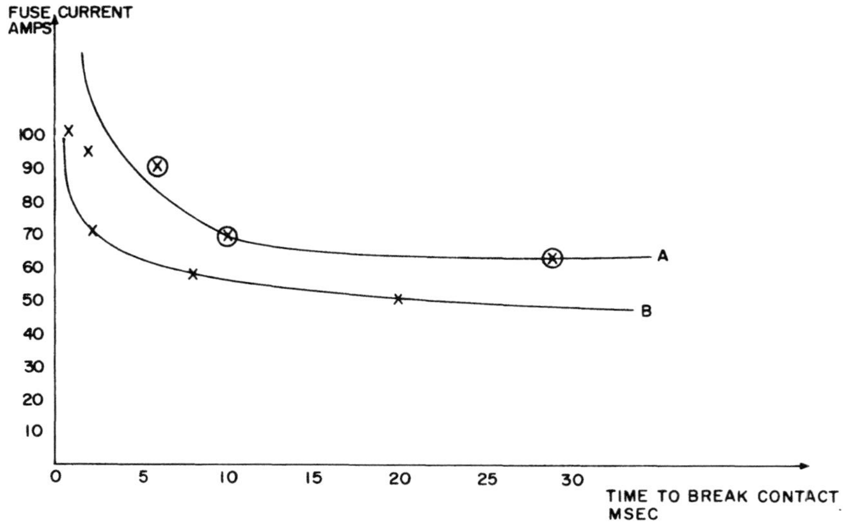

Fig. 3. Fusing characteristics of constricted alumina tube fuse. Curve A shows delay time to interrupt current when initially there is no current flowing in fuse. Curve B shows shortened delay time when fuse initially carries 20 A.

some hope that the use of paralleled fine channels may aid here by cutting short the plasma discharge phase because higher voltages are required to sustain a discharge in a fine bore tube [2]. Investigations are underway to gain a quantitative understanding of the design criteria for fuses of a prespecified rating.

REFERENCES

1. A. Kelly, Strong Solids, Chap. 5, Oxford University Press (1966), p. 121.
2. A. von Engel, Ionized Gases, 2nd ed., Oxford University Press (1965), p. 236.

DISCUSSION

Question by G.J. Woffinden, Aerojet-General Corp.: Have flexible materials such as Teflon or RTV been considered to form the capillary to eliminate the problem of breakage? The flexible material could be confined by a metal tube.

Answer by Author: I don't know of any that will withstand the temperature achieved by boiling Hg and the plasma.

Question by D.V. Turnquist, E.G. & G., Inc.: Have you considered quartz?

Answer by Author: Yes. Unfortunately it proved weaker than pyrex.

Question by J. D. Harrell, Mason-Hanger: What diameters are used to conduct the mercury?

Answer by Author: In the neighborhood of 20 mils.

Question by D. M. Wetstone, United Aircraft Research Laboratories: Do you not have the problem of column recombination in the capillary region such as in the column of a Hg thermometer that has separated?

Answer by Author: No, because the system is completely filled with Hg and is under pressure.

Question by F. D. Bennett, Ballistic Research Laboratories, Aberdeen Proving Ground, Md.: Is there any safety hazard from escaping Hg or its vapor as a fuse ages and suffers erosion or damage?

Answer by Author: We have not investigated this point yet.

Deflagration of Secondary Explosives by Slowly Exploding Wires*

Robert J. Reithel

University of California
Los Alamos Scientific Laboratory
Los Alamos, New Mexico

Small wires, identical to those exploded rapidly to initiate detonation in secondary high explosives, have been slowly exploded to ignite self-sustaining deflagrations in PETN, RDX, HMX, and tetryl. Producing deflagrations required circuits with relatively larger inductances and capacitances and much lower voltages than those used for exploding wire detonators. Choice of electrical parameters allowed the slow explosion of the wire without ignition of the explosive, but provided sufficient energy storage in the inductance to effect an arc between the terminals subsequent to the wire vaporization. This arc supplied the heat to start the deflagration. While the wire characteristics and current configurations used differ markedly from those employed in hot-wire ignition of secondary explosives, the required conditions of explosive confinements and densities and the qualitative description of the ignition mechanism are quite similar for deflagrations produced by the two methods.

INTRODUCTION

The use of exploding wires to initiate the detonation of secondary explosives such as PETN (pentaerythritol tetranitrate) and RDX (cyclotrimethylenetrinitramine) was described at the Third Conference on the Exploding Wire Phenomenon [1–4] and in the basic patent on an exploding wire detonator [5]. This paper will described a method by which explosive devices resembling those used for the exploding wire initiation of detonation may also be used to ignite deflagrations in several secondary explosives [6]. The advantages of such a device include the safety features inherent in the use of secondary explosives in ordnance items, such as pressure cartridges and other types of squibs. Also, lower capacitor voltages are re-

*This work was performed under the auspices of the U.S. Atomic Energy Commission.

quired for these slowly exploding bridgewire ignition systems [7]. Furthermore, methods exist [8] for converting the deflagration device into one which detonates.

While bridgewires identical to those commonly used in detonators were employed in the study described here, both the mode of wire explosion and the process by which chemical reaction is completed were found to be quite different from those in the firing of detonators. In the latter case, the shock wave transmitted from the fast explosion of a bridgewire can be seen to develop into a detonation, with the entire process occurring in a few microseconds [1]. While the attractively simple identity of shock initiation of explosives with exploding wire initiation remains unproven [1], a vast amount of data indicates that the mechanism of the phenomenon has an intimate relation to the fluid dynamic configuration generated by the fast expansion of the bridgewire. These considerations appear relatively unimportant for the exploding wire ignition of deflagrations. Rather, the bridgewire is fired more slowly with a circuit of relatively large capacitance and inductance and functions primarily to cause an arc to be established between the assembly electrodes subsequent to its explosion. This arc discharge is the heat source for the ignition.

Experimental evidence supporting this assertion of the role of the bridgewire in igniting deflagrations and representative sets of bridgewire and circuit parameters to achieve the cited mode of explosion will be presented in the latter part of the following section. This discussion will be preceded by a presentation of observations demonstrating that deflagrations, rather than detonations, were the explosive burning mode produced with slowly exploded wires, and by an examination of the conditions of explosive density and confinement necessary to establish self-sustaining deflagrations. This detailed discussion will be limited to studies with PETN; some similar results with other secondary explosives will be presented briefly in a subsequent section.

EXPERIMENTAL PROCEDURES AND RESULTS

The investigation was initially planned to study the behavior of PETN exploding bridgewire (EBW) detonators when fired by circuits containing large capacitors; e.g., of the order of 1000 μF as compared to the 1.0 μF more commonly used. Figure 1 shows the details of the test assembly, including the pin switch used both to confine the explosive and to measure the time at which ionization occurred at the face of the explosive charge opposite the bridgewire.

The electrical firing circuit and the associated instrumentation for measuring bridgewire voltage and current as functions of time have been described at a previous Exploding Wire Conference [9]. This circuit was modified to provide a large capacitor and an external inductance. A mercury switch was substituted for the triggered spark-gap switch for many of the measurements.

Experimental techniques included the measurement of transit time (the time between the spike in the voltage trace and the arrival of an ionization front at the face of the explosive opposite the bridgewire) by means of the pin switch, the examination of damage to recovered explosive containers, from which the type of explosion process could be inferred, and a determination of the voltage threshold for each test assembly configuration. The latter determination was of the voltage required to explode 50% of the detonators in a given population and was obtained by applying the Bruceton up-and-down method of attempting to fire the detonator [10]. In this test, attempts are made to fire the detonator with the initial voltage determined by the results of the preceding test. If a detonator fails to fire, the succeeding one is tested at a voltage which has been increased by a fixed increment. This procedure is continued until a detonator fires. Succeeding initial voltages are decreased by the same increment until a failure occurs. A statistical analysis applied to the results of the Bruceton test on the selected sample of a population yields an estimate of the voltage at which 50% of the population will fire.

The Effect of Confinement on the Explosion Process in a PETN–EBW Detonator

The determination of the voltage threshold was the first step of our study of the behavior of EBW detonators fired at high capacitance. The detonators in these initial tests were as shown in Fig. 1, except the pin switches were not included in the assemblies. Two lots of these assemblies were used; one had aluminum slugs placed against the open surface of the PETN and held in place by masking tape, while the other had no retaining slugs. The detonators of both lots contained PETN loaded to a density of 0.88 g/cc and a depth of 6.23 mm. A firing circuit capacitance and inductance of 1330 μF and 19 μH, respectively, were employed. The detonators whose explosive charges were unconfined by the aluminum slugs

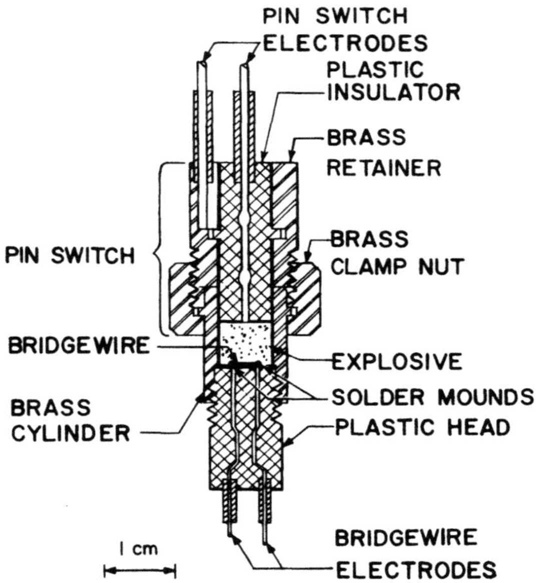

Fig. 1. Test assembly for exploding wire ignition of deflagration. All components are shown to scale except bridgewire. Gold wires of 0.038-mm diameter by 1.0-mm length were used for all work reported here.

could not be fired at the 550-V maximum permitted by the capacitor. The detonators containing the slugs yielded a threshold value of 230 V. A 50-μF capacitor with a higher operating voltage was substituted in the circuit. Detonators without the confinement of the aluminum slugs fired at a voltage threshold of 1075 V. Assemblies with aluminum slugs were not tested with this circuit, but detonators tested with the PETN confined by the pin switch fired at test voltages down to 450 V.

The transit time measurements provided by the pin switches are shown in Fig. 2, relating the firing voltage and the transit time of the detonator. Detonators fired at voltages at or above 1075 V exhibited transit times of a few microseconds, of the order expected of EBW detonators. The detonators fired at voltages less than 1075 V exhibited transit times of the order of hundreds of microseconds. The long transit times strongly suggested that deflagration rather than detonation had occurred.

Further evidence to support the conjecture that deflagrations had occurred was obtained by substituting a brass cylinder of

21.6-mm wall thickness for the test assembly of Fig. 1. A plastic head bearing the bridgewire and the electrical leads was placed in the well of the cylinder. A charge of PETN was loaded to a density of 0.88 g/cc and a depth of 6.23 mm into the well. A rod threaded at the end opposite the explosive was placed in the well and screwed into a short section of thread at the top of the well until it touched the explosive, thus providing confinement. When these assemblies were fired at a voltage greater than the threshold

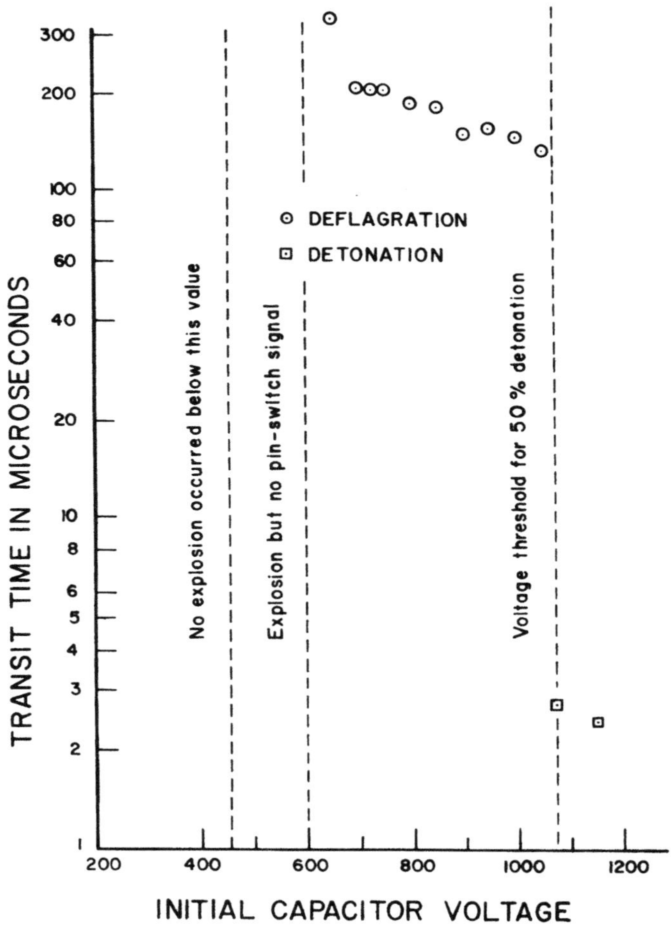

Fig. 2. Effect of firing voltage on transit time (time between the bridgewire voltage spike and pin-switch signal). The capacitance and inductance of firing circuit, 50 μF and 19 μH, respectively.

voltage for an unconfined assembly (1075 V) and then sectioned on a plane including the axis of the well, sharply defined recesses of the cylinder wall at the position of the explosive charge were observed, indicating the occurrence of a detonation. Assemblies fired with less than 1075 V on the firing set did not exhibit the recesses resulting from the shock wave pressure generated by a detonating explosive. Instead, the gas pressure generated by the deflagrating explosive charge frequently was sufficient to strip the brass threads and to expel the confining rod. The destruction of the test assemblies which exhibited the long transit times shown in Fig. 2 was, therefore, effected by the pressure of the gaseous products of the deflagrating PETN.

The fact that deflagrations developed only in the confined assemblies in the tests described above was not surprising. The requirement of an imposed ambient pressure for the ignition of self-sustained deflagrations is well known, and was described in detail by Kistiakowski in 1944 [11]. This condition was quite evident in a study of hot-wire ignition of secondary explosives conducted by the author [12]. The experiments discussed in the next two subsections demonstrate that some means of imposing a pressure in the region of ignition is also necessary to achieve self-sustaining deflagrations with slowly exploded wires.

Effect of PETN Density on the Voltage Threshold

Detonators were secured in several lots containing PETN loaded to densities of 0.75, 0.88, 1.0, 1.2, 1.4, and 1.6 g/cc. These were tested by the Bruceton method to determine the voltage firing threshold, using a circuit of 1330-μF capacitance and 19-μH inductance. Figure 3 shows the results. The marked reduction of the voltage threshold as the density of the PETN increased was ascribed to the greatly reduced volume of the interstitial space in the PETN and, also, to the greater degree of confinement of the decomposition products provided by the increased mechanical strength and reduced permeability of the higher density explosive charges. These effects all facilitated the development of a high pressure in the region near the bridgewire.

Degree of Confinement Required to Ensure Deflagration of the Explosive Charge

Three additional experiments were devised to examine the degree of confinement required for deflagration. The first con-

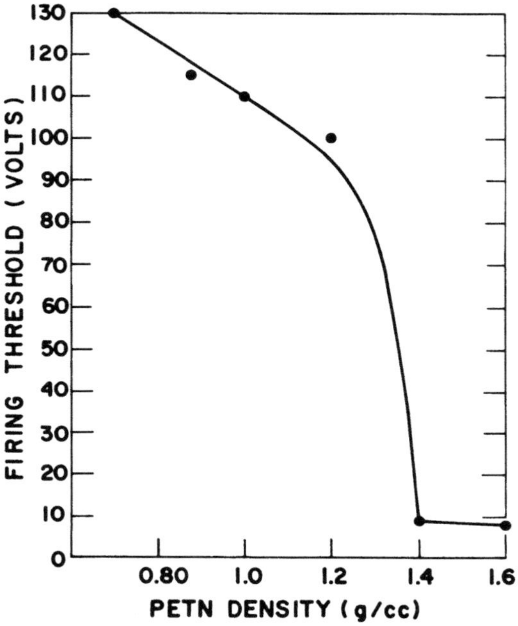

Fig. 3. Effect of PETN density on firing voltage threshold. Capacitance and inductance of firing circuit, 1330 μF and 19 μH, respectively.

sisted in increasing the length of the charge in the Fig. 1 test assembly cylinders. These cylinders were left open rather than being closed by aluminum slugs or pin switches. Progressively longer charges of PETN loaded to a density of 1.4 g/cc were tested at 25 V, a value greater than the voltage threshold for the same charge confined by a pin switch. Charges of 0.90-mm length fired while those of 0.76-mm length failed to fire.

The second experiment on the degree of confinement required for firing consisted in boring progressively larger axial holes through the plastic insulator of a group of pin switches. The modified switches were then used to support 0.13-mm thick aluminum foil disks against the surface of a 1.4 g/cc PETN charge 6.23 mm in length. These items were also tested with 25 V on a 1330-μF capacitor. The explosive charges confined by aluminum disks 6.2-mm unsupported diameter failed while those confined by disks of 5.0-mm unsupported diameter fired. As with the first test, this experiment did not yield quantitative data on the pressures attained in the explosive charge.

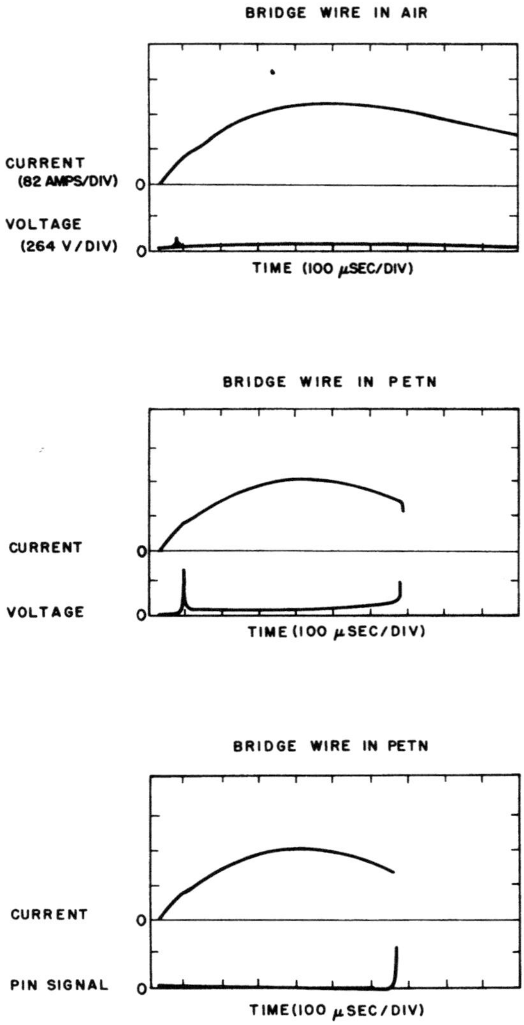

Fig. 4. Current, voltage, and pin signal oscillograms illustrating slowly exploding-wire ignition of deflagration in low-density PETN. Capacitance and inductance of firing circuit, 1380 μF and 19 μH, respectively. Capacitor charged initially to 175 V.

A third experiment was performed by substituting a pipe nipple for the pin switch and imposing a pressure on the explosive charge with dry nitrogen. It was determined that a pressure threshold did, indeed, exist. It was noted that in a certain range of imposed pressures all of the PETN charge was consumed; the assembly was not noticeably damaged but the copper tube used to connect the assembly to the nitrogen tank was heated for about 10 cm. The pressure threshold for igniting a self-sustaining deflagration of 0.88 g/cc PETN was determined. With the capacitor voltage held constant at 25 V, a Bruceton up-and-down test was conducted in which the nitrogen pressure was varied in fixed steps. A charge 0.86 mm in length gave a pressure threshold of 45.0 bars gauge. A threshold value of 41.5 bars gauge was yielded by a PETN load of the same density but of 6.23-mm length, showing the self-confinement effect of the longer explosive charge even at low density.

The Role of the Slowly Exploding Wire in Igniting Deflagration

A study of oscilloscope records of the bridgewire current and voltage and of the signal from the pin switch showed that the energy of explosion of the bridgewire had little significance in the deflagration of the explosive charge. Figures 4 and 5 display typical oscillograms for bridgewires igniting deflagration in low density and high density PETN charges, respectively. The top and center photographs in each figure include current and voltage traces for bridgewires slowly exploded in air and in PETN by a circuit of 152 μH inductance containing a capacitor of 1380 μF. The bottom photograph shows the same current trace as the center photograph, this time in conjunction with the electrical signal from the pin switch. This signal was generated when an ionization front in the deflagrating explosive shorted the electrode of the pin switch to the case, thus discharging the pin circuit capacitor.

Records of the electrical parameters of the slowly exploding bridgewire displayed in Figs. 4 and 5 show that the existence of an electrical arc is a necessary condition for the explosion of the detonators. Although the PETN density of 0.88 g/cc used in the detonator of Fig. 4 and that of 1.4 g/cc used in the detonator of Fig. 5 required currents of different amplitudes and durations to achieve explosions, both show the same sequence of events. Upon the closing of the firing circuit switch, the bridgewire first served as a low-resistance circuit element, which permitted the capacitor to discharge through the circuit. During this time, a portion of the electrical energy initially stored in the capacitor was transferred to

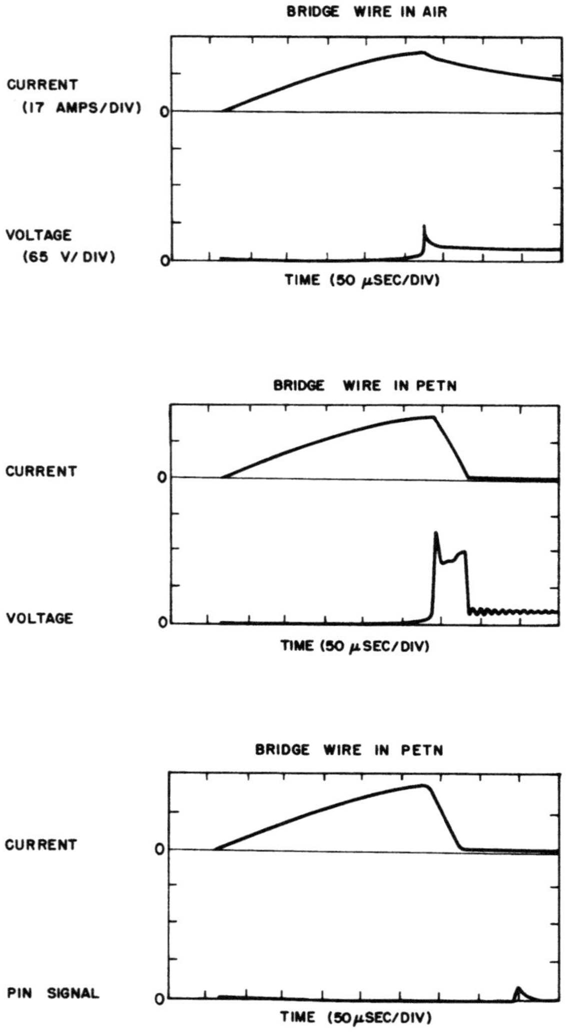

Fig. 5. Current, voltage, and pin signal oscillograms illustrating slowly exploding wire ignition of deflagration in high-density PETN. Circuit capacitance and inductance were same as Fig. 4. Initial voltage on the capacitor, 25 V.

the magnetic field of the inductor. As the bridgewire heated and slowly exploded, its resistance increased and acted to limit and to reduce the current. Both the ohmic resistance of the bridgewire and the consequent negative rate of change of the current in the inductance resulted in the development of a voltage across the slowly exploding wire of sufficient magnitude to strike an arc between the terminals of the bridgewire. Since the initial capacitor voltage used to ignite the low density PETN was 175 V while the voltage in the higher-density case was only 25 V, the two arcs exhibited quite different current and voltage characteristics. A consideration of the effect of the different environments provided by the two explosive densities is necessary for an understanding of the lowering of the voltage threshold which results from increasing the explosive densities, shown in Fig. 3. For this purpose, the current and voltage records of bridgewires exploded in air will be compared separately with those of bridgewires exploded in PETN at each of the two densities.

As is evident in Fig. 4, the arcs resulting from the wire exploded in air and from the wire exploded in low-density PETN differ only slightly. A somewhat higher voltage was required to strike an arc in the explosive assembly as the PETN between the terminals provided some resistance to the formation of an arc. Therefore, a greater voltage developed across the terminals. The arc in the explosive environment continued to exhibit a higher resistance than that in air throughout its duration, as becomes particularly evident approximately 180 μsec before the appearance of the pin switch signal. This may be ascribed to the increasing pressure of the decomposition products of the heated PETN. As deflagration progressed, the rapidly increasing pressure, as indicated by the increasing resistance of the arc, finally explosively destroyed the test assembly, thus abruptly terminating the current. The pin signal indicated the arrival of an ionization front at the end of the explosive charge at about the same time that the arc was terminated. Thus, the slowly exploding bridgewire was the controlling circuit element in the establishment of the arc which supplied the heat required to decompose a sufficient quantity of explosive to attain the necessary pressure for the self-sustaining deflagration of the remainder of the explosive.

The sequence of events described above may also be seen in the oscillograms of Fig. 5. Since the capacitor was initially charged to only 25 V, the bridgewire explosion was even slower than that shown in Fig. 4. Current and voltage traces recorded for the

Table I

Explosive	Specific surface, cm^2/g	Density, g/cc	Firing threshold, V
PETN	4000	1.4	9
RDX	1970	1.4	50
RDX	6861	1.4	130
Tetryl	597	1.4	150
HMX	696	1.4	200
DATB	–	1.4	270
TNT	2302	1.4	175
Nitroguanidine	20,000	1.1	150
Nitromethane	(liquid)	1.1	150

bridgewire exploded in air show the slow heating of the bridgewire followed by its eventual failure as a conducting element. While an arc was formed at this time, the arc current decreased continually thereafter. The bridgewire slowly exploded in the test assembly containing a PETN charge of 1.4 g/cc density developed a considerably higher voltage before the arc was established. The reduced interstitial volume and the reduced rate of diffusion in the denser charge resulted in a faster increase in pressure after the inception of the arc than that which occurred in the low-density explosive charge. The rapidly increasing pressure terminated the arc before the arrival of an ionization front at the surface of the pin switch. Altough the termination of the arc ended the deposition of energy by the current, the necessary condition for a self-sustaining deflagration had been established.

DEFLAGRATION OF OTHER SECONDARY EXPLOSIVES

The voltage thresholds were determined for the deflagration of a number of secondary explosives. Gold bridgewires of 0.038-mm diameter and 1.0-mm length were exploded by the 1330-μF and 19-μH firing circuit. Table I lists the results including the threshold for PETN under similar conditions.

CONVERSION OF DEFLAGRATOR TO DETONATOR

An electroexplosive device containing a deflagrating secondary explosive charge may be converted into a device to detonate another secondary explosive charge by the method of Savitt, Stresau, and Weber [8]. The gas pressure in the first explosive container is used to drive a flying plate against the second charge and, thus, to initiate detonation of the second charge. A device of this type was tested, using a hot bridgewire to ignite a charge of RDX in a steel container of considerably greater strength than the test assembly of Fig. 1 [12]. The flying plate, initially a part of the container, impacted against the end of a length of Mild Detonating Fuze (MDF). The PETN core of the MDF was detonated by the flying plate.

SUMMARY

A slowly exploding wire can be used to ignite deflagrations in PETN and other secondary explosives if the necessary degree of confinement of the explosive charge is present and if the elements of the electrical circuit used to explode the wire are chosen to ensure the formation of an arc subsequent to the explosion of the wire. The voltage to which the capacitor must be initially charged to effect a deflagration is a function of the particular explosive and of the density of the explosive charge. Experiments on the degree of confinement required for the deflagration of the explosive charge indicate that a threshold value of pressure in the charge must be attained before the deflagration becomes self-sustaining. The density of the explosive charge affects the rate at which pressure of the decomposition products can reach the threshold pressure. Therefore the voltage to which the capacitor must be charged to result in a bridgewire explosion and an arc sufficient to effect a deflagration decreases as the density of the explosive charge increases. As a result of these factors, electroexplosive devices containing only secondary explosives may be fired at much lower voltages than those characterizing the initiation to detonation of the same explosives by rapidly exploding wires.

REFERENCES

1. J.H. Blackburn and R.J. Reithel, "Exploding Wire Detonators: Sweeping-Image Photographs of the Exploding Bridgewire Initiation of PETN," in Exploding Wires, Vol. 3, W.G. Chace and H.K. Moore, eds., Plenum Press, New York, (1964), pp. 153–173.
2. H.S. Leopold, "Effect of Bridgewire Parameters on Explosive Initiation," Exploding Wires, Vol. 3, W.G. Chace and H.K. Moore, eds., Plenum Press, New York (1964), pp. 125–152.

3. T.J. Tucker, "Exploding Wire Detonators: The Burst Current Criterion of Detonator Performance," in Exploding Wires, Vol. 3, W.G. Chace and H.K. Moore, eds., Plenum Press, New York (1964), pp. 175–184.
4. E.C. Cnare, "Exploding Wire Detonators: An Approximate Method of Predicting Exploding Wire Detonator–Capacitor Discharge System Performance," in Exploding Wires, Vol. 3, W.G. Chace and H.K. Moore, eds., Plenum Press, New York (1964), pp. 185–192.
5. U.S. Patent Number 3,040,660, L.H. Johnston, "Electric Initiation with Exploding Bridge Wire," June 26, 1962.
6. U.S. Patent Number 3,158,098, R.J. Reithel, "Low-Voltage Detonator System," November 24, 1964.
7. W.G. Chace, "Introduction," in Exploding Wires, Vol. 2, W.G. Chace and H.K. Moore, eds., Plenum Press, New York (1962), pp. 1-4.
8. U.S. Patent Number 3,062,143, J. Savitt, R.H. Stresau, and J.P. Weber, November 6, 1962.
9. R.J. Reithel and J.H. Blackburn, "A Hydrodynamic Explanation of the Anomalous Resistance of Exploding Wires," in Exploding Wires, Vol. 2, W.G. Chace and H.K. Moore, eds., Plenum Press, New York (1962), pp. 21-31.
10. W.J. Dixon and A.M. Mood, J. Am. Stat. Assoc. 48:109 (1953).
11. G.B. Kistiakowski, in Third Symposium on Combustion, Flame, and Explosion Phenomena, Williams and Wilkins Company, Baltimore (1949), pp. 560-565.
12. R.J. Reithel, "The Hot-Wire Ignition of Secondary Explosives in Squibs," Los Alamos Scientific Laboratory Report, LA-3668 (1967).

Capacitance—Voltage Relationships for the Initiation of PETN by Exploding Wires

Howard S. Leopold

U. S. Naval Ordnance Laboratory
White Oak, Maryland

Practical usage of exploding bridgewire systems in explosive ordnance requires optimization of the circuit parameters in order to conserve available space and energy and also decrease insulation problems. The 50% firing voltage levels of PETN for 1-, 2-, and 3-mil-diameter gold bridgewires were studied (as a function of capacitance) over a range of 0.05 to 9.9 μF. There is an optimum capacitance value where a minimum of stored energy is necessary to initiate PETN high order for a specific size bridgewire. This optimum capacitance value depends upon the energy transfer efficiency to the wire at the time of burst. A concurrent energy and voltage minimum occurs with the 2-mil-diameter wire at approximately 1 μF. Smaller bridgewires can be used to decrease the required initiation energy down to a limiting wire size.

INTRODUCTION

In exploding bridgewire systems used in missiles, as in most other exploding wire arrangements, the designer or investigator is initially faced with a choice of what wire and circuit parameter magnitudes to employ out of the myriad possible combinations. The desired results can usually be effected by many combinations. However, factors such as voltage and energy limitations, size, weight, etc., quickly place limits on the magnitude ranges which can be employed. Within these limits, there remains the search for an optimized design.

U.S. Patent 3,040,660 [1] appears to be the original patent on exploding bridgewire initiators (EBW's). The patent mentions the use of such diverse values as a 2-μF capacitor charged to 1500 V to explode a 1-mil-diameter, 125-mil-length tungsten wire and a

50-μF capacitor charged to 500 V to explode a 6-mil-diameter, 125-mil-length nichrome wire. In both cases, PETN (pentaerythritol tetranitrate) was initiated. The dimensions of the bridgewire are partially influenced by two extremes. It is difficult to initiate the critical volume of secondary explosive necessary to ensure propagation with a bridgewire too small in diameter or too short in length. The use of bridgewires much larger than necessary can be extremely wasteful of energy, especially when employed in systems where only limited amounts of energy are available. EBW firing units currently used in missiles employ a 1-μF capacitor charged to 2000 V or a 0.5-μF capacitor charged to 2600 V to explode bridgewire sizes falling within the aforementioned limits. This paper describes the results of a laboratory investigation with bridgewire sizes which conceivably might be used to initiate PETN in EBW devices. Various voltage–capacitance relationships were explored with the test bridgewire sizes to determine how to design an optimized system.

EXPERIMENTAL APPARATUS

The test circuit used in the investigation is shown in Fig. 1. It can be adjusted to fire over a range of 300 to 5000 V and has a capacitor bank adjustable from 0.05 to 9.95 μF. General Electric paper pyranol capacitors were used. Except for the 0.05-μF capacitor, the capacitor bank operates on a 1-2-2-4 combination basis, which by paralleling capacitors, provides for 0.1 to 9.9 μF in 0.1 μF steps. Changes in the number of capacitors used in the bank cause changes in the circuit inductance and resistance. To keep the circuit inductance approximately constant, a series compensating inductance was varied as the capacitor bank values were varied. Slope measurements (di/dt) showed that the circuit inductance was held between 0.9 and 1.0 μH for the various test capacitance values employed. No compensation was made for the circuit resistance variation (0.38 to 0.42 Ω) which was considered negligible.

An EG&G KN-6 Kryton tube was used as the discharge switch. Although the typical operating limits of the tube are between 700 and 5000 V, it is possible to use the tube at lower voltages. Potentials as low as 550 V were employed in this investigation.

A Tektronix Type 555 dual-beam oscilloscope was used to record the current and voltage waveforms obtained, respectively, from a

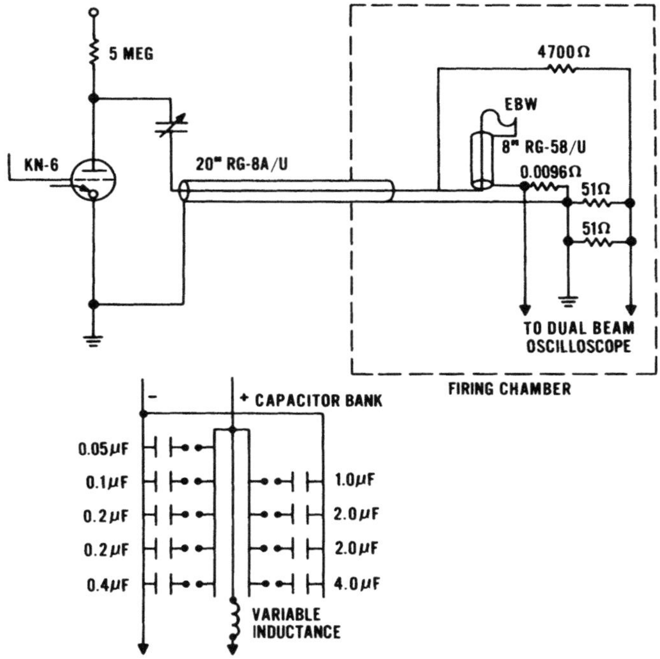

Fig. 1. Test circuit with variable capacitance.

coaxial current shunt and a carbon resistor voltage divider. The voltage was corrected for the inductive component. Instantaneous resistance, instantaneous power, action, and cumulative energy values were then calculated. The calculated quantities have an inherent error, but are within ±10% of actual values.

Explosive testing was performed with the fixture shown in Fig. 2. The bridgewire was mounted flush on a transparent plastic plate. The mounting was typical of that used in electroexplosive devices. The transparent plastic permitted photographic observation of the bridgewire in contact with the explosive. Streak camera photography was used to observe the growth or failure mode of explosion along the surface of the PETN in contact with the transparent plastic plate. The PETN used was from a single lot. It was loaded at a density of 1.0 g/cm^3 (56.5% of crystal density) in the test fixture for all the testing.

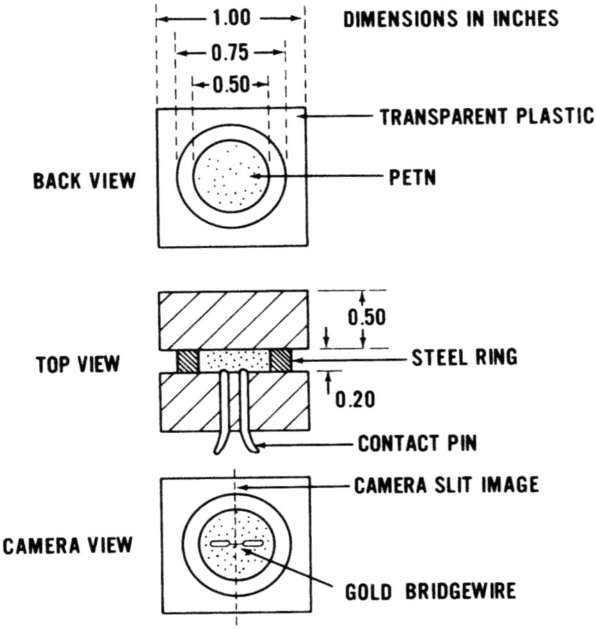

Fig. 2. Test fixture.

TEST PROCEDURE

Bruceton-type tests [2] were conducted to determine the capacitances and voltages necessary to cause high-order initiation of PETN by each of three gold bridgewires, viz., 1-mil diameter, 50-mil length; 2-mil diameter, 75-mil length; and 3-mil diameter, 100-mil length. The Bruceton statistical technique [2] was used at each test capacitance to determine the 50% firing voltage. The range of test capacitances covered approximately two orders of magnitude, one order of magnitude in each direction from the frequently used 1-μF value. The circuit inductance was increased to 3.95 μH for an additional test run with the 2-mil-diameter bridgewire to determine the effect on the 50% firing voltage levels.

TEST RESULTS

The 50% firing voltage for PETN over the capacitor test range for each of the three bridgewire sizes is shown in Fig. 3. It can be observed that a minimum stored energy requirement occurs in the 2-μF region for the 3-mil-diameter bridgewire, in the 1-μF region for the 2-mil-diameter bridgewire, and in the 0.05-μF region for the 1-mil-diameter bridgewire. The minimum energy does not occur at a sharply defined point, but at the bottom of a broad trough. As the bridgewire length and diameter are decreased, the minimum energy requirement for 50% firing decreases. This indicates that the energy required for initiation can be reduced by using smaller bridgewires and lower capacitance values, perhaps until a limit is reached where too small a bridgewire precludes initiation of a critical volume of explosive or is unable to transfer sufficient energy to the explosive. It can also be seen in Fig. 3 that the smallest wire will not necessarily always require the least amount of stored energy. At the highest test capacitor values, the 1-mil-diameter bridgewire requires more stored energy for 50% firing than both the 2- and 3-mil-diameter bridgewires which have, respectively, 6 and 18 times the mass of the 1-mil bridgewire.

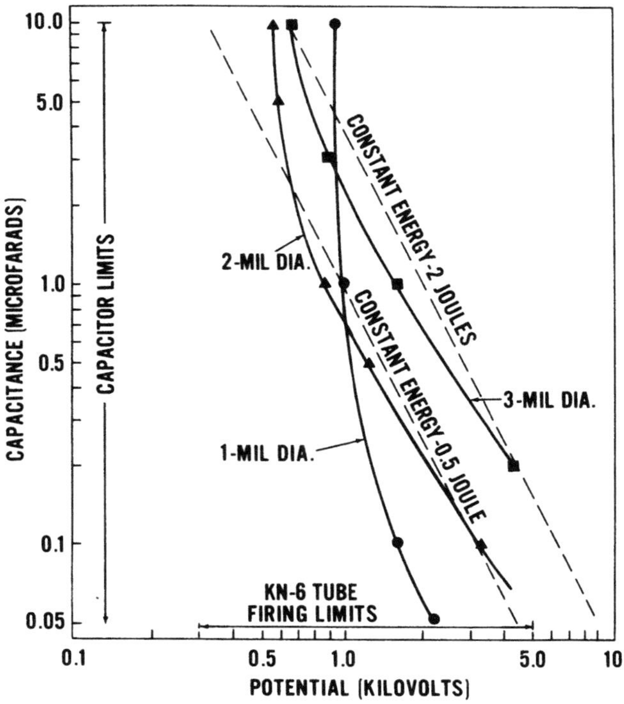

Fig. 3. Capacitance–voltage relationships for 50% firing.

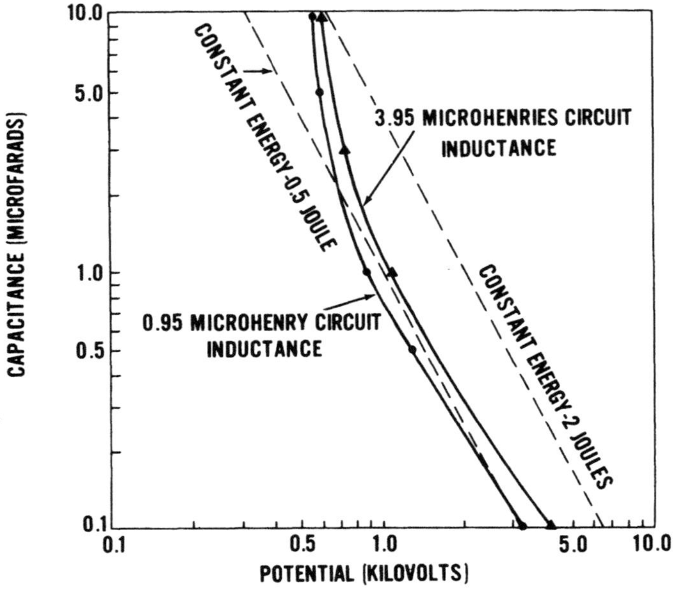

Fig. 4. Effect of inductance on capacitance–voltage relationships for 2-mil-diameter, 75-mil-length bridgewire.

Additional Bruceton-type tests with the 2-mil-diameter bridgewire were run with the circuit inductance increased approximately fourfold from 0.95 to 3.95 μH. The results are shown in Fig. 4. The firing energy curve at the higher inductance paralleled the curve obtained with the lower inductance and indicates the increase in stored energy necessary for 50% firing. The energy minimum was unexpectedly found to occur at the same capacitance for both inductance values.

The relationships between the bridgewire and the current pulse at the 50% firing point differ for each of the bridgewire sizes. The 1-mil-diameter bridgewire appears to burst close to the current peak at the 0.05-μF optimum capacitance (least stored energy required). This results in a single pulse giving what is considered to be a well-matched wire explosion. As the capacitance is increased, the bridgewire bursts well forward of the current sine wave peak resulting in a current resurge (see Figs. 5a and 5b). The 2-mil-diameter bridgewire also gave a single pulse in the low inductance circuit when exploded at the critical firing levels with the various test capacitances (see Fig. 5c). When the 2-mil-diameter bridgewire was fired in the higher inductance circuit, current continuance was generally observed (see Fig. 5d). Only single pulses were observed with the 3-mil-diameter bridgewires at the 50% firing levels. They were similar to that illustrated in Fig. 5c. If high-order initiation is not effected during the first half sine wave pulse, it appears highly unlikely that it can be effected by a wire burst at a later period, since there will be a period of low power flow to the bridgewire when the current direction changes. As the current decreases during the latter part of the first pulse, power input (I^2R) will decrease even though the bridgewire resistance is increasing due to heating. The power input will drop to zero as the current passes through zero, and remain low during the initial part of the second half sine wave pulse. The lower amplitude of the second pulse due to damping will also further decrease the probability of high-order initiation.

When firing with a fixed capacitance, the peak current before burst has been found to be a good indicator of whether or not detonation will be produced by the various size bridgewires [3].

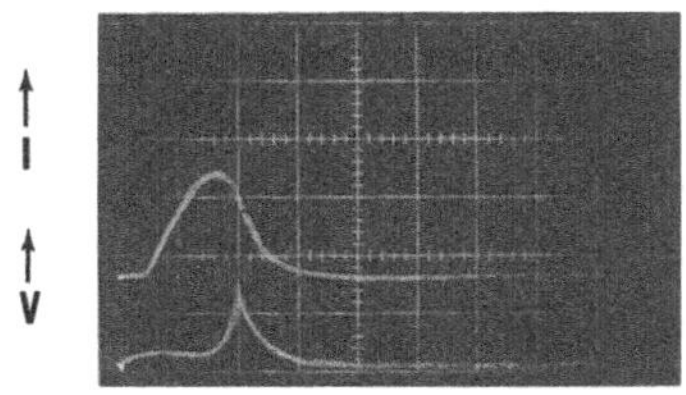

(a) 1-MIL DIA., 0.05μ F-SWEEP-0.2μ SEC/DIV CURRENT=208 AMPS/DIV, POTENTIAL=1840 VOLTS/DIV

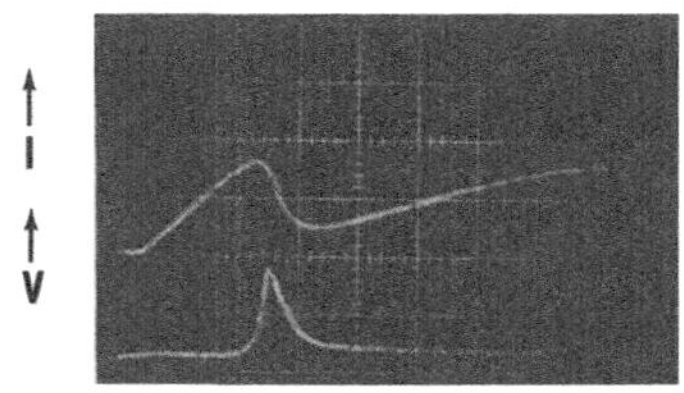

(b) 1-MIL DIA., 1.0μ F-SWEEP=0.2 μ SEC/DIV CURRENT=208 AMPS/DIV, POTENTIAL=1840 VOLTS/DIV

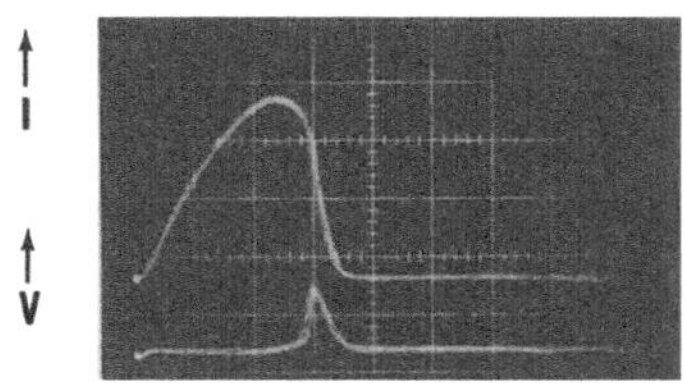

c) 2-MIL DIA., 1.0μ F-SWEEP=0.5μ SEC/DIV CURRENT=208 AMPS/DIV, POTENTIAL=1840 VOLTS/DIV

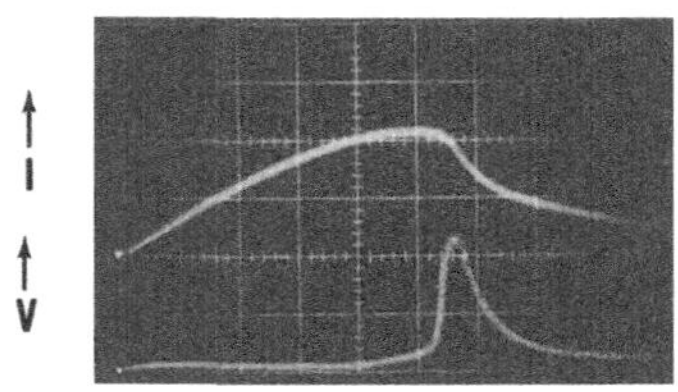

(d) 2-MIL DIA., 1.0μ F- SWEEP=0.5μ SEC/DIV CURRENT=208 AMPS/DIV, POTENTIAL=920 VOLTS/DIV (HIGH INDUCTANCE CIRCUIT)

Fig. 5. Typical oscillograms.

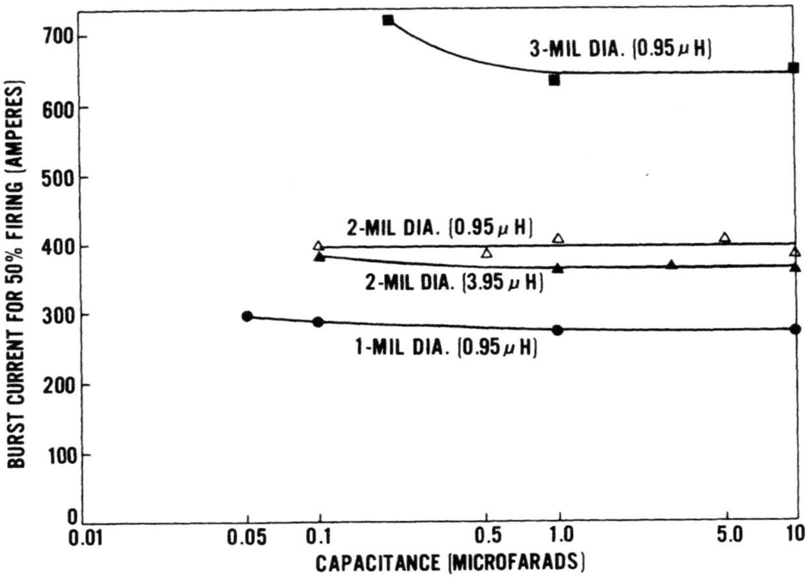

Fig. 6. Effect of capacitance on burst current for 50% firing.

However, the peak current value before burst needed to effect detonation varies with the capacitance. Consequently, the peak current before wire burst cannot be used alone as a guide to explosive detonation by the bridgewire. The peak current before burst decreases as the capacitance increases. The time to wire burst (peak voltage time) increases as the capacitance increases. If it is desired to control the time of wire burst by variation of the capacitance–voltage relationship, larger wires and high inductance circuits permit the largest time variation.

Whereas the peak current value for 50% firing was not constant over a capacitance range, the current value at the time of peak voltage (burst current) remains fairly constant for 50% firing as reported by Tucker [4], except at low capacitance values. Three of the four runs indicate a possible increase in the 50% burst current at low capacitance values (see Fig. 6). The 2-mil-diameter bridgewire data show a slightly lower burst current with increased circuit inductance. It is not known whether this lower current is due to statistical scatter or possibly to the sustained current flow after burst in the high inductance circuit. Previous work using pulse termination methods had indicated that under marginal conditions, a sustained energy input after burst can be helpful in effecting detonation [5].

DISCUSSION

Anderson and Smith in early studies on exploding wires proposed that maximum power will be dissipated when the circuit resistance approaches $(L/C)^{1/2}$ in an $R-L-C$ circuit [6]. Bennett [7], in a mathematical examination of the optimum resistance for exploding wires found that a more precise value for an underdamped $R-L-C$ circuit would be in the range

$$1.1(L/C)^{1/2} < R_{opt} < 1.3(L/C)^{1/2} \quad (1)$$

where R_{opt} is the mean resistance for the entire circuit including the bridgewire, L is the circuit inductance, and C is the capacitance. The lower resistance bound is for maximum power dissipation during burst at the current peak and the upper resistance bound is for burst during the latter part of the first half sine wave. The shaded areas in Fig. 7 show these wire resistance bounds for the two test inductance values employed over the range of test capacitances. The circuit resistance, excluding the bridgewire, is 0.4 Ω, so the upper bound of each shaded area is delineated by a 1.3 $(L/C)^{1/2}$ -0.4 value and the lower bound by a 1.1 $(L/C)^{1/2}$ -0.4 value. Bennett's criterion illustrates that, in general, for maximum energy transfer efficiency: 1) A larger mean wire resistance at burst is desirable as the capacitance is decreased, and 2) A larger mean wire resistance at burst is desirable as the inductance is increased.

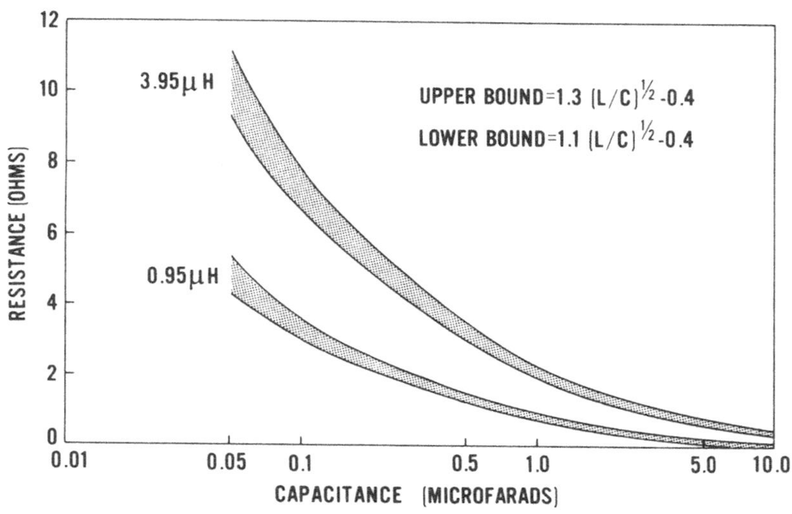

Fig. 7. Determination of optimum bridgewire resistance (Bennett criterion) for maximum energy transfer with a series circuit resistance of 0.4 Ω.

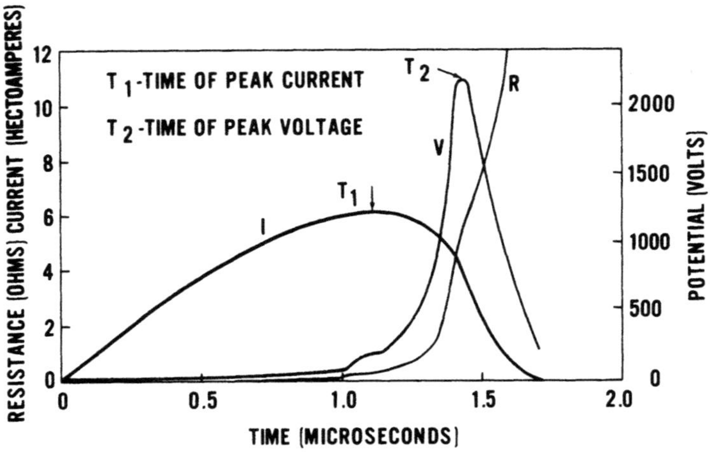

Fig. 8. Current, voltage, and resistance curves for typical 2-mil-diameter gold bridgewires shot with a 1-μF capacitor charged to 850 V.

In the initiation of secondary explosives by exploding bridgewires, the most important period of energy deposition occurs at the time of wire burst. The dynamic variables (burst current, peak power) that can indicate whether or not detonation will ensue, occur at this time. It is near the current peak that the resistance of the gold bridgewire begins to increase significantly (see Fig. 8). Therefore, to make comparisons with Bennett's criterion, the mean resistance of the bridgewire for the finite interval between the current peak and the time of peak voltage was used. The mean resistance for this interval was calculated by dividing the energy input by the action.

$$\bar{R} = \frac{\int_{T_1}^{T_2} RI^2 dt}{\int_{T_1}^{T_2} I^2 dt} \tag{2}$$

where T_1 is time of current peak and T_2 is time of peak voltage. When the mean resistance of this interval is compared to Bennett's criterion, it can be seen that the experimentally determined minimum energy regions occur when the calculated resistance falls within the Bennett criterion limits (see Figs. 3 and 9). For example, Fig. 3 shows that the minimum energy requirement occurs in the 0.05-μF region for the 1-mil-diameter bridgewire. Figure 9 shows that the 1-mil-diameter bridgewire mean resistance (4.6 Ω)

at the 0.05-μF capacitance value falls within the resistance limits indicated by Bennett's criterion. This indicates that an optimum transfer efficiency is responsible for the minimum energy region.

The resistance of the wire burst is not determined uniquely by the amount of energy supplied (which is fairly constant over the capacitor test range for 50% firing), but also by the rate at which the energy is supplied [8]. This phenomenon manifests itself in two places in the results. First, it can be seen in Fig. 9 that the calculated mean resistance for each wire size tends to increase as the capacitance is increased and the charge potential lowered. Secondly, one would expect that when the circuit inductance was increased fourfold with the bridgewire size kept constant, a definite shift to a higher optimum capacitance would occur. Experimentally, almost the same optimum capacitance region was found with the 2-mil-diameter bridgewire for both test inductance values. This is due to the higher resistance exhibited by the 2-mil-diameter bridgewire in the higher inductance circuit. The various experimental mean resistances over the time range (T_2 - T_1) were found to increase from 30 to 90 times over the original room-temperature resistances for the gold bridgewire dimensions used. Since the resistance increase depends upon the rate and amount of energy supplied, the wire dimensions and material, and the circuit parameters, it is difficult to employ Bennett's criterion other than as a

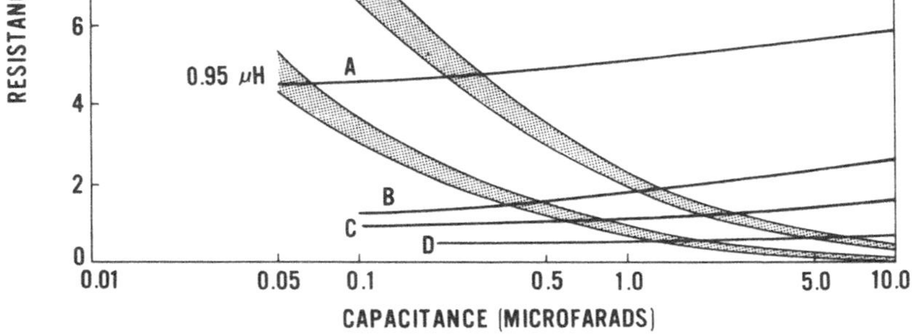

Fig. 9. Comparison of mean exploding bridgewire resistance with Bennett's criterion.

rough rule of thumb for predicting an optimum bridgewire circuit match.

To initiate PETN high order with a 50% probability, it was found that at the time of peak voltage the 1-mil-diameter bridgewire requires more delivered energy than necessary for vaporization of the wire (163%), the 2-mil-diameter bridgewire requires approximately the wire vaporization energy, and the 3-mil-diameter bridgewire requires less energy than necessary for vaporization (93%) (see Fig. 10). The 2-mil-diameter wire in the higher inductance circuit was found to also require approximately the same vaporization energy for 50% firing as in the lower inductance circuit. This indicates that the extra capacitor energy is stored temporarily in the magnetic field of the circuit and released during the discharge at a later time when it is no longer useful for effecting detonation. The trend of the graph in Fig. 10 indicates that as the wire size is further decreased, greater amounts of delivered energy per cylindrical mil of wire will be required to initiate the explosive high order. Russian investigators, however, have shown that the energy input into an exploding wire apparently cannot be increased indefinitely when the voltage is increased [8]. There appears, therefore, to be a lower limit on the bridgewire dimensions that

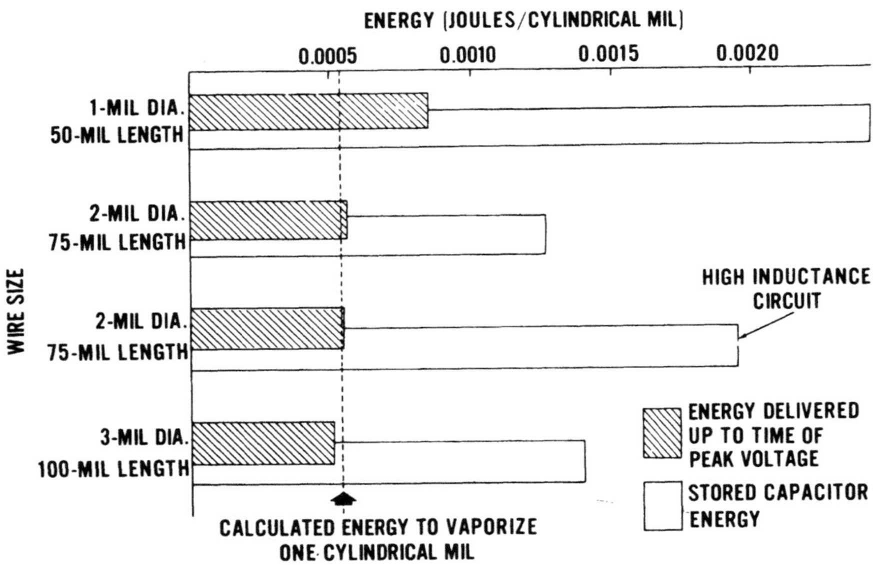

Fig. 10. Comparison of energy per cylindrical mil required for 50% firing at experimental minimum stored energy points.

can be used due to electrical considerations (aside from explosive critical volume considerations). Although the 1-mil-diameter bridgewire requires more energy than necessary for vaporization to initiate the PETN, it requires the least total energy of the test wire sizes.

For initiating PETN under the conditions employed, the 2-mil-diameter, 75-mil-length bridgewire requires the least stored energy per cylindrical mil and requires the lowest voltage in the minimum stored energy region, making this size the most logical choice for systems use unless other factors are important to the design engineer. If minimum stored energy is an important design consideration, smaller bridgewires can be employed. However, as finer wires are employed, the EBW device will become more susceptible to stray current dudding.* Larger wires can make the EBW device quite unsusceptible to dudding, but will tend to require larger capacitors, higher voltage levels, and a greater amount of stored energy.

SUMMARY

There is an optimum capacitance value where a minimum of stored energy is necessary to initiate PETN high order for a specific size bridgewire. The optimum capacitance value depends upon the energy transfer efficiency to the wire at the time of burst. Finer bridgewires can be used to decrease the required initiation energy for PETN apparently down to some limiting size where the wire size fails to ignite a critical volume of PETN or insufficient energy is transferred to the PETN.

ACKNOWLEDGMENT

The author wishes to thank Prof. L. Rosenthal of Rutgers University for construction of the test circuit.

REFERENCES

1. U.S. Patent 3,040,660 by Lawrence H. Johnston, dated June 26, 1962, filed November 8, 1944.
2. "Statistical Analysis for a New Procedure in Sensitivity," AMP Report No. 101.1R SRG-P No. 40, July 1944 (Report submitted by the Statistical Research Group, Princeton University).

*Assuming no use of a protective spark gap.

3. H. Leopold, "Initiation of Explosives by Exploding Wires II Effect of Circuit Resistance on the Initiation of PETN by Exploding Wires," U.S. Naval Ordnance Laboratory, White Oak, Md., Technical Report 63-244, May 1963.
4. T.J. Tucker, "Exploding Wire Detonators: The Burst Current Criterion of Detonator Performance" in Exploding Wires, Vol. 3, W.G. Chace and H.K. Moore, eds. Plenum Press, New York (1964) p. 175.
5. H. Leopold, "Initiation of Explosives by Exploding Wires VII Effect of Energy Termination on the Initiation of PETN by Exploding Wires," U.S. Naval Ordnance Laboratory, White Oak, Md., Technical Report 65-56, June 1965.
6. J.A. Anderson and S. Smith, "General Characteristics of Electrically Exploded Wires," Astrophys. J. 64:295 (1926).
7. F.D. Bennett, H.S. Burden, and D.C. Shear, "Correlated Electrical and Optical Measurements of Exploding Wires," Phys. Fluids 5:102 (1962).
8. I.F. Kvartskhava, V.V. Bondarenko, A.A. Pliutto, and A.A. Chernov, "Oscillographic Determination of Energy of Electric Explosion of Wires," Soviet Physics—JETP 4, p. 623, June 1957.
9. Assuming no use of a protective spark gap.

DISCUSSION

Question by F.D. Bennett, Ballistic Research Laboratories, Aberdeen Proving Ground, Md.: Did the smaller wires function efficiently according to the matching criterion limits on average resistance?

Answer by Author: The minimum energy requirement for all the wire sizes tested occurred very close to where the mean resistance fell within the criterion limits.

Apparatus with Electronic Crowbar for Chemical Investigations of Exploding Wire Phenomena*

M. J. Joncich, S. K. Lott, C. W. Spangler, and H. B. Giddings

Michael Faraday Laboratories
Northern Illinois University
DeKalb, Illinois

An apparatus suitable for investigations of chemical phenomena during the explosions of wires in various gaseous atmospheres has been developed. Electrical energy of 6900 J maximum was used to explode a variety of metals, as well as graphite filaments. A new "crowbar" system is included which will reroute the normal restrike current of an exploding wire through a circuit not containing the sample wire. Time to the initiation of the "crowbar" may be varied from 4 to 200 μsec. This new technique provides an additional variable for investigating the various mechanisms of exploding wire phenomena. The extent of chemical reaction at any point in the explosion process can now be determined.

INTRODUCTION

Comparatively few fundamental investigations of the chemical aspects of the exploding wire phenomena (EWP) have been carried out. The major emphasis appears to involve specific product synthesis. Baker and Warchal [1] studied metal–water reactions using the exploding wire technique. Joncich and Reu [2] have prepared sulfides, carbides, and iodides of aluminum and magnesium by exploding metal wires coated with the corresponding nonmetal. Mahieux [3] prepared $PtXeF_6$ by exploding platinum wire in an atmosphere of xenon and fluorine. The exploding wire synthesis

*Supported by the donors of the Petroleum Research Fund, administered by the American Chemical Society.

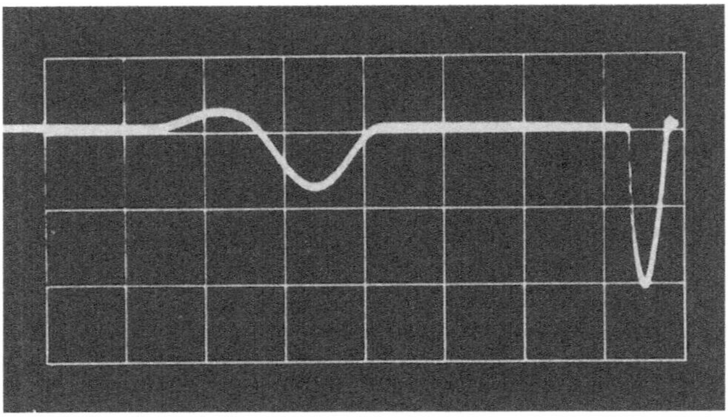

Fig. 1. Typical oscilloscope tracing of current (20,000 A per division) versus time (10 μsec per division) for copper wire explosion without "crowbarring."

of metal nitrides has been reported [4], as well as some organic reaction studies [5] and the preparation of metal fluorides [6]. In these and related investigations [7], a charged capacitor bank was completely discharged through the sample wire, causing it to explode. No attempts were made to regulate or "crowbar" the current flow during the explosion. The term "crowbarring" refers to the process by which current is rerouted at some instant during the course of the explosion through a path not containing the sample wire.

The amount of current flowing through a copper wire exploded by a capacitor discharge varies with time as shown in Fig. 1. An initial period of large current flow (approximately 50,000 A) is usually followed by a period of low current (approximately 500 A) known as the "dwell." This, in turn, is followed by an abrupt "restrike." The current then flows in a damped oscillation until the capacitor bank is discharged. Although temperatures achieved during the explosion are too high to be measured by conventional means, it is likely that the temperature is related to the magnitude of the current. From the chemical viewpoint, many reactions which take place at high temperature may be instantly "quenched" by a rapid drop in the temperature, and reaction intermediates may be isolated which normally would react further or decompose if the temperature were decreased more slowly. Knowledge of the intermediates formed is an important step in determining chemical reaction mechanisms. A rapid temperature drop gives better quenching and raises the possibility of isolating reactive intermediates. In exploding wires, the rapid current decrease leading to the corresponding rapid decrease in temperature occurs during the time between the initial current maximum and the onset of the dwell. The system described here is particularly suitable for chemical studies in which the exploding wire is one of the reactants; in addition, the reaction may be quenched at any point by use of the electronic crowbar which is described in some detail.

APPARATUS DESCRIPTION

The basic components of the EWP crowbar system include (1) capacitor banks, (2) a high-voltage power supply, (3) low-noise switches capable of rapid closing of high currents, (4) special trigger generators necessary for operating the switches, (5) a time delay circuit, (6) devices for measuring electrical parameters, and (7) the reaction vessel, including the sample wire. A block diagram of the system is shown in Fig. 2. Capacitor bank C_1 provides the energy for the sample wire explosion and capacitor bank C_2, the energy source to explode a fine wire activating the crowbar switch.

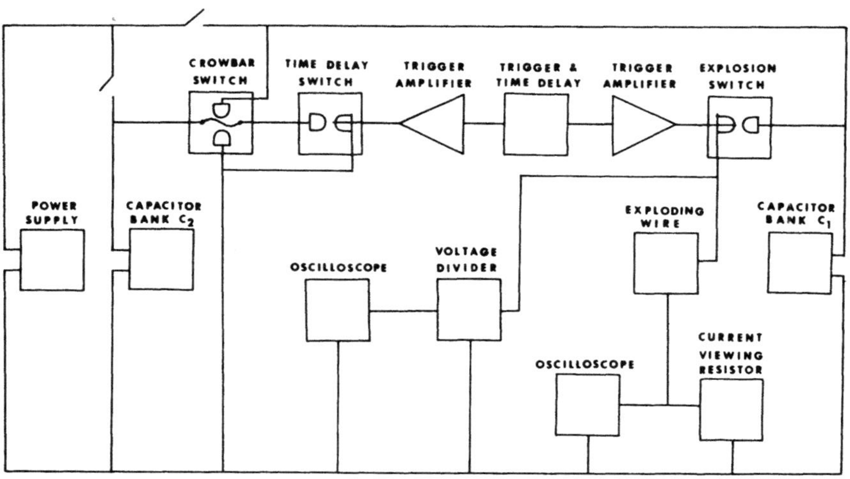

Fig. 2. Block diagram of exploding wire system with provisions for "crowbarring."

In our system, each capacitor bank consists of five 31.25-μF capacitors which can be charged by the 0 to 17,000-V, 5-mA power supply to a maximum of 10 kV (6900 J). The system was designed so the two capacitor banks could be connected to provide twice the energy for wire explosions at times when the crowbar system was not used. If this is not required, capacitor bank C_2 may be of lower energy. The capacitors in each bank were connected together in parallel by a copper plate sandwich consisting of two sheets of ⅛-in. copper separated by two sheets of thin (0.030-in.) polyethylene which in turn were separated by one sheet of (0.005-in.) Mylar. The three insulation sheets extended about 2 in. beyond the copper plates to prevent arcing at the edges from plate to plate. The polyethylene prevents surface corona and the Mylar prevents breakdown through the insulation. To prevent arcing at the capacitor terminals, a 4-in.-diameter hole was drilled in the copper sheet at each terminal not connected to that particular sheet. Through the use of the copper plate sandwich, the leads were held very close together, thus reducing the inductance to a reasonable level (0.39 μH). Short flat copper or aluminum leads were used exclusively for connecting the various components of the system to keep inductance at a minimum.

During the course of our experimental work reaction vessels of several types were constructed in which the sample wire was exploded and the products collected. A Teflon vessel gave the most satisfactory results, due to its excellent electrical insulating qualities and its ability to withstand shock and fairly high temperatures. This vessel was cylindrical with a 2-in. inside diameter and an 8-in. length. The two copper electrodes to which the sample wire was attached were introduced through the side of the vessel and were 5 in. apart. The ends of the vessel contained valves for evacuation and filling with the desired gaseous atmosphere.

A Lovotron [8] switch was found to be vastly superior to both mechanical and vacuum spark gap switches for general EWP experiments. This switch consists of two hemispherical copper electrodes; one solid, and one with a hole along the axis joining the two electrodes. A piece of RG 58/u coaxial cable with the shielding removed was passed through the pierced electrode and served as the trigger electrode. The end of the trigger electrode formed a smooth surface with the pierced electrode. The spheres were mounted solidly so that the force of the discharge through the spark gap did not lengthen the spacing between the electrodes.

For best results it was found that the electrode gap should be such that a 10% increase in voltage across the gap caused arcing without the trigger being applied. Jitter and delay were kept at a minimum when the switch was connected in the circuit with the pierced electrode positive with respect to the solid electrode. Lovotron-type switches were utilized for both the explosion switch and the time delay switch. The two pulses used to deliver the sparks between the end of the trigger electrode and pierced electrode of each switch originated in the trigger and time delay and each was amplified by a trigger generator.

A schematic diagram of the trigger and time delay is shown in Fig. 3. Transistors Q1 and Q2 comprised a common Schmitt trigger which provided an output pulse with a very fast rise time. The output of the Schmitt trigger, amplified slightly by Q3 and inverted by Q4, provided the negative pulse necessary for the trigger amplifier and the time delay. The time delay, composed of transistors Q5 and Q6, was a modification of a conventional monostable multivibrator [9]. The delay output pulse was found to be stable and sharp, even with fairly large variations in the power supply and trigger pulse amplitude. The timing network consisting of R1, R2, and C1 was isolated from the base of Q6 during the timing interval by diode D3. Therefore, the collector base leakage current of Q6 did not affect the time delay. D3 was a silicon diode with an extremely high resistance in the reverse direction. In the quiescent state, Q6 was saturated while Q5 was cut off. Zener diode D2 acted as open circuit, leaving the base of Q5 near ground. The insertion of a negative pulse caused Q5 to start conducting.

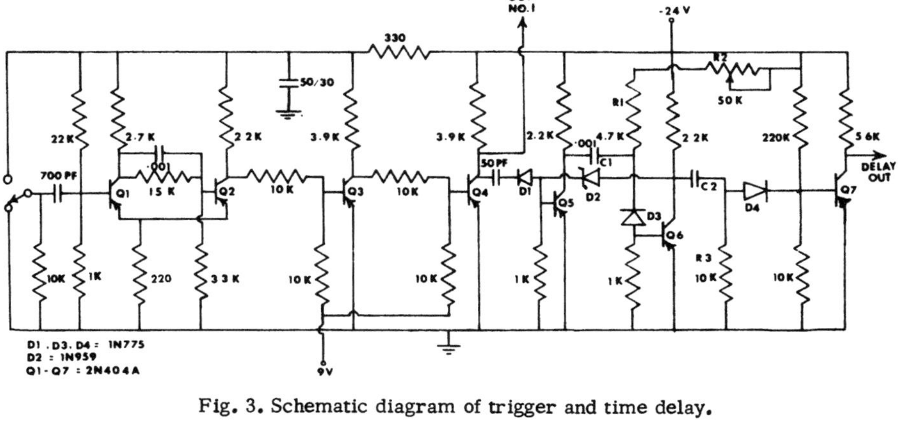

Fig. 3. Schematic diagram of trigger and time delay.

As the collector of Q5 rose toward ground potential, the positive pulse passed through C1 and reverse biases D3. This reduced the base and collector currents in Q6 to zero. As the collector of Q6 rapidly approached −24 V, zener diode D2 broke down producing sufficient current to the base of Q5 to hold it in conduction until the charging circuit allowed D3 to once again become forward biased causing Q6 to turn on again. The 8.2 V output pulse was differentiated by C2 and R3, where the trailing edge triggered Q7 and gave a −24 V output pulse. The time delay could be varied from 4 to 200 μsec by adjusting R2.

Each of the trigger generators used was of the Theophanis [10] type, as shown in Fig. 4. The 5C22 used in this trigger generator was a hydrogen thyratron tube capable of withstanding 16 kV peak forward and inverse anode voltages. A trigger of +200 V (minimum) with a rise time (26 to 70%) of 0.5 μsec (maximum) was required to trigger the 5C22.

The RG 58/u coaxial cable used in the trigger generator was arranged in a small coil to minimize spatial requirements. The fast-rise positive pulse required to trigger the 5C22 was obtained by amplifying the pulse from the trigger and time delay, thereby firing a 2D21 thyratron into a pulse transformer with a 1:1 turn ratio. This gave a fast-rise-time, positive 300 V pulse.

The grid resistor at pin 7 of the 12AU7 was reduced from the value suggested in the tube manual to 47 K. This lowered noise at the grid and prevented the trigger generator from firing when the switch between the high-voltage supply and the capacitor bank was opened and closed. The negative potential at the grid of the 2D21 was varied by adjusting the bias control. In this manner it was possible to prevent small noise pulses from triggering the

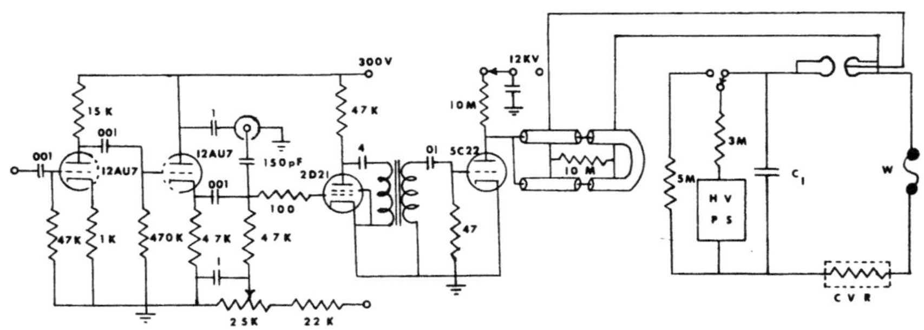

Fig. 4. Schematic diagram of trigger generator, Theophanis type.

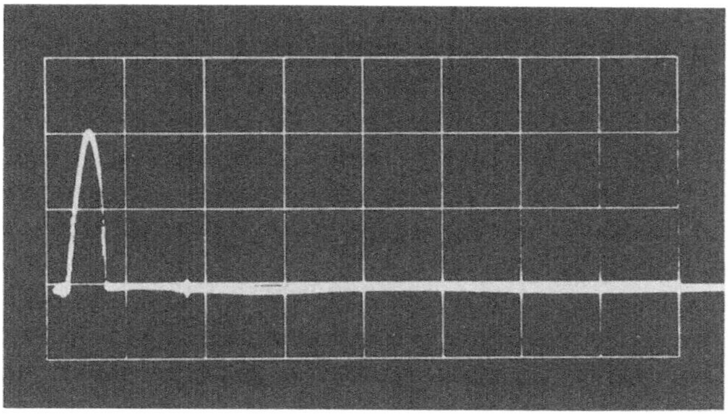

Fig. 5. Oscilloscope tracing of current for explosion of copper wire. Identical conditions as in Fig. 1, except current was "crowbarred" during dwell period.

thyratron. After charging the capacitors using a 0 to 17 kV power supply, a SPDT knife switch was opened to remove the power supply from the circuit. The knife switch was an antenna switch to which a larger insulated handle was installed. The current-viewing resistor is commercially available and has a resistance of 0.0025 Ω and a rise time of 36 nsec. Care must be taken when connecting the circuit to be sure the surge current passes in the proper direction through the CVR. If the current is passed in the wrong direction, a high-voltage pulse may be fed to the chassis of the oscilloscope, which could damage the scope or injure the operator. A type 517A Tektronix oscilloscope was used because of its ability to show fast rise times clearly.

DISCUSSION

The oscilloscopic trace for the current during a typical explosion of a copper wire with crowbarring is shown in Fig. 5. All experimental parameters except for crowbarring during the "dwell" were the same as for the experiment used to obtain Fig. 1. In each case, a 5-in. length of copper wire (B & S gauge 22) was exploded with an initial charge of 8 kV on the capacitor bank.

The results shown in Figs. 1 and 5 are perfectly reproducible. Also, in our laboratories most wire explosions have been found to correlate with the general model. Upon exploding wires of different metals, some individual differences in the shape of the initial current pulse are observed.

Numerous experiments were carried out on the explosion of copper wires in a nitrogen atmosphere. Previous attempts [4] were unsuccessful in preparing copper nitride. When crowbarring was used, no nitride formation could be detected. Although the synthesis was unsuccessful under crowbarring conditions, we know now that the difficulty is not due to the formation of nitride during the initial current surge with subsequent decomposition during the "restrike" period.

Additional experiments with metals known to yield nitrides by EWP [4] were carried out to determine the effect, if any, of crowbarring the current during the dwell period. Aluminum and tantalum metal wires were exploded in a closed vessel containing nitrogen at 1 atm pressure. Using the analytical procedures described previously [4], it was determined that the yields of the nitrides formed (approximately 20%) were very similar to the yields obtained when no crowbarring was used. The results obtained, however, were much more reproducible under crowbarring conditions. These results indicate strongly that the chemical reaction is complete prior to the restrike period, and that the reaction products do not remain in the immediate proximity of the wire after the initial current surge. Additional experiments are planned to make use of the crowbarring technique to determine the point at which the reaction does take place.

REFERENCES

1. L. Baker, Jr., and R.L. Warchal, Exploding Wires, Vol. 2, W.G. Chace and H.K. Moore (eds.), Plenum Press, New York (1962), pp. 207–223.
2. M. J. Joncich and D. Reu, Exploding Wires, Vol. 3, W.G. Chace and H.K. Moore (eds.), Plenum Press, New York (1964), p. 353.
3. M. F. Mahieux, Compt. Rend. 257:1083 (1963); 258:3497 (1964).
4. M. J. Joncich, J.W. Vaughn, and B.F. Knutsen, Can. J. Chem. 44:137 (1966).
5. C.W. Spangler, S.K. Lott, and M.J. Joncich, Chem. Communications 842 (1966); Electrochem. Tech. 5:214 (1967).
6. E. Cook and B. Siegel, J. Inorg. Nucl. Chem. 29:2739 (1967).
7. W.G. Chace and H.K. Moore (eds.), Exploding Wires, Vols. 1–3, Plenum Press, New York (1959, 1962, 1964).
8. E.H. Cullington, W.G. Chace, and R.L. Morgan, "Lovotron — A Low Voltage Triggered Gap Switch," Instrumentation for Geophysical Research No. 5, FCRC-TR-55-227, September, 1955.
9. John M. Carroll, "Design Manual for Transistor Circuits," McGraw-Hill Book Co., New York, p. 147.
10. G.A. Theophanis, Rev. Sci. Instr. 31:427 (1960).

DISCUSSION

Question by C. C. Gallagher, AFCRL: For your crowbar system, why did you use an electronic delay with a second exploding wire rather than an LC delay system with a spark gap?

Answer by Author: The system we use is a high-energy system which would be more difficult to control using an LC circuit instead of the explosion of a second wire for crowbarring.

Author Index

Subject Index

GPSR Compliance

The European Union's (EU) General Product Safety Regulation (GPSR) is a set of rules that requires consumer products to be safe and our obligations to ensure this.

If you have any concerns about our products, you can contact us on ProductSafety@springernature.com

In case Publisher is established outside the EU, the EU authorized representative is:

Springer Nature Customer Service Center GmbH
Europaplatz 3
69115 Heidelberg, Germany

Batch number: 10620163

Printed by Printforce, the Netherlands